MASS LOSS FROM RED GIANTS

ASTROPHYSICS AND SPACE SCIENCE LIBRARY

A SERIES OF BOOKS ON THE RECENT DEVELOPMENTS
OF SPACE SCIENCE AND OF GENERAL GEOPHYSICS AND ASTROPHYSICS
PUBLISHED IN CONNECTION WITH THE JOURNAL
SPACE SCIENCE REVIEWS

VOLUME 117
PROCEEDINGS

MASS LOSS FROM RED GIANTS

PROCEEDINGS OF A CONFERENCE HELD AT
THE UNIVERSITY OF CALIFORNIA
AT LOS ANGELES, U.S.A., JUNE 20–21, 1984

Edited by

MARK MORRIS

and

BEN ZUCKERMAN

Department of Astronomy, UCLA,
Los Angeles, California, U.S.A.

D. REIDEL PUBLISHING COMPANY

A MEMBER OF THE KLUWER ACADEMIC PUBLISHERS GROUP

DORDRECHT / BOSTON / LANCASTER

Library of Congress Cataloging in Publication Data

CIP

Main entry under title:

Mass loss from red giants.

(Astrophysics and space science library; v. 117)
Includes index.
1. Red giants–Congresses. 2. Mass loss (Astrophysics)–Congresses.
I. Morris, Mark, 1947– II. Zuckerman, Benjamin, 1943–
III. Title. IV. Series.
QB843.R42M37 1985 523.8'1 85-11724
ISBN 90-277-2075-4

Published by D. Reidel Publishing Company,
P.O. Box 17, 3300 AA Dordrecht, Holland.

Sold and distributed in the U.S.A. and Canada
by Kluwer Academic Publishers,
190 Old Derby Street, Hingham, MA 02043, U.S.A.

In all other countries, sold and distributed
by Kluwer Academic Publishers Group,
P.O. Box 322, 3300 AH Dordrecht, Holland.

Printed in The Netherlands

TABLE OF CONTENTS

1. Linsky
2. Roddier
3. Beckwith
4. Lucy
5. Golreich
6. Leger
7. Jewell
8. Scoville
9. Sargent
10. Cohen
11. Bowers
12. Knapp
13. Claussen
14. Sahai
15. Lane
16. Goldberg
17. Rieu
18. Sreenivasan
19. Iben
20. Wannier
21. Masson
22. Hagen
23. Carpenter
24. Matthews
25. Werner
26. Wood
27. Trimble
28. Deguchi
29. Sackmann
30. Tielens
31. Jura
32. Huggins
33. Heiligman
34. ?
35. Fix
36. ?
37. Augason
38. ?
39. Beckers
40. ?
41. ?
42. Sutton
43. ?
44. Morris
45. Carsenty
46. Drake
47. Lo
48. Dupree
49. Johansson
50. Olofsson
51. Alcock
52. Willems
53. Omont
54. Herman
55. de Jong
56. Kuiper
57. Aller
58. Leighton
59. ?
60. Olnon
61. Keady
62. Thronson
63. Betz
64. Bieging
65. Holzer
66. ?
67. Baud
68. Zuckerman

LIST OF PARTICIPANTS

Alcock, C.	MIT, Boston, MA, U.S.A.
Aller, L.H.	UCLA, Los Angeles, CA, U.S.A.
Augason, G.C.	NASA Ames Research Center, Moffett Field, CA, U.S.A.
Baud, B.	Space Research Groningen, Groningen, The Netherlands
Beckers, J.M.	NOAO, Tucson, AZ, U.S.A.
Beckwith, S.	Cornell University, Ithaca, NY, U.S.A.
Betz, A.L.	University of California, Berkeley, CA, U.S.A.
Bieging, J.H.	University of California, Berkeley, CA, U.S.A.
Boothroyd, A.	Caltech, Pasadena, CA, U.S.A.
Bowers, P.F.	Naval Research Lab., Washington D.C., U.S.A.
Carpenter, K.G.	University of Colorado, Boulder, CO, U.S.A.
Carsenty, U.	UCLA, Los Angeles, CA, U.S.A.
Claussen, M.J.	University of Massachusetts, Amherst, MA, U.S.A.
Cohen, M.	NASA Ames Research Center, Moffett Field, CA, U.S.A
Danchi, W.C.	University of California, Berkeley, CA, U.S.A.
de Jong, T.	Astronomy Institute, University of Amsterdam, Amsterdam, The Netherlands
Deguchi, S.	University of Illinois, Urbana, IL, U.S.A.

Drake, S.A.	University of Colorado, Boulder, CO, U.S.A.
Dupree, A.K.	Smithsonian Astrophysical Observatory, Cambridge, MA, U.S.A.
Ellis, H.B.	NASA Ames Research Center, Moffett Field, CA, U.S.A.
Engels, D.	Universitats-Sternwarte, Bonn, F.R.G.
Etzel, P.B.	UCLA, Los Angeles, CA, U.S.A.
Fix, J.D.	University of Iowa, Iowa City, IA, U.S.A.
Goldberg, L.	Kitt Peak National Observatory, Tucson, AZ, U.S.A.
Goldreich, P.	Caltech, Pasadena, CA, U.S.A.
Hagen, W.	University of New Mexico, Albuquerque, NM, U.S.A.
Hawkins, I.	UCLA, Los Angeles, CA, U.S.A.
Heiligman, G.	Caltech, Pasadena, CA, U.S.A.
Herman, J.	European Space Agency, Noordwijk, The Netherlands
Hinkle, K.H.	Kitt Peak National Observatory, Tucson, AZ, U.S.A.
Holzer, T.E	NCAR/HAO, Boulder, CO, U.S.A.
Huggins, P.J.	New York University, New York, N.Y., U.S.A.
Iben Jr., I.	University of Illinois, IL, Urbana, U.S.A.
Jewell, P.R.	NRAO, Tucson, AZ, U.S.A.
Johansson, L.E.B.	Onsala Space Observatory, Onsala, Sweden
Jura, M.	UCLA, Los Angeles, CA, U.S.A.
Kaiser, M.E.	UCLA, Los Angeles, CA, U.S.A.
Keady, J.	Los Alamos National Observatory, Los Alamos NM, U.S.A.
Keene, J.B.	Caltech, Pasadena, CA, U.S.A.

Knapp, G.R.	Princeton University, Princeton, NJ, U.S.A.
Kuiper, T.	JPL, Pasadena, CA, U.S.A.
Lane, A.P.	NRAO, Charlottesville, VA, U.S.A.
Leighton, R.B.	Caltech, Pasadena, CA, U.S.A.
Léger, A.	Université de Paris VII, Paris, France
Likkel, L.	UCLA, Los Angeles, CA, U.S.A.
Linsky, J.	University of Colorado, Boulder, CO, U.S.A.
Lo, K.Y.	Caltech, Pasadena, CA, U.S.A.
Lucy, L.B.	Columbia University, New York, NY, U.S.A.
Masson, C.R.	Caltech, Pasadena, CA, U.S.A.
Matthews, H.E.	Herzberg Institute of Astrophysics, Ottawa, Ontario, Canada
Morris, M.	UCLA, Los Angeles, CA, U.S.A.
Olnon, F.M.	Huygens Laboratorium, Leiden, The Netherlands
Olofsson, H.	Onsala Space Observatory, Onsala, Sweden
Omont, A.	Université Scientifique et Médicale de Grenoble, St. Martin d'Hères, France
Rieu, Nguyen-Quang	Observatoire de Paris, Meudon, France
Roddier, F.	Université de Nice, Nice, France
Sackman, J.	Caltech, Pasadena, CA, U.S.A.
Sahai, R.	Owens Valley Radio Observatory, Caltech, Pasadena, CA, U.S.A.
Sargent, A.	Caltech, Pasadena, CA, U.S.A.
Scoville, N.	Caltech, Pasadena, CA, U.S.A.
Sedwick, K.	UCLA, Los Angeles, CA, U.S.A.
Sreenivasan, S.R.	University of Calgary, Calgary, Alberta, Canada

Sutton, E.	University of California, Berkeley, CA, U.S.A.
Thronson, H.A.	University of Wyoming, Laramie, WY, U.S.A.
Tielens, A.G.G.M.	NASA Ames Research Center, Moffett Field, CA, U.S.A.
Toth, J.	UCLA, Los Angeles, CA, U.S.A.
Trimble, V.	University of California, Irvine, CA, U.S.A.
Wannier, P.	JPL, Pasadena, CA, U.S.A.
Werner, M.W.	NASA Ames Research Center, Moffett Field, CA, U.S.A.
Willems, F.J.	JPL, Pasadena, CA, U.S.A.
Wood, P.R.	Mount Stromlo and Siding Spring Observatories, Woden, Australia
Zuckerman, B.	UCLA, Los Angeles, CA, U.S.A.

PREFACE

Red giant and supergiant stars have long been favorites of professional and amateur astronomers. These enormous stars emit up to 10^6 times more energy than the Sun and, so, are easy to study. Some of them, specifically the pulsating long-period variables, significantly change their size, brightness, and color within about a year, a time scale of interest to a single human being.

Some aspects of the study of red giant stars are similar to the study of pre-main-sequence stars. For example, optical astronomy gives us a tantalizing glimpse of star forming regions but to really investigate young stars and protostars requires infrared and radio astronomy. The same is true of post-main-sequence stars that are losing mass. Optical astronomers can measure the atomic component of winds from red giant stars that are undergoing mass loss at modest rates ($\dot{M} \lesssim 10^{-6}$ $M_\odot$/yr.). But to see dust grains and molecules properly, especially in stars with truly large mass loss rates, $\gtrsim 10^{-5}$ $M_\odot$/yr, one requires IR and radio astronomy. As this stage of copious mass loss only lasts for ~10^5 years one might be tempted to ask, "who cares?".

But these few years near the tip of the asymptotic giant branch (AGB) have a number of profound implications for astronomy. It is now generally believed that most stars with masses between 1.4 and about 8 $M_\odot$ shed sufficient mass on the AGB to die peacefully as white dwarfs rather than violently as supernovae. Such stars constitute about half of all dying stars in our galaxy. White dwarfs have average masses of ~0.6 $M_\odot$ but their progenitors had main sequence masses that, on the average, were about one solar mass larger. If the white dwarf birth rate is ~1/year and if mass loss occurs primarily on the AGB, as appears very likely, then the total rate of return of stellar material from AGB stars is ~ 1 $M_\odot$/yr in the galaxy. Thus AGB stars have returned ~10^{10} $M_\odot$ to the interstellar medium during the course of galactic evolution. The average rate of return from supernovae is likely to have been an order of magnitude smaller.

Astronomy is currently in the midst of a "building boom" of ground and space based telescopes, both single mirrors and interferometers, that have or will cover essentially all of the electromagnetic spectrum from ultraviolet to radio wavelengths. Along with Dr. Michael Jura, we felt that, before these instruments deluge the astronomical community with a great rush of new data pertaining to red giant stars, it would be worthwhile to synthesize current understanding of the observational and theoretical aspects of mass loss in the upper right hand portion of the Hertzsprung-Russell diagram (encompassing M supergiants as well as AGB stars). It also seemed to be an appropriate time to confront investigators involved with diverse wavelength regimes since, too often, high and low frequency astronomers studying mass loss communicate exclusively with members of their own camp. Hence we organized a meeting that was held at the UCLA faculty center on June 20 and 21, 1984 to which we invited essentially all of the most active researchers in this field. We were delighted that so many of them could come so far for such a short meeting and we are gratified that, almost without exception, they were able to supply us with written versions of their oral presentations. We believe that the reviews presented in this volume represent the most comprehensive account of the outermost layers of the coolest and most luminous stars in our galaxy. In addition, the contributed papers give a representative flavor of the cutting edge of research in this field. The results described in many of these papers are not being published elsewhere.

The editors gratefully acknowledge the assistance of many members of the astronomy department at UCLA without which our meeting could not have taken place. The California Space Institute and the National Science Foundation kindly supplied financial aid, both for the conference and for these proceedings. We thank Ms. Françoise Quéval for her help in preparing the camera-ready manuscript. Most of all, we wish to thank Dr. Michael Jura for his help in planning and organizing the conference and for many stimulating discussions on mass loss from stars.

Ben Zuckerman
Mark Morris
Los Angeles, California
March 1985

ON THE SIGNIFICANCE OF MASS LOSS FROM BRIGHT RED GIANTS

Icko Iben, Jr.
Departments of Astronomy and of Physics
University of Illinois
Champaign - Urbana, Illinois
USA

ABSTRACT. Mass loss plays a dominant role in the lives of main-sequence stars of initial mass larger than 25 M_o and in the lives of asymptotic giant branch stars which are descendants of main-sequence stars less massive than 10 M_o. Mass loss is also of decisive importance in the lives of close binary stars which interact as a consequence of Roche-lobe overflow. In many instances, mass loss leads to exposure of layers in which matter has experienced significant nuclear processing. Because of wind mass loss, asymptotic giant branch stars become white dwarfs rather than supernovae. Because of Roche-lobe overflow, most intermediate-mass stars in close binaries become white dwarfs; perhaps 10 percent of all white dwarf pairs formed in this way are close enough that gravitational wave radiation will lead to merging of the white dwarfs; 10 percent of the merging dwarfs become type I supernovae.

1. Introduction

It is only in recent years that it has become clear that almost all stars lose a significant fraction of their initial main-sequence mass while still burning nuclear fuel quiescently. Stars more massive than about 25 M_o lose as much or more than half of their initial mass while still on the main sequence, burning hydrogen at their centers. As a consequence of this mass loss, interior layers in which extensive nuclear processing has occurred may be exposed.

If matter in exposed layers has experienced extensive hydrogen burning (initial oxygen and carbon have been converted almost completely into nitrogen), the mass losing star is known as an N-type Wolf-Rayet star. When matter in exposed layers has experienced even more extensive processing (hydrogen totally exhausted, all N converted into Ne and some He converted into C and O), the mass-losing star is called a C-type Wolf-Rayet star. All such stars go on to become supernovae of a variety that leaves a neutron star (or perhaps black hole) remnant.

M. Morris and B. Zuckerman (eds.), Mass Loss from Red Giants, 1–10.

Single stars less massive than about 10 $M_\odot$ do not lose much mass while on the main sequence, but develop strong winds after having exhausted first hydrogen and then helium at their centers. They are then known as asymptotic giant branch (AGB) stars. They consist of a compact, electron-degenerate core composed primarily of carbon and oxygen and an extended hydrogen-rich envelope. Between the envelope and the core is a helium-rich region which grows in mass as a consequence of hydrogen burning at the base of the envelope. When the mass of the helium-rich zone exceeds a critical value, a helium-burning thermonuclear runaway (or helium shell flash) occurs. Following the runaway, helium continues to burn quiescently until the amount of helium fuel is reduced by the amount produced through hydrogen burning between flashes.

The rate at which an AGB star loses mass increases with its luminosity. In turn, the average luminosity increases with the mass of the inert carbon and oxygen core. As a star evolves upward in the Hertzsprung-Russell diagram along the AGB, the mass-loss rate intensifies to such an extent that, at some point, almost all of the matter remaining in the hydrogen-rich envelope is stripped from the star. The remnant contracts rapidly at nearly constant luminosity while the ejected nebular material drifts outward. When its surface temperature becomes large enough, the remnant emits photons of sufficiently large energy to cause the matter in the nebula to fluoresce. The nebula is then called a planetary nebula (PN) and the central star is called a planetary nebula nucleus (PNN). Eventually the nebula disperses and the remnant becomes a white dwarf. Were it not for mass loss during the AGB phase, every star of initial mass in the range (1.4-10 $M_\odot$) would become a supernova when the mass of its CO cores reached the Chandresekhar limit of about 1.4 $M_\odot$. As it is, mass loss causes the ejection of almost all of the hydrogen-rich envelope before a central explosion occurs. Thus, all single stars of initial mass less than 10 $M_\odot$ become white dwarfs composed of carbon and oxygen.

A much wider variety of white dwarf compositions is produced by stars in close binaries as a consequence of mass loss which is made possible by the gravitational field configuration around the binary pair. Passing an imaginary plane through the line joining the centers of the stars in a binary, one may imagine a zero gravitational potential energy curve which forms a lopsided figure eight in this plane. Spinning this figure eight through 360 degrees about the line joining the stellar centers generates a zero-potential surface. Each portion of the double lobed structure so produced is called a Roche lobe. If and when, in consequence of expansion brought about by composition changes in its interior, either star fills its lobe, matter from that star spills over onto its companion. Depending on the circumstances, the companion may either accrete the transferred matter or reject it. In the latter case, both Roche lobes become filled and matter and angular momentum leave the system. In consequence of frictional drag forces between the stars and the outwardly expanding common envelope about them, the remnant stars are drawn closer together. The final result is, in most cases, a pair of white dwarfs in a binary of period much shorter than the period of the parent binary system. In contrast to the situation for single stars, one or both remnant degenerate dwarfs may be composed primarily of helium, or of oxygen and neon, as well as of carbon and oxygen.

2. Mass Loss from AGB Stars

After exhausing first hydrogen and then helium at their centers, single stars initially less massive than about 10 M_o (the precise value depends on the initial composition and on the input physics) develop a very compact core composed primarily of carbon and oxygen and a very extended envelope in which the element abundances are close to the original main-sequence abundances --- that is, primarily hydrogen, helium, and CNO elements in roughly solar proportions. The dimensions of the core are similar to those of a white dwarf of the same mass and, therefore, the electrons in the core are Fermi-degenerate. Conduction by electrons at the top of the Fermi Sea maintains the core at nearly uniform temperature (only factor of two variations from the mean of one to two hundred million degrees K). The dimensions of the envelope are typically on the order of the distance between the Earth and the Sun.

For ninety percent of the life of an AGB star, hydrogen burning at the base of the hydrogen-rich envelope provides most of the energy leaving the stellar surface. Neutrino losses from the core keep the core too cool for carbon to ignite. When the mass of the helium layer laid down by the hydrogen-burning shell exceeds a critical mass (e.g., $\sim 10^{-2}$ M_o for M_{core} $\sim$0.6 M_o and $\sim 10^{-3}$ M_o for M_{core} $\sim$1.0 M_o), helium-burning reactions are ignited. Because of the high temperature sensitivity of these reactions, nuclear energy is released more rapidly than it can be carried outward from the energy-producing region. Energy fluxes become so intense that the entire region between the base of the helium-burning region and the base of the hydrogen-burning shell becomes convective. The nuclear energy released is converted rapidly into particle kinetic energy and the consequent overpressure causes matter in and on either side of the convective shell to expand. Hydrogen-rich material is pushed outward and cools to such an extent that hydrogen ceases to burn. We say that the AGB star is experiencing a convective helium shell flash (or thermal pulse).

Expansion and cooling of matter in the helium-burning region eventually quenches the flash and the star embarks on a quiescent helium-burning phase which lasts for about 10 percent of the total time between successive flashes. Recontraction and heating of outer layers eventually reignites hydrogen whereas continued cooling in the helium-rich region leads to the effective cessation of helium burning. The whole cycle then repeats.

During the peak of a shell flash, approximately 20 percent of the helium in the convective shell is converted into ^{12}C and into ^{16}O. The precise ratio of ^{16}O to ^{12}C remaining after burning ceases depends upon the as-yet uncertain cross section for the $^{12}C + \alpha \rightarrow {}^{16}O + \gamma$ reaction. In all cases, the ^{14}N engulfed by the convective shell is converted almost immediately into ^{22}Ne. In AGB stars of small core mass, some ^{13}C is incorporated into the convective shell during a shell flash. Neutron capture reactions convert ^{56}Fe and its progeny into the so called s-process isotopes. The source of neutrons in AGB stars of core mass less than $M_{core} \sim 0.7\ M_{\odot}$ is probably the $^{13}C(\alpha,n)^{16}O$ reaction and the source in AGB stars of core mass larger than $M_{core} \sim 0.9\ M_{\odot}$ is the $^{22}Ne(\alpha,n)^{25}Mg$ reaction. Theoretical analysis to date suggests that only in AGB stars in which the ^{22}Ne source of neutrons is important will s-process isotopes be produced in the solar-system distribution.

After the peak of a flash, and before the quiescent helium-burning phase, some matter that was earlier in the helium-burning convective shell is "dredged" to the surface by convection that extends all the way from the outer layers of the freshly processed region to the surface. Thus, both carbon and s-process isotopes are brought to the surface. Some stars may in this way become carbon stars (defined as those exhibiting a surface carbon-to-oxygen ratio greater than unity).

As the mass of the CO core grows, temperatures and densities in the core continue to increase. If the mass of the AGB star is initially as large or larger than about 1.4 $M_{\odot}$ and if the core were to grow as large as 1.4 $M_{\odot}$, carbon would be ignited and the star would experience a star-disrupting supernova explosion. If all stars of initial main-sequence mass in the range 1.4 $M_{\odot} \rightarrow 10\ M_{\odot}$ were to behave in this fashion, then the supernova rate in a galaxy like our own would be about 20 times larger than is observed. The standard interpretation of this arithmetic is that AGB stars lose mass on average much more rapidly than they process nuclear fuel. And, indeed, direct observations of winds from many "normal" cool supergiants of appropriate luminosity have led to the parameterized "Reimers" mass-loss rate which, within the uncertainties, predicts that mass is lost from the surface at a rate which is between 3 and 10 times the rate at which the core grows by nuclear processing. But, even losing mass at this rate, AGB stars initially more massive than about 4-5 $M_{\odot}$ would go on to become supernovae. That they do not is brought home by the paucity of AGB stars brighter than about $M_{Bol} = -6$ both in Magellanic Cloud globular clusters of intermediate age and in open clusters of intermediate age in our Galaxy.

Particularly clear cut is the story told by clusters which contain Cepheids. The ages of stars in a typical Cepheid-containing cluster are of the order of 10^8 years and the masses of the main-sequence progenitors of the Cepheids are of the order of 4-6 M_o. The lifetime of the Cepheid phase is of the order of 10^6 years and this, too, would be the lifetime of an AGB star in the cluster if its core could grow from, say, 0.8 M_o to 1.4 M_o. There should therefore be as many AGB stars of magnitude between M_{BOL} ~-6 and M_{BOL} ~-7.3 (luminosity increases with core mass and this magnitude range corresponds to the chosen core mass range) as there are Cepheids. In fact, both in the Cloud clusters containing Cepheids and in the 12 Cepheid-containing Galactic clusters which are used to normalize the cosmic distance scale, there are essentially no AGB stars in this magnitude range.

The full story is, of course, more complex than this simple argument suggests. Nevertheless, the evidence is very compelling that, when AGB stars become brighter than about M_{BOL} ~-6 (corresponding to M_{core} ~0.8 M_o-0.85 M_o), they lose mass much faster than the Reimers rate implies. The inferred phase of rapid mass loss is variously referred to as a "superwind" phase or a "planetary nebula ejection" phase. The fact that planetary nebulae appear to be formed at a rate which is comparable to the birthrate of stars of initial mass between 1 M_o and 10 M_o is a strong argument supporting this interpretation. There are also good arguments to suggest that OH/IR sources contain stars which are in the final portion of the superwind phase; some of these sources are known to support wind mass loss rates as large as 10^{-4} M_o yr^{-1}, which is several orders of magnitude larger than the rate of core growth due to nuclear processing.

Thus, the evidence is almost overwhelming that, because of rapid mass loss in the AGB configuration, single stars initially less massive than about 10 M_o become white dwarfs. The mass of the most massive white dwarf produced by a single star is expected to be about 1.1 M_o, a limit which is consistent with the scanty information we have about the number-mass distribution of white dwarfs in the solar neighborhood (e.g., M (Sirius B) ~ 1.05 M_o).

Mass loss has other far reaching consequences. The carbon that is produced in the convective helium-burning shells and dredged to the surfaces of low mass AGB stars is quite possibly the origin of most of the visible carbon in the Universe. The s-process isotopes produced in the convective helium-burning shell of a massive AGB star may possibly account for the distribution of s-process isotopes in solar-system material. It remains to be determined whether an AGB star of large core mass can live long enough, despite the occurrence of a superwind, to permit a quantitatively satisfactory explanation.

3. The Effects of Mass Loss in Close Binaries of Intermediate Mass

Stars in binary systems in which the primary fills its Roche lobe and attempts to transfer mass to its companion may bypass many of the phases experienced by single stars of the same initial mass. This is especially true of a component that fills its Roche lobe before leaving the main sequence or just after it leaves the main sequence but before it ignites helium at its center. In the latter case, after losing most of its hydrogen-rich envelope through its Roche lobe, the remnant star burns helium as a compact object, dimmer and hotter than a main-sequence star, rather than brighter and cooler, as in the case of a single star. Thus, it does not become a Cepheid or an RR Lyrae star during core helium burning. If it is massive enough (main-sequence mass between about 4.8 M_o and 10 M_o), the remnant (of mass between about 0.75 M_o and 2.2 M_o) expands again during shell helium burning and, in a second stage of Roche-lobe filling, loses the remainder of the hydrogen-containing material at its surface and even some of the helium-rich matter above the burning shell. Thus, such a star bypasses the AGB double shell-burning phase and goes on directly to become a CO degenerate dwarf (of mass between about 0.75 M_o and 1.1 M_o).

Stars of initial mass between about 2.3 M_o and 4.8 M_o that fill their Roche-lobes during the transit between core hydrogen burning and core helium burning also lose all but a thin skin of hydrogen and burn helium as hot, compact objects of mass between about 0.32 M_o and 0.7 M_o. However, they do not expand again during shell helium burning and cease nuclear burning while still containing significant quantities of helium near the surface, (e.g., M_{He} = ~0.04 M_o in the 0.5 M_o remnant of a component of initial mass M_{MS} ~4 M_o and M_{He} ~0.1 M_o in the 0.4 M_o remnant of a component of initial mass M_{MS} ~3 M_o). Stars that are initially less massive than about 2.3 M_o and that fill their Roche lobes during the transit between the two major core nuclear-burning stages do not ignite helium. They become degenerate dwarfs which are composed primarily of helium and are of mass less than 0.5 M_o. After filling their Roche lobes twice and losing mass each time, components of initial mass in a very narrow interval of about 0.3 M_o centered on about 10 M_o become degenerate dwarfs composed predominantly of oxygen and neon. In each case, whether it is composed of helium, of carbon and oxygen, or of oxygen and neon, the mass of the final degenerate dwarf remnant is smaller than the mass of the remnant produced by a single star of the same initial mass (a CO dwarf if M_{MS} <10 M_o; a neutron star if M_{MS} ~10 ± 0.15 M_o).

If a component fills its Roche lobe for the first time only after it has exhausted helium at its center, it loses essentially all of its surface hydrogen and burns up most of its helium fuel, becoming a CO dwarf of mass larger than would be the case had it filled its Roche lobe during the transit between the two major nuclear burning phases. The reason for this is that hydrogen continues to burn and the mass of the hydrogen-exhausted central region continues to grow at the same time that helium burns at the center. Thus, when helium is exhausted at the center, the mass of the hydrogen-exhausted (or helium) core is larger than is the case had Roche-lobe filling occurred prior to the onset of helium burning. When the core expands during shell helium burning and Roche-lobe filling finally occurs, most of the hydrogen-rich surface layers are lost in all cases. If the hydrogen-exhausted core is larger than about 0.85 M_o when Roche-lobe filling occurs, some helium-rich matter is also lost. In every case, most of the helium remaining in the star after the Roche-lobe filling experience is converted into carbon and oxygen. The final remnant is a degenerate CO dwarf with either a thin helium-rich layer at the surface (initial main-sequence mass larger than about 5 M_o and remnant mass between about 0.85 M_o and 1.1 M_o) or a CO dwarf with a thin hydrogen-rich layer at the surface (main-sequence mass between 1 M_o and 5 M_o and remnant mass between about 0.5 M_o and 0.85 M_o).

In summary, if the first Roche-lobe filling episode occurs during the shell helium-burning stage, the final remnant is always a CO degenerate dwarf of nearly the same mass (slightly larger mass if initial mass is between 5 M_o and 10 M_o) as that of the CO dwarf produced by a single star of the same initial mass. In contrast, the degenerate dwarf end product of Roche-lobe overflow which first occurs after core hydrogen burning but before core helium burning is smaller than the degenerate dwarf produced by a single star of the same initial mass and is often of a different composition: nearly pure helium (but also ^{14}N since all initial CNO elements are converted almost completely into ^{14}N during complete hydrogen burning) if initial main-sequence mass is less than 2.3 M_o; a CO core with a sizeable helium- and ^{14}N-rich envelope and a thin hydrogen-rich surface layer if initial mass is between 2.3 M_o and about 5 M_o; a CO core and a thin helium-and ^{14}N-rich surface layer if initial mass is between about 5 M_o and 10 M_o; an ONe core plus a thin helium- and ^{14}N-rich layer if initial mass is about 10±0.15 M_o.

Mass loss by Roche-lobe overflow has a pronounced impact on the orbital characteristics of the binary as well as on the evolution of the mass-losing star. If the mass-losing component is a giant or supergiant with a deep convective envelope, attempted mass transfer proceeds at a rate faster than the companion can accept the proffered matter. The companion rejects most of the proffered matter into a "common envelope" that overflows both Roche lobes. Pressure waves are generated by friction between the two stellar cores and the matter in the common envelope, forcing matter away from the system at essentially the same rate as the primary (or "driver") attempts to transfer mass onto its companion. The same frictional forces lead to a loss of angular momentum from the system and the binary components drift closer together. This situation persists until the mass of the primary decreases to below about 60 percent of the mass of the companion. Thereafter, mass transfer occurs at a much slower rate and, if it is a main-sequence star, the companion accretes most of the proferred matter. In first approximation, system angular momentum is conserved and the components again drift apart. However, if the companion is a compact helium-burning remnant produced in an earlier mass transfer event or is a degenerate dwarf, mass continues to be rejected by this companion and the components continue to approach one another.

The observational evidence for common envelope action which leads to orbital shrinkage is manifold. For example, one may cite the existence of planetary nebulae containing close binary pairs, the degenerate component of which must at one time have occupied a volume much larger than the volume defined by the current orbital separation. Another example is the existence of main-sequence and subgiant Barium and CH stars in binary orbits with degenerate dwarf companions. The evolutionary stage of the optically luminous component is such that this component cannot be expected to have, of itself, produced its peculiar composition characteristics. It is highly likely that the degenerate dwarf companion was once a thermally pulsing AGB star which manufactured and dredged to its surface carbon and s-process elements. The size of such an AGB precursor is larger than the size of the Roche lobe about the degenerate dwarf in the current system. Thus, the mass, composition, and dynamical characteristics of many current binary systems containing at least one degenerate component are most easily explained by the supposition of a prior phase or phases of attempted mass transfer that lead to a common envelope, to mass and angular momentum loss from the system, and to a pronounced shrinkage of orbital dimensions.

During the common envelope stage it is probable that the system will appear as a single, very cool giant star which is losing mass at an extremely rapid rate. Thus, the system may well masquerade as a single star in the planetary ejection or superwind stage. It would be worthwhile to monitor systems exhibiting very large mass-loss rates for possible periodic variations (on time scales of hours to a few years), although such systems might be so inherently "noisy" that it is difficult to pick the signal out of the noise. Further, the total duration of the mass-loss event is expected to be so short (typically only hundreds to thousands of years) that the probability of catching a system in the act is very small.

One final possible outcome of mass transfer/mass loss in close binary systems is worth mentioning. If the initial primary in a typical binary system experiences Roche-lobe overflow, its companion will in general also experience Roche-lobe overflow. If the components are initially of comparable mass and if these masses are in the (5-10) M_{o} range, then four Roche-lobe filling episodes will occur as the system evolves into a pair of CO degenerate dwarfs. If the orbital separation of a newly formed pair of degenerate dwarfs is less than about 3 R_{o}, gravitational waves will carry away system angular momentum fast enough, and cause the pair to drift close enough in less than the Hubble time (10^{10}years), that the lighter of the pair will overflow its Roche lobe for one final time. The net result is a merging of the two erstwhile degenerate dwarfs into one. If the sum of the masses of the dwarfs prior to merging exceeds the Chandresekhar limit, then the result of the merging process is likely to be a very explosive event. This scenario is currently thought to be a very likely one for accounting for type I supernova explosions. It would be fascinating if observers could catch a precursor in the act of merging. The energy released and the rate of release are expected to be quite large: 10^{50} erg in 100s!

4. A Guide to Evolution in the Hertzsprung Russell Diagram

In Figure 1 are evolutionary tracks of representative nuclear burning models of single stars that experience mass loss at a large rate during the main-sequence phase ($M_{MS} \gtrsim 25\ M_o$) or during the AGB phase ($1\ M_o \lesssim M_{MS} \lesssim 10\ M_o$). In Figure 2 is the track of a close binary component of intermediate mass which first fills its Roche lobe just prior to helium ignition in the core.

I apologize for a lack of literature references, but circumstances have not permitted me both to meet the submission deadline for this paper and to do justice to the many scientists who have helped construct the picture painted in these pages. The interested reader may find several appropriate references in the March, 1985 issue of the Quarterly Journal of the Royal Astronomical Society.

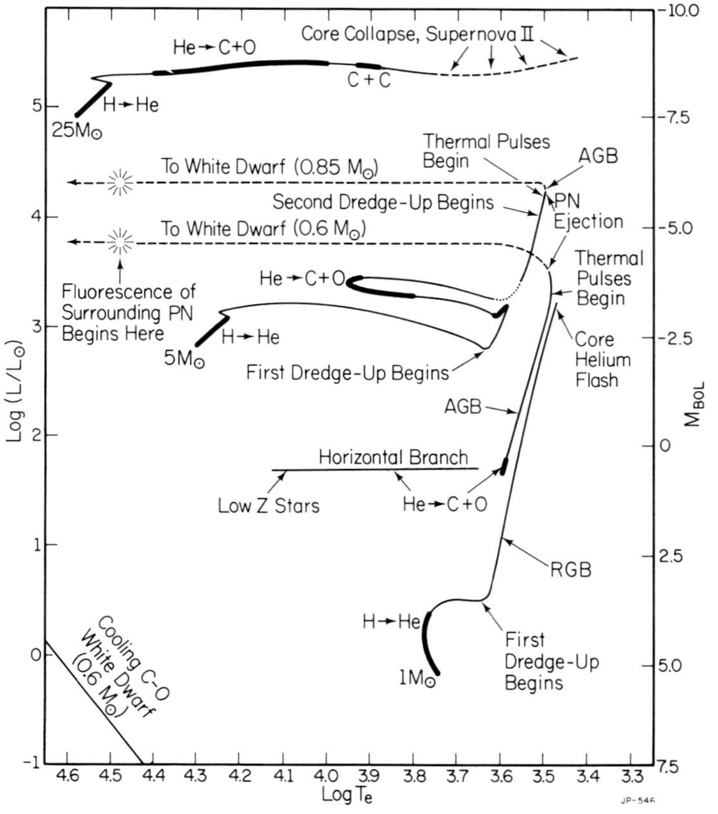

Figure 1

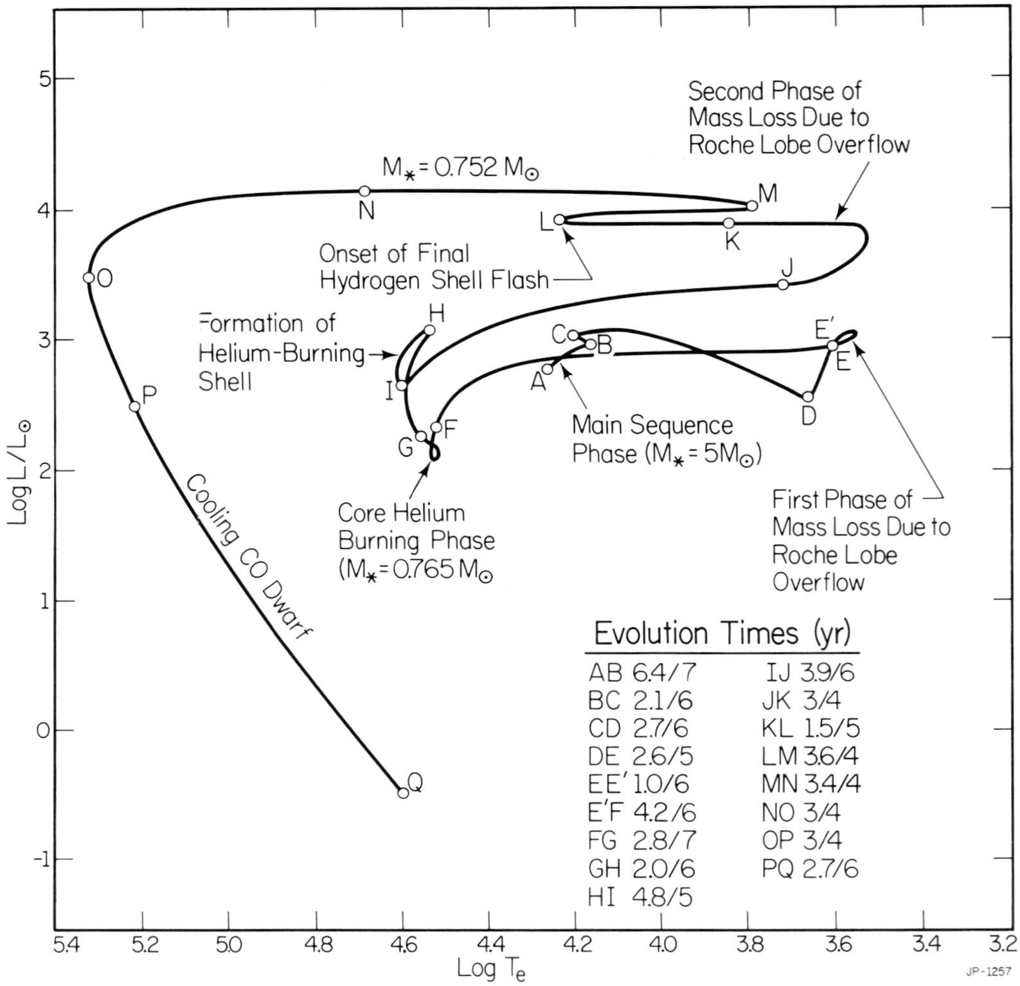
Second Phase of
Mass Loss Due to
Roche Lobe Overflow
$M_* = 0.752\ M_\odot$
Onset of Final
Hydrogen Shell Flash
Formation of
Helium-Burning
Shell
Main Sequence
Phase ($M_* = 5M_\odot$)
Core Helium
Burning Phase
($M_* = 0.765\ M_\odot$
First Phase of
Mass Loss Due to
Roche Lobe
Overflow
Cooling CO Dwarf
Evolution Times (yr)
AB 6.4/7
BC 2.1/6
CD 2.7/6
DE 2.6/5
EE′ 1.0/6
E′F 4.2/6
FG 2.8/7
GH 2.0/6
HI 4.8/5
IJ 3.9/6
JK 3/4
KL 1.5/5
LM 3.6/4
MN 3.4/4
NO 3/4
OP 3/4
PQ 2.7/6
Log L/L⊙
Log T_e
JP-1257

Figure 2

MASSES OF WHITE DWARFS AND OTHER REMNANTS: THE ULTIMATE CONSTRAINT ON $\dot{M}$

Virginia Trimble
Astronomy Program, Univ. of Maryland
College Park, MD 20742, USA
and
Dept. of Physics, Univ. of California, Irvine, CA 92717, USA

ABSTRACT. Direct measurements of stellar masses during and after post-main-sequence evolution allow us to probe the amount and effect of losses. The main conclusions are that non-explosive loss of about half the initial main sequence mass is fairly universal, that close binaries normally lose more than single stars (though not always much more), and that massive stars lose most during helium burning, while low mass stars (at least in Pop II) lose most during hydrogen shell burning. Only this last is at all surprising. The large amounts of mass loss generally claimed for massive stars (yielding Wolf-Rayets etc.) are somewhat difficult to reconcile with the requirements for producing Type II supernovae.

1. MASSIVE STARS

1.1. Cepheids

Cepheid variables represent a subset of (initially) 4-20 $M_\odot$ stars during the phase of core helium burning. Mass loss $\gtrsim$ 20% during shell hydrogen burning will prevent the loop back to the blue in the HR diagram needed to bring these stars into the instability strip. Until recently, this seemed to present a serious problem, because the masses of the variables determined from their pulsation properties (Cox 1980; Iben 1983; Burki 1984) were 30-40% lower than those of the main sequence stars whose evolutionary tracks passed through the Cepheid colors and luminosities.

Changes in the distance scale, color-temperature calibration, estimated surface composition, and opacity calculations have largely removed this discrepancy in the last few years (Cox 1980; Iben 1983; Simon and Davis 1983; Carson and Stothers 1984), at least for stars of 4-10 $M_\odot$. More massive ones may have lost about 30% of their initial mass before becoming Cepheids.

The lower-mass Cepheids are the stars that ought to evolve into electron-capture, Type II supernovae, like the 1054 event (Davidson et al. 1982). We can tentatively conclude, from comparing Cepheids to the Crab Nebula, that they lose almost nothing before core He burning, but 50% or more during the double shell phase, perhaps appearing as type II

M. Morris and B. Zuckerman (eds.), Mass Loss from Red Giants, 11–19.

OH/IR stars. Willson and Bowen (1984) suggest that a large part of the loss might occur during the Cepheid phase itself, though Hendon et al. (1984) note that TU Cas, at least, currently has little or no wind.

Cepheids in the Magellanic Clouds have evolutionary and pulsational masses in essential agreement over the range 9-20 $M_\odot$. Undoubtedly selection effects have discriminated against our seeing low-mass objects, but van Genderen (1983) suggests that the lower metal abundance may lead to less mass loss than in the Milky Way Cepheids.

1.2. Progenitors of Type II Supernovae

Type II supernovae are defined by their spectra showing hydrogen lines (Minkowski 1941). A few tenths of a solar mass of H, with roughly solar abundances of other elements, suffice to produce the observed spectra (Branch 1984). But modelling a Type II light curve requires at least a few solar masses of H (e.g. a 5 $M_\odot$ envelope with 40% helium; Weaver 1984) extended over a region 1-3 X 10^{13} cm in radius. The extended H envelope is required, first, to let energy leak out as light rather than going into kinetic energy of layers that have to be dragged out of a deep potential well, and, second, to produce a recombination wave responsible for the plateau phase of the light curve.

This means that the average SN II progenitor (a 10-15 $M_\odot$ star, cf. Trimble 1975, 1982) can lose, at most, about 15% of its initial mass at all stages prior to the supernova event. Litvanova and Nadyozhin (1983) have suggested that the rarer sort of Type II without a plateau phase may come from stars that have lost most of their hydrogen. Flamsteed's star of 1685, if it gave rise to Cas A, may have been of this type (Chevalier 1976).

The Crab Nebula apparently reflects an intermediate case. Assuming the ionized material we now see represents the outer layers of the star just before explosion, then the envelope contained about 1 $M_\odot$ each of H and He, with a solar admixture of heavies (some of these now in dust), and we are not surprised that the event may have been rather fainter than the average SN II (Davidson et al. 1982; Glaccum 1982; Trimble 1983; Marsden et al. 1984).

The need to retain a hydrogen rich envelope to make SN II's from massive stars is at variance with current popular views (Conti et al. 1983 and many others) in which O stars evolve to Of's and then to WR's by losing essentially all their H. I do not know how to resolve this discrepancy, but suspect it may be a serious one. Maeder (1981) has advocated WR's as the immediate predecessors of SN II's but without addressing the light curve problem.

1.3. Massive Binaries

Masses measured for Cepheids in binary systems show that, like single Cepheids, they have not yet undergone much mass loss (Cox 1980; Iben 1983), though they presumably will later.

Production of short period (close) binary X-ray sources and binary pulsars requires very extensive loss of both mass and angular momentum, via expulsion of a common envelope, prior to neutron star formation (van den Heuvel 1981; van den Heuvel and Taam 1984). Both kinds of systems are rare enough that producing them this way will not significantly reduce the supply of progenitors for Type II supernovae.

The presence of a close companion raises the initial main sequence mass needed to develop an iron core to something like 12-15 $M_\odot$ (Webbink 1979). Primaries almost this massive give rise to degenerate dwarfs very close to the Chandrasekhar limit. They can be driven above 1.4 $M_\odot$ relatively easily by eventual mass transfer back from the secondary. In some cases, the dwarf may undergo electron-capture collapse and so give rise to a close binary pulsar in due course (van den Heuval and Taam 1984). In other cases it may detonate, yielding a Type I supernova. Sramek et al. (1984) and Chevalier (1984) suggest that SN 1983n, the radio-emitting event in M83, arose from such a system. In either case, both components of a close binary typically lose more mass earlier in their evolution that do comparable single stars, further hindering the production of SN II's.

2. STARS OF LOW AND INTERMEDIATE MASS

2.1. RR Lyrae Variables

The low-mass, population II pulsational variables called RR Lyrae's present a significant contrast to the population I Cepheids. We know from the age of our galaxy and from globular cluster HR diagrams that their initial masses must have been at least 0.85 $M_\odot$ (and probably not much larger). But their current luminosities, temperatures, and pulsational properties are best matched by masses near 0.5 for field halo stars, 0.55 for globular clusters of Oosterhoff Type II and 0.65 for Oosterhoff Type I's (Cox et al. 1983; van Albada and Baker 1973; Stellingwerf 1984; Rood 1973; the types are distinguished by the pulsation periods of the RR Lyraes and result from differences in metal abundance). Thus at least 25-35% of the MS mass must have been lost prior to or during core helium burning. Willson and Bowen (1984) suggest that also in this case most of the loss occurs during the phase of pulsational instability.

Field stars that become RR Lyraes, though still retaining hydrogen atmospheres deep enough to include the ionization zone, have already shed so much material that they are unlikely to expel significant planetary nebulae or be able to ionize them if they did (Schoenberner 1983; Sabbadin et al. 1983). Some may never complete helium burning and will leave mostly-helium white dwarfs. Even globular cluster RR Lyrae's must not lose much more if they are to keep the 0.55 $M_\odot$ needed to ionize a planetary.

2.2. The Planetary Nebula in M15

We can check the last point because one (and apparently only one) PN has been found in the Oosterhoff I cluster M15. The progenitor star, then, was about 0.85 $M_\odot$ and lost about 0.2 $M_\odot$ as a red giant. The planetary contains 0.01-0.04 $M_\odot$ of ionized gas (Adams et al. 1984; Turner and Terzian 1984). The luminosity of the central star implies $M = 0.53 \pm 0.02\ M_\odot$; and we conclude that only about $0.1 M_\odot$ can have been lost during the asymptotic giant phase. Low mass Pop II stars apparently lose more during the first ascent of the giant branch than during the second.

2.3. Other Planetary Nebulae

Other planetaries are less informative. We know their progenitor masses only on average from statistical and dynamical arguments. The progenitor range is at least 1-6 M_o and the median 1.2-1.5. The average ionized mass is only about 0.05 M_o (to within a factor of two, depending on your choice of distance scale; Gathier et al. 1983; Phillips and Pottasch 1984), and so not a significant fraction of the total loss over the stellar lifetime. But 8 of 55 objects in the Gathier et al. sample have ionized masses in excess of 1 M_o. And an increasing number show direct or indirect evidence for 0.1-1.0 M_o of neutral matter (Pottasch 1983), up to as much as 5 M_o for NGC 7027 (Gee et al. 1984; Ellis and Werner 1985).

One is led tentatively to the conclusion that the main epoch of mass loss occurs very late in these star's lives, through a superwind persisting over 10^{3-4} yr or episodic mass loss triggered by shell flashes or both. If this is so, then 1-6 M_o disc population stars behave more like massive population I ones than like globular cluster objects.

Irrelevant but interesting are the facts that the bright, famous planetaries tend to have unusually large ionized masses as well as being close to us, and that we would expect, on statistical grounds, to see slightly less than one PN in an open cluster. None is known.

2.4. OB Subdwarfs

OB subdwarfs are simply hot stars too faint for the main sequence but too bright for white dwarfs. They are, thus, a heterogeneous group, including some nuclei of planetary nebulae and cataclysmic variables discussed later. I have always found the single, isolated ones, and their globular cluster analcgs (the "extended horizontal branch stars") something of an evolutionary puzzle. As a result of recently-computed evolutionary tracks for asymptotic giant stars with rapid mass loss, it has become possible (and moderately fashionable) to interpret sd OB's as ordinary pre-white-dwarfs that have been stripped down to such low masses ($\lesssim 0.5$ M_o) that they either do not eject PNe (Bergeron et al. 1984) or are unable to ionize them (Schoenberner and Drilling 1984).

Kilkenny and Lynas-Gray (1984) provide some confirmation for this view via a pulsation analysis of the helium subdwarf V652 Her, which yields a core mass of 0.475 M_o. These stars apparently lose comparable amounts of mass during shell hydrogen and shell helium burning. Stars stripped this far will not have the extended hydrogen envelope needed to ascend to the tip of the AGB. Thus the scenario could be checked by looking for a correlation between the presence of extended horizontal branches and deficiencies of bright AGB stars in globular clusters with good HR diagrams. Correlation with composition or any other property that might be causal would also be interesting.

2.5. Nuclei of Planetary Nebulae

A handful of planetary nuclei are spectroscopic and/or eclipsing binaries, none, however, sufficiently clean and well-studied to give a useful mass for the hot star (cf. tables in Patterson 1984; Ritter 1984). Our ideas on the subject therefore come largely from fitting evolutionary tracks through positions on the HR diagram, with or without an additional constraint from the expansion times of the nebulae and fading times of the models. The choice of distance scale enters in a rather complex way, because most PNe nuclei are near the knee of the evolutionary tracks where more massive objects switch from being brighter but shorter lived (via effects of nuclear reaction rates) to being fainter and longer lived (via the mass-radius relation for degenerate objects).

Despite all this, there is considerable recent agreement that virtually all PNe nuclei are between 0.5 and 0.75 $M_{\odot}$, with quite a sharp peak at 0.6 or 0.65 (Cahn 1984; Schoenberner 1983; Pottasch 1983; Schoenberner and Weidemann 1983).

Cahn's (1984) figures show a handful of stars in the 1.0-1.3 $M_{\odot}$ range, all with error bars permitting much smaller values. Tylenda (1983) has suggested NGC 7027 and 2440 and Lutz (1984) HDE 330036 as likely to have come from massive stars and to have nuclei in excess of 1.0 $M_{\odot}$. Kaler (1983) has entered a caveat that the popular masses are smaller than the core masses needed to produce N/O ratios as large as those seen in some nebulae. Serrano (1983) implicitly makes the same point. The current majority view is that a 2 $M_{\odot}$ MS star gives rise to a 1 $M_{\odot}$ AGB star and a 0.6 $M_{\odot}$ nucleus and white dwarf. My prejudices are on the side of rather larger remnant masses (Trimble and Sackmann 1977).

2.6. The DB Accretion Problem

White dwarfs that show metals but no hydrogen in their spectra are a problem. The metals they were born with will sink out of sight in less than the cooling time (Muchmore 1984). Carbon can be dredged up from the interior (Fontaine et al. 1984; Wegner and Yackovich 1983), but other metals seemingly must be accreted from the interstellar medium and should carry hydrogen with them (Shipman and Greenstein 1983).

At one time, it appeared that adopting as mass as low as 0.3 $M_{\odot}$ for DB stars would resolve the problem by slowing accretion and mixing the H into a deep convective envelope. Then one might conclude that helium atmosphere white dwarfs with metal lines came from low mass progenitors that were unusually efficient mass shedders. But the metals would get mixed down, too, and the approach has essentially been abandoned (Alcock 1984). For what it is worth, vMa 2, with iron lines but no H in its spectrum, has $M = 0.7\ M_{\odot}$ by conventional techniques (Gatewood and Russell 1974). A possible alternative is that accreted H diffuses downward and burns as fast as it arrives (Michaud et al. 1984). Metals don't burn.

2.7. Cataclysmic Variables and Other Close Binaries

Qualitatively, these systems of white dwarf plus normal companion are straightforward. They arise from relatively wide systems that lose mass and angular momentum during a common envelope phase, producing systems like V 471 Tau (Paczyński 1976) that later spiral together and come into contact as further angular momentum is lost in gravitational radiation and magnetic winds. Quantitatively, there are surprises both in the current WD masses and in what seem to be their final states.

White dwarfs in CV's are more massive on average than single ones, and this is not just an observational selection effect (Livio and Stoker 1984). According to the tables given by Patterson (1984), Ritter (1984), and Sion et al. (1984, for V471 Tau stars), the mean WD mass is 0.75 $M_{\odot}$. The dispersion in also larger, a quarter of the systems having M_{WD} in excess of 1.0 $M_{\odot}$. This is a surprise because the presence of a close companion considerably increases the initial mass needed to produce a particular WD (Webbink 1979). Comparing the present systems to likely progenitors indicates that systems with initial total masses around 2 $M_{\odot}$ have so far lost about 60% of what they started with, while systems beginning with 6-12 $M_{\odot}$ have lost 80-90%. Much of the loss occurred during back transfer.

What about the final state? There are two possibilities: either the WD gradually gains accreted matter, in which case it is likely to become a Type I supernova (Iben and Tutukov 1984), or it is gradually eroded by CO getting mixed into accreted H and being expelled with it in nova explosions, so that the system evolves into a very close pair of very low mass degenerate dwarfs like AM CVn, GP Com, and PG 1346+082 (Nather 1985; Solheim et al. 1984). In either case, essentially 100% of the initial system mass is returned to the ISM in diffuse form, including 0.5-1.0 $M_{\odot}$ of heavy elements.

2.8. Isolated White Dwarfs (Single and Wide Binaries)

The average white dwarf mass and its dispersion affects determinations of the total mass density in the galactic plane and the amount of recycled gas that must be included in calculations of galactic evolution as well as constraining mass loss at earlier stages.

2.8.1. Wide Binaries. These systems are exceedingly informative because we can determine both the current WD mass and at least a limit on its progenitor (larger than the secondary by an amount sufficient to allow for the WD cooling time). Table I gives data on six such systems. The Sanduleak-Pesch object and G 107-70AB are double degenerates; the others have main sequence secondaries. M_1^o is the estimated initial primary mass, dM the fraction of it lost, and q the initial mass ratio. The implication of these numbers is that some wide or single WD's have lost as much mass as the interacting ones, and that there may not be an absolutely one-to-one relation between main sequence mass and white dwarf mass.

2.8.2. Cluster Members. A handful of young clusters each contain one or more white dwarfs (Reimers and Koester 1982; Anthony-Twarog 1984). Combining data for the Pleiades, Hyades, Praesepe, and NGC 2168, 2287, 2516, and 2422 tells us that main sequence stars of at least 5$\pm$1 to 8 $M_{\odot}$ can form WD's at least some of the time (Weidemann and Koester 1983; Anthony-Twarog 1984), which is clearly relevant to galactic evolution, supernova models, and so forth. Most classic models of galactic evolution have put this cut-off at rather lower masses, and at least the higher numbers push rather hard on estimated Type II supernova rates.

TABLE I: PARAMETERS OF NON-INTERACTING WHITE DWARF BINARIES

OBJECT	M_{WD}	M_2	M_1^o	dM	q	Reference
Sanduleak-Pesch	0.8	0.4	$\geq$5.0	$\geq$80%	0.5-0.7	Greenstein et al. 1983
Sirius AB	1.05	2.14	2.25	55%	0.95	Gatewood & Gatewood 1978
Procyon	0.63	1.77	2-3	65%	$\leq$0.90	
40 Eri AB	0.43	0.16	$\geq$0.9	$\geq$50%	$\leq$0.18	Heintz 1974
Stein 2051B	0.5	0.25	$\geq$0.9	$\geq$44%	$\leq$0.3	Strand 1977
G107-70AB	0.67	0.25	$\geq$0.9	$\geq$26%	$\leq$0.28	Harrington et. al. 1980

In addition, cluster members can help calibrate the relation between initial and final masses. All workers on the subject (Shipman 1979; Weidemann and Koester 1983; Greenstein and Trimble 1967; etc.) agree that WD's in young clusters are more massive than average, and that clusters with the largest turn-off masses tend to have the most massive degenerates. There is less agreement on precise numbers, the discrepancies mirroring those found for isolated stars, discussed in the next section.

2.8.3. Single White Dwarfs. Masses for isolated WD's can be derived by combining a model mass-radius relation with a spectroscopic log g, a radius from the Barnes-Evans relation (plus parallax), or a gravitational redshift, which gives M/R. There is general agreement that WD masses are rather tightly peaked around a median value, the one-sigma dispersion being only 0.1-0.15 M_o (Greenstein 1984), but less agreement on what that median value is.

The Barnes-Evans value is the smallest, 0.50-0.55 M_o (Moffett et al. 1978). Measurements of gravitational redshift (Greenstein and Trimble 1967; Trimble and Greenstein 1972; Wegner 1974) give the largest values, 0.75-0.80 M_o for DA's and more like 0.6 M_o for other types. Some of the spectroscopic determinations roughly concur, with Shipman (1979) and Guseinov et al. (1983, 1984) finding 0.75 M_o for DA's and 0.5-0.6 M_o for the helium atmosphere types. Given these relatively large values, white dwarfs make a significant contribution to the local mass density, nearly supplying the previously missing mass in the galactic plane; and a typical 1-2 M_o population I star seems to be losing at most about half its mass during pre-WD evolution.

Other atmospheric analyses (Wickramasinghe et al. 1983; Weidemann and Koester 1983, 1984; Oke et al. 1984) yield 0.55-0.60 M_o as the mean mass for all spectral types, including helium-rich atmospheres, the pulsationally variable ZZ Ceti stars, and members of Magellanic Cloud clusters (Weidemann 1984). All methods, naturally, give the right answer when applied to Sir B and 40 Eri B. Koester (1984) has considered differences between the Kiel and Soviet scales and remains on the Kiel side.

Small amounts of supporting evidence can be advanced on each side. A high mass, 0.8 M_o, comes from a pulsational analysis of one ZZ Ceti star (Dolez et al. 1983). On the other hand, Starrfield et al.'s (1984) analysis of the pulsation of the hydrogen-deficient PG1159-035 requires both C and O in the outer layers, implying that the star has been stripped deep into the helium-burning zone, and favoring small remnant masses. But this star can be roughly counterbalanced by the WC8 nucleus of NGC 40, a planetary nebula of normal composition (Clegg et al. 1983), which suggests that only the hydrogen-rich envelope is removed prior to WD formation. I remain prejudiced in favor of the larger WD mass values (which have been used in most galactic evolution models), but it is a prejudice, and I am prepared to be crushed by a low-mass bandwagon.

REFERENCES

Adams, S. et al. 1984. MNRAS 207, 471.
van Albada, B. & N. Baker 1973. Ap.J. 185, 477.
Alcock, C. 1984. Priv. Comm. at this meeting.
Anthony-Twarog, B.J. 1984. Astr. J. 89, 267.
Bergeron, P. et al. 1984. Astr. J. 89, 374.
Branch, D. 1984. Priv. Comm.
Burki, G. 1984. Astr. Ap. 133, 185.
Cahn, J. 1984. Ap.J. 279, 304.
Carson, T.R. & R. Stothers 1984. Ap.J. 276, 593.
Chevalier, R.A. 1976. Ap.J. 207, 872 & 208, 826.
..... 1984. Ap.J. Lett. (submitted).
Clegg, R.S. et al. 1983. MNRAS 205, 417.
Conti, P.J., D.S. Dearborn & P.L. Massey 1981. MNRAS 195, 165
Conti, P.J., C. Garmany, C. de Loore, & D. Vanbeveren 1983. Ap.J. 274, 302.
Cox, A.N. 1980. Ann. Rev. Astr. Ap. 18, 15
Cox, A.N., S.W. Hudson & S.P. Clancy 1983. Ap.J. 266, 94.
Davidson, K. et al. 1982. Ap.J. 253, 429.
Dolez, N., G. Vauclair & M. Chevreton 1983. Astr. Ap. 121, L23.
Ellis, H.B. & M. Werner 1985. Poster paper at this meeting.
Fontaine, G. et al. 1984. Ap.J. 277, L61.
Gatewood, G.D. & C.V. Gatewood 1978. Ap.J. 225, 191.
Gatewood, G. & J. Russell 1974. Astr. J. 79, 815.
Gathier, R. et al. 1983. Astr. Ap. 128, 325.
van Genderen, A.M. 1983. Astr. Ap. 124, 223.
Gee, G. et al. 1984. MNRAS 208, 517.
Greenstein, J.L. 1984. Ap.J. 276, 602.
Greenstein, J.L., N. Dolez, & G. Vauclair 1983. Astr. Ap. 127, 25.
Greenstein, J.L. & V. Trimble 1967. Ap.J. 149, 283.
Guseinov, O.H., H.I. Novruzova, & Yu.S. Rustamov 1983. Ap.Sp. Sci. 96, 1.
..... 1984. Ap. Sp.Sci. 97, 309.
Harrington R.S., J.W. Christy, & Ka. A. Strand 1981. Astr.J. 86, 909.
Heintz, W.D. 1974. Astr. J. 79, 819.
Hendon, A.A. et al. 1984. Publ. Astr. Soc. Pac. 96, 310.
van den Heuvel, E.P.J. 1981. IAU Symp. No. 93, 155 and No. 95, 379.
van den Heuvel, E.P.J. & R.E. Taam 1984. Nature 309, 598.
Iben, I. 1983. Sol. Phys. 82, 457.
Iben, I. & A.V. Tutukov 1984. Ap.J. Sup. 54, 335.
Kaler, J.B. 1983. Ap.J. 271, 188.

Glaccum, W. 1982. Priv. Comm.
Kilkenny, D. & A.E. Lynas-Gray 1984. MNRAS 208, 613.
Koester, D. 1984. Ap. Sp. Sci. 100, 471.
Litvanova, I. Yu. & D.K. Nadyozhin 1983. Ap. Sp. Sci. 89, 89.
Livio, M. & N. Soker 1984. MNRAS 208, 613.
Lutz, J.H. 1984. Ap.J. 279, 714.
Maeder, A. 1981. Astr. Ap. 102, 401.
Marsden, P.L. et al. 1984. Ap.J. 278, L29.
Michaud, G., G. Fontaine, & Y. Charland 1984. Ap.J. 280, 247.
Minkowski, R. 1941. Publ. Astr. Soc. Pac. 53, 224.
Moffett, J., T.G. Barnes, & D.S. Evans 1978. Astr. J. 83, 820.
Muchmore, D. 1984. Ap.J. 278, 769.
Nather, R.E. 1985. in P.P. Eggleton & J.P. Pringle (eds.) Interacting Binaries (Dordrecht: D. Reidel).
Oke, J.B., V. Weidemann, & D. Koester 1984. Ap.J. 281, 267.
Paczyński, B. 1976. IAU Symp. No. 73, 65.
Patterson, J. 1984. Ap.J. Sup. 54, 443.
Phillips, J.P. & S.R. Pottasch 1984. Astr. Ap. 130, 91.
Pottasch, S.R. 1983. IAU Symp. No. 103, 391.
Reimers, D. & D. Koester 1982. Astr. Ap. 116, 341.
Ritter, H. 1984. Astr. Ap. (in press).
Rood, R.T. 1973. Ap.J. 184, 477.
Sabbadin, F. et al. 1983. Astr. Ap. 123, 147.
Schoenberner, D. 1983. Ap.J. 272, 708.
Schoenberner, D. & J.S. Drilling 1984. Ap.J. 278, 702.
Schoenberner, D. & V. Weidemann 1983. IAU Symp. No. 103, 359.
Serrano, A. 1983. IAU Symp. No. 103, 463.
Shipman, H.L. 1979. Ap.J. 228, 240.
Shipman, H.L. & J.L. Greenstein 1983. Ap.J. 263, 761.
Simon, N.R. & C.G. Davis 1983. Ap.J. 266, 787.
Sion, E.M., F. Wesemael, & E.F. Guinan 1984. Ap.J. 279, 758.
Solheim, J. et al. 1984. Astr. Ap. 135, 1.
Sramek, R.A., N. Panagia, & K.W. Weiler 1984. Ap.J. Lett. (submitted)
Starrfield, S. et al. 1984. Ap.J. 281, 800.
Stellingwerf, R.F. 1984. Ap.J. 277, 322.
Strand, K. Aa. 1977. Astr. J. 82, 745.
Trimble, V. 1975. Rev. Mod. Phys. 47, 877.
..... 1982. Rev. Mod. Phys. 54, 1183.
..... 1983. Rev. Mod. Phys. 55, 811
Trimble, V. & J.L. Greenstein 1972. Ap.J. 177, 441.
Trimble, V. & I.-J. Sackmann 1977. MNRAS 182, 97.
Turner, K.L. & Y. Terzian 1984. Astr. J. 89, 501.
Tylenda, R. 1983. Astr.Ap. 126, 299 and IAU Symp. No. 103, 230.
Weaver, T.A. 1984. Priv. Comm.
Webbink. R.F. 1979. in H. Van Horn & V. Weidemann (eds.) White Dwarfs and Variable Degenerate Stars (U. Rochester) p. 443.
Wegner, G.A. 1974. MNRAS 166, 271.
Wegner, G. & F. Yackovich 1983. Ap.J. 275, 240.
Weidemann, V. 1984. Astr. Ap. 134, L1.
Weidemann, V. & D. Koester 1983. Astr. Ap. 121, 77; 1984 Astr.Ap. 132,195.
Wickramasinghe, D.T. et al. 1983. MNRAS 203, 887.
Willson, L.A. & G.H. Bowen 1984. Nature (submitted).

OPTICAL SPECTROSCOPY OF RED GIANTS

Leo Goldberg
Kitt Peak National Observatory
P.O. Box 26732
Tucson, AZ 85726-6732
U.S.A.

ABSTRACT. Accomplishments and problems in the study of mass loss from cool stars in the optical spectrum are reviewed. Accurate mass loss rates remain elusive principally because of inadequate knowledge of density, velocity and ionization structure in the circumstellar shells. Improved rates may be expected from binary star systems. Empirical rates derived for stars in globular clusters are found to be of doubtful validity. It is shown that optical observations can furnish important clues to the understanding of mass loss mechanisms.

1. INTRODUCTION

Nearly 30 years have elapsed since A.J. Deutsch (1956,60) employed the techniques of optical spectroscopy to prove the existence of massive stellar winds in red giants and supergiants. Even the first rough estimates of the rates of mass loss indicated that the winds played a major role in the scenario of the late stages of stellar evolution. Weymann (1962) then undertook a detailed study of the blue-displaced components of atomic resonance absorption lines in α Ori, using curves of growth to derive ion column densities and solar abundances to obtain a mass loss rate on the order of 10^{-6} $M_{\odot}$/yr. He also showed that the mean heliocentric velocity of the expanding shell, 11 km/sec, had not varied by as much as 2 km/sec in several decades and, moreover, that the velocity of the expanding matter was not affected by the variable motion of the photosphere below. This led to the concept of a detached, circumstellar shell, expanding at a constant velocity and having a sharply defined inner radius of at least a few stellar radii.

Weymann (1962) also found that the Balmer lines in α Ori were considerably stronger than expected from its spectral type, implying a contribution from the circumstellar shell, which was also suggested by the asymmetric shape of the line core. Not only was the center of the Hα line found to be blue-shifted by about 8 km/sec relative to the wings of the central core, but the velocities appeared to be relatively constant and unaffected by the photospheric movements.

M. Morris and B. Zuckerman (eds.), Mass Loss from Red Giants, 21–27.

Following the passage of about a dozen years, during which the 10 μm silicate emission feature became the principal focus of studies of mass loss from red giants, interest in the optical spectrum was revived by Reimers' (1975) survey of circumstellar absorption lines in the spectra of 120 K and M giants, and G, K, M supergiants. Reimers was able to show that the stars in which CS lines were observed are confined to the upper-right portion of the H-R diagram, and that expansion velocities tended to vary from about 5-100 km/sec, increasing towards the boundary between stars that showed CS lines and those that did not. The flow also tended to become time-dependent as the boundary was approached.

Using the curve of growth to infer column densities and mass loss rates in selected stars, Reimers summarized his results in the form of an interpolation formula:

$$\dot{M} = 4 \times 10^{-13} \, L/gR \; (M_\odot/yr), \qquad (1)$$

where the luminosity, L, the surface gravity, g, and the radius, R, are all in solar units. This parametric formula expresses the ad hoc physical assumption that the kinetic energy carried away by mass loss is always the same fraction of a star's luminosity. Theorists have made extensive use of the formula in evolutionary calculations, introducing a multiplicative factor $0.3 \leqslant \eta \leqslant 3$ to represent the uncertainties in empirical mass loss rates (e.g. Renzini, 1981).

2. TECHNIQUES AND PROBLEMS

In recent years, treatment of the line formation problem in the context of plane-parallel atmospheres and the curve of growth has been largely abandoned in favor of exact solutions of the radiative transfer equation for expanding, spherical atmospheres and non-LTE treatment of ionization equilibrium (Sanner, 1976; Bernat, 1977). Improved, high-resolution spectra have also revealed the P-Cygni type profiles (emission plus blue-displaced absorption) predicted by the theory to be formed in extended CS shells. Nevertheless, accurate mass loss rates from optical observations continue to elude us, principally for two reasons. First, almost nothing is known of the density structure of the shells. Unanswered questions include the degree of departure from spherical symmetry and the height dependence of the flow velocity. The chief uncertainty in rate determinations is generally said to be the poorly known inner radius of the shell, but the real question is whether or not the shell is shaped like a doughnut, as is commonly assumed. Second, the ionization structure of the shells is also beset by many contradictions (Hagen, 1978; Bernat, 1982). Since the most abundant metals are represented in the optical spectrum by neutral lines, they are apparently predominantly singly-ionized in the shell, and consequently knowledge of their column densities relies on accurate calculations of ionization equilibrium. Discrepancies of up to four orders of magnitude in such calculations have been attributed by

Hagen (1978) to uncertainties in cool-star radiation fields, which fall steeply below 2000 Å, where the metal-ionizing radiation occurs.

One way to avoid the use of ionization calculations is to base the rate of mass loss on the line of Sr^+ 4077 Å, to make the not unreasonable assumption that Sr is mostly singly-ionized, and to guess that the Sr/H abundance ratio has the solar value. A second stratagem is to observe lines of singly-ionized metals in the spectrum of a companion star to a red giant in a visual binary system. If the geometry of the orbit is known, no assumption need be made about the inner radius of the shell (Reimers, 1977; Kudritski and Reimers, 1978). Since most resonance lines of singly-ionized metals lie below 3000 Å, the technique is best applied to space observations with balloons (van der Hucht et al., 1980), or with the IUE satellite (Bernat, 1982). Bernat's (1982) analysis of lines of 10 different ions in the spectral region 1190 Å - 2631 Å of α Sco gave hydrogen column densities differing by as much as a factor of 30, when solar-like abundances were assumed. Blends with interstellar lines and element depletion in grains account for some but not all of the discrepancies.

Few stars are accessible to the binary star approach unless one uses spectroscopic binaries consisting of a red giant and a B-type companion. Observation in the UV, e.g., with IUE, effectively removes the continuous spectrum of the M giant, but the separation is so small that radiation from both the B star and the CS shell pass through the slit. The resulting P Cygni profiles must be analysed in the context of the non-spherical, three-dimensional line transfer problem, which has been solved by Hempe (1982) and applied to ζ Aur, 32 Cyg and 31 Cyg (Che et al., 1983), δ Sge (Reimers and Schröder, 1983) and Boss 1985 (Che and Reimers, 1983).

3. RESULTS

Table I gives mass loss rates and terminal expansion velocities derived from the optical spectrum for selected stars. The optical techniques are best suited to giants and supergiants of relatively early spectral type, in which most of the shell material is in gaseous form. The rates for the early M supergiants exceed those of the M giants by 1-2 orders of magnitude and the K supergiants by still another factor of ten. Still higher rates are found for late M, S and C stars (Morris, this volume) and for OH/IR stars (Bowers, this volume).

4. STARS IN GLOBULAR CLUSTERS

Optical spectroscopy has provided the only estimates of mass loss rates for stars in globular clusters, although their validity is open to question. Empirical determinations of rates for such stars are of high importance in the light of very narrow limits on their values apparently imposed by evolutionary requirements (Renzini, 1981).

Observational evidence for mass loss in globular cluster stars is of two kinds: 1) emission in either the blue or red wing, or both, of the Hα line (Cohen, 1976), and 2) blue-shifted, asymmetric cores of Hα and Na D absorption lines in some stars (Peterson, 1981, 1982). Cohen (1976) first postulated that the Hα emission arises in a circumstellar shell of recombining, ionized hydrogen and used this model to estimate rates of mass loss on the order of 10^{-9} $M_\odot$/yr. The same method has been applied to well over a hundred stars by Mallia and Pagel (1978) and by Cacciari and Freeman (1983), both of whom found values more than 10 times greater, primarily because they used larger stellar radii. Both of these discordant estimates can be made compatible with evolutionary requirements by adjustment of the fraction of the time spent on the red giant branch during which the star is permitted to lose mass.

The reality of the evidence for mass loss in globular cluster stars has been questioned by Reimers (1981), who pointed out that Hα emission by itself is not evidence of mass loss and by Dupree et al. (1984), who show by model calculations that emission wings can arise in warm chromospheres with mass loss rates as low as $\ll 2 \times 10^{-9}$ $M_\odot$/yr. The assumption that the emission is formed by recombination in a fully-ionized gas is also purely arbitrary.

Table I. Mass Loss Rates, $\dot{M}$, and Expansion Velocities, V_{exp}, derived for early-type red giants and supergiants from optical spectra.

Star	Spectral Class	$\dot{M}$ (10^{-6} $M_\odot$/yr)	V_{exp} (km/sec)	Ref
ζ Aur	K4 Ib	0.006	40	6
32 Cyg	K5 Iab	0.003	60	6
R Lyr	M4-5 III	0.014	5	1
ρ Per	M4 II-III	0.012	7	1
δ Sge	M2 II	0.02	28	2
δ^2 Lyr	M4 II	0.05	7	1
α Her	M5 Ib-II	0.11	8	1,3
119 Tau	M1 Ib	0.24	7	1
α Ori	M2 Iab	4	10	4,5
6 Gem	M1-2 Ia-Iab	2.1	8	1
α Sco	M1.5 Iab	5.4	9	1
μ Cep	M2 Ia	1.0	9	1,5

References to Table I. 1. Sanner (1976). 2. Reimers and Schröder (1983). 3. Reimers (1977). 4. Bernat et al. (1979). Hagen et al. (1983). 6. Che et al. (1983).

5. MECHANISMS

Apart from their utility in mass loss rate determinations, optical observations can provide important clues to the understanding of mass loss mechanisms, by exposing phenomena that occur in the deepest parts of the atmosphere. Because of its brightness and fascinating behavior, the red supergiant α Ori has been the best studied of the highly-evolved stars and promises to be even more intensively observed in the future. The star has already provided a number of clues that may be relevant to the mass loss process (Goldberg, 1984):

1) The star undergoes semi-regular variations in brightness and radial velocity in a mean period of 5.78 years, with amplitudes of about half a magnitude and 6 km/sec, respectively. The phase relations between the brightness and velocity curves suggest pulsation as the cause of these changes.
2) Superimposed on the 6-year cycle are irregular short-period fluctuations in brightness and radial velocity on a timescale of about 100 days. The irregular fluctuations in radial velocity, and possibly in visual brightness as well, are sometimes so pronounced as to mask the 6-year period altogether; at other times, the longer period is clearly in evidence.
3) Observations of linear polarization in the B-band (Hayes, 1984) show ordered changes in amount and angle, the degree of polarization varying from zero up to nearly 1%. The amount but not the angle of polarization is roughly correlated with the brightness (Guinan, 1984).
4) The scale height of the photosphere, as implied by TiO-related variations of the interferometrically-measured diameter (Balega et al., 1982), is on the order of a stellar radius, which may or may not be connected with (5).
5) The widths of the photospheric Fraunhofer lines at half intensity are about 27 km/sec, a grossly non-thermal situation that has received very little attention and study.
6) Occasional large outbursts have been observed, in which ejection of matter has been followed by light-dimming of about half a magnitude. The most spectacular event of this kind was observed in August, 1925 (Goldberg, 1984). The possibility that these are dust-forming episodes makes it important that similar events in the future be monitored in the infrared at 10 μm.

One of the pressing questions requiring attention is whether or not the short-period brightness fluctuations are caused by local temperature changes, as would be expected, for example, from very large convection cells (Schwarzschild, 1975), which Hayes (1984) believes are consistent with the polarization changes. Guinan (1984) finds, however, that broadband indicators of TiO band strength show no evidence of temperature changes associated with the brightness fluctuations.

Recent model calculations (Schwarz and Clarke, 1984; Clarke and Schwarz, 1984) indicate that the wavelength dependence and time variation of the polarization are qualitatively consistent with the growth of a hot spot, or large convection cell, occupying a few percent of the area of the stellar disk. A weakness of the calculations is that both the undisturbed photosphere and the hot spots are assumed to radiate like black bodies, whereas in fact the visible spectra of red giants are dominated by TiO absorption bands (Collins, 1975). High-resolution spectra should settle this question in the near future.

Finally, I wish to draw attention to the interesting behavior of the CaII triplet lines 8498 Å, 8542 Å, and 8662 Å in the spectrum of α Ori (Goldberg, 1979). The cores of these lines are strongly asymmetric, implying that they are formed in an atmospheric region, the bottom of which has approximately the center-of-mass, or systemic velocity, and in which the flow is accelerated outward to nearly its terminal velocity. Since this layer seems not to share in the pulsation of the photosphere, it must lie at least several radii above it. A similar stationary layer has been identified in the Mira star χ Cyg, from observations of the 5 μm CO bands by Hinkle et al. (1982), who suggest that matter is carried up to the stationary layer by shock waves connected with the pulsation. The subsequent formation of dust grains in the stationary layer facilitates mass loss by radiation pressure. It is intriguing to speculate that a similar process is at work in α Ori, with matter being elevated to a stationary layer either by the 6-year pulsation or by ejections associated with the short-term changes in brightness and polarization.

REFERENCES

Balega, Y.A.,Blazit, A., Bonneau, D.L., Koechlin, L., Foy, R., Labeyrie, A. 1982, Astr. Ap., **115**, 253.
Bernat, A. 1977, Ap. J., **213**, 756.
Bernat, A. 1982, Ap. J., **252**, 644.
Bernat, A., Hall, D.N.B., and Ridgway, S.T. 1979, Ap. J., **233**, L135.
Cacciari, A. and Freeman, K.C. 1983, Ap.J., **268**, 185.
Che, A. and Reimers, D. 1983, Astr. Ap., **127**, 227.
Che, A., Hempe, K. and Reimers, D. 1983, Astr. Ap., **126**, 225.
Clarke, D. and Schwarz, H.E. 1984, Astr. Ap., **132**, 375.
Cohen, J.G. 1976, Ap. J., **203**, L127.
Collins, J.G. 1975, Ph.D. thesis, Indiana University.
Deutsch, A.J. 1956, Ap. J., **123**,210.
Deutsch, A.J. 1960 in Stars and Stellar Systems VI, ed. J.L. Greenstein, Univ. Chicago, p. 453.
Dupree, A.K., Hartmann, L. and Avrett, E.H. 1984, preprint.
Goldberg, L. 1979, Quart. J. Roy. Astr. Soc., **20**, 361.
Goldberg, L. 1984, Pub. A.S.P., **96**, 366.
Guinan. E.F. 1984, to be published in Third Cambridge Workshop on Cool Stars, Stellar Systems and the Sun, eds. Hartmann L. and Baliunas, S.
Hagen, W. 1978, Ap. J. Suppl., **38**, 1.
Hagen, W., Stencel, R.E and Dickinson, D.F. 1983, Ap.J., **274**, 286.
Hayes, D.P. 1984, Ap. J. Suppl., in press.
Hempe, K. 1982, Astr. Ap., **115**, 133.
Hinkle, K.H., Hall, D.N.B. and Ridgway, S.T. 1982, Ap. J., **252**, 697.

Kudritski, R.P. and Reimers, D. 1978, Astr. Ap., 70, 227.
Mallia, E.A. and Pagel, B.E. 1978, Mon. Not. Roy. Astr. Soc., 184, 55P.
Peterson, R.C. 1981, Ap. J., 248, L31.
Peterson, R.C. 1982, Ap. J., 258, 499.
Reimers, D. 1975, Mem. Soc. Roy. Soc. Liege, 6th Ser. 8, 369.
Reimers, D. 1977, Astr. Ap., 61, 217 (Erratum 67, 161).
Reimers, D. 1981, in Physical Processes in Red Giants, eds. Iben, I. and Renzini A., Reidel, Dordrecht, p.269.
Reimers, D. and Schröder, K.-P. 1983, Astr. Ap., 124, 241.
Renzini, A. 1981, in Effects of Mass Loss on Stellar Evolution, I.A.U., Colloq. No. 59, eds. Chiosi, C. and Stalio, R., Reidel, Dordrecht, p.319.
Sanner, F. 1976, Ap. J. Suppl., 32, 115.
Schwarz, H.E. and Clarke, D. 1984, Astr. Ap., 132, 370.
Schwarzschild, M. 1975, Ap. J., 195, 137.
van der Hucht, K.A., Bernat, A.P. and Kondo, Y. 1980, Astr. Ap., 82, 14.
Weymann, R.J. 1962, Ap. J., 136, 844.

MASS LOSS IN METAL DEFICIENT RED GIANTS

A. K. Dupree, L. Hartmann, G. H. Smith, and E. H. Avrett
Harvard-Smithsonian Center for Astrophysics
60 Garden Street
Cambridge, MA 02138

The occurrence and rate of mass loss from metal deficient red giant field stars have consequences for the theory of evolution of low mass stars. This paper reports several of the results from an observational and theoretical program to assess the presence and character of chromospheric line profiles and their variations, to obtain quantitative measurements of radiative losses from chromospheres, and to construct semi-empirical model chromospheres and extended atmospheres in order to provide limits on the mass loss rates.

Observations of three metal deficient field stars in the ultraviolet and optical spectral regions (Dupree, Hartmann, and Smith 1984) show that Hα profiles can exhibit strong asymmetric emission that is quite different from luminous Population I stars. Moreover, the Hα emission and core asymmetry can vary on a time scale of two months or less. A full report of the Hα survey will be published by Smith and Dupree (1985). Our observations substantially reduce the previous upper limit (Ramsey 1979) of one year or less for variability in field giants, and are not inconsistent with the time scale of a few days (Cacciari and Freeman 1983) for red giants in globular clusters. Moreover such variations suggest that the Hα emission arises in the chromosphere of the star (Dupree, Hartmann, and Avrett 1984) and does not necessarily imply mass loss as previous authors have conjectured (Cohen 1976, Mallia and Pagel 1978).

The Hα emission profiles can be fit (see Figure 1) with extended atmosphere models driven by Alfven waves (Hartmann and MacGregor 1980) to derive steady mass loss rates $<< 2 \times 10^{-9}$ $M_{\odot}$ yr^{-1} (Dupree, Hartmann, and Avrett 1984). This value is a factor of 20 to 50 less than the relation proposed by Reimers (1975) for Population I stars, and 5 to 40 times less than the mass loss rate inferred from a circumstellar interpretation. This does not rule out a massive cool wind typical of

M. Morris and B. Zuckerman (eds.), Mass Loss from Red Giants, 29–30.

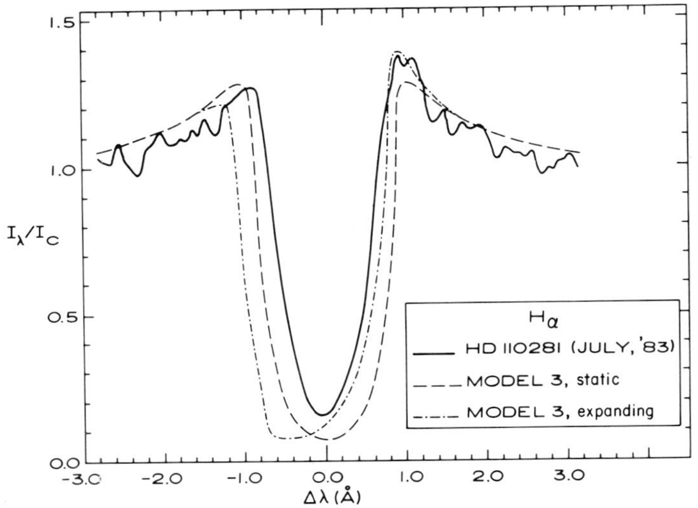

Figure 1. Calculated Hα profiles for an extended atmosphere model in static and expanding versions. The calculated shift of the line core to shorter wavelengths corresponds to ~ 21 km s^{-1}, the terminal wind velocity. Parameters of the observed profile of HD 110281, a metal deficient field giant, that are well matched by this calculation include: the core depth, the peak separation, the wing shape and the peak height. The observed line is, however, narrower than the calculated profile, and the observed position of the line core is not shifted and agrees with the photospheric velocity to within 1 km s^{-1}. (Figure from Dupree, Hartmann, and Avrett 1984.)

Population I giants (Reimers 1975), for which high resolution Mg II or Ca II line profiles are required. Existing observations also do not eliminate the possibility of substantial mass loss by transient events perhaps associated with pulsation.

REFERENCES

Cacciari, C., and Freeman, K.: 1983, Astrophys. J., **268**, p. 185.
Cohen, J.G.: 1976, Astrophys. J., **203**, p. L127.
Dupree, A.K., Hartmann, L., and Avrett, E.H.: 1984, Astrophys. J., **281**, p. L37.
Dupree, A.K., Hartmann, L., and Smith, G.: 1984, in Proc. Third Cambridge Workshop on Cool Stars, Stellar Systems, and the Sun, Baliunas, S. and Hartmann, L., eds., Lecture Notes in Physics, **No. 193**, Berlin, Springer-Verlag, p. 326.
Hartmann, L., and MacGregor, K.B.: 1980, Astrophys. J., **242**, p. 260.
Mallia, E.A., and Pagel, B.E.J.: 1978, Mon. Not. Roy. Astron. Soc., **184**, p. 55p.
Ramsey, L.W.: 1979, Pub. Astron. Soc. Pacific, **91**, p.252.
Reimers, D.: 1975, Mem. Soc. Roy. des Sci. de Liege, **8**, p. 369.
Smith, G. H., and Dupree, A.K.: 1985, Astrophys. J., in preparation.

MASS LOSS FROM RED GIANTS: RESULTS FROM ULTRAVIOLET SPECTROSCOPY

Jeffrey L. Linsky*
Joint Institute for Laboratory Astrophysics, National Bureau of Standards and University of Colorado, Campus Box 440, Boulder, Colorado 80309, USA

ABSTRACT. New instrumentation in space, primarily the IUE spacecraft, has enabled the application of ultraviolet spectroscopic techniques to the determination of physical properties and reliable mass loss rates for red giant winds. One important result is the determination of where in the H-R diagram are found stars with hot outer atmospheres and with cool winds. So far it appears that single cool stars, except perhaps for the so-called hybrid stars, have either hot outer atmospheres or cool winds but not both. The C II resonance (1335 Å) and intersystem (2325 Å) mutliplets have been used to derive temperatures, densities, and geometrical extents for the chromospheric portions of red giant winds, with the result that the red giants and the earlier giants with hot coronae have qualitatively different chromospheres. Mass loss rates can now be derived accurately from the analysis of asymmetric emission lines, such as the Mg II resonance lines, and from P Cygni profile lines of atoms in the dominant ionization stage when a hot star is available to probe the wind of a red giant. The ζ Aur systems, consisting of a K-M supergiant and a main sequence B star are important systems for which reliable mass loss rates for the red supergiant components are becoming available.

1. INTRODUCTION

The bulk of the radiative output from red giants lies in the infrared. Nevertheless, ultraviolet spectroscopy has been and will continue to be a powerful tool for probing the physical properties of their extended outer atmospheres and for measuring their mass loss rates. In this review I will attempt to summarize the rapidly expanding literature in this field by describing the ultraviolet spectra of these stars and what has been inferred from these data.

---

*Staff Member, Quantum Physics Division, National Bureau of Standards.

M. Morris and B. Zuckerman (eds.), Mass Loss from Red Giants, 31–54.

In analogy with the solar atmosphere, I refer to the region that lies immediately above the photosphere of a late-type star as a chromosphere. By this term I mean (see Linsky 1980) a region extending over at least several pressure scale heights for which the temperature structure is hotter than predicted by radiative equilibrium and which is heated by nonradiative processes. Chromospheres as I have defined them are ubiquitous for main sequence stars of spectral type F-M and for giants and supergiants of spectral type F-K. The term chromosphere as I have defined it may not be valid for the M giants and supergiants, however, if the chromospheric emission lines observed are not formed in an extended relatively homogeneous region but instead behind isolated shocks that are propagating outward through an otherwise cool, extended envelope. We cannot determine which of these two pictures is valid empirically at this time.

1.1. Instrumentation and Observational Techniques

The available evidence for the existence of chromospheric (4000-10,000 K) plasma in the outer atmospheres of late-type giants consists primarily of emission lines observed in the ultraviolet. These lines are formed either by collisional excitation, recombination and subsequent cascade, or fluorescence. The first mechanism is generally thermal in character, the second and third are also thermal but nonlocal in the sense that the ionizing or stimulating radiation originates elsewhere. The Ca II H and K lines and hydrogen Hα lines are among the few lines in the visible that indicate chromospheric plasma. In addition to spectral lines, chromospheric plasma can be observable by microwave emission and ultraviolet continuum emission -- the former has been detected so far only from K and M supergiants, and the latter so far only from dMe stars during flares.

The accumulation of evidence for the existence of nonradiatively heated chromospheres has been limited by our observational capability. Prior to observations from space, we could observe only relatively cool chromospheric plasma using the Ca II H and K lines and could obtain indirect evidence for hot coronae from the He I 10830 Å line. Linsky and Avrett (1970) and Linsky (1977) have summarized these data and the usefulness of various spectroscopic diagnostics. The first few space observations by rockets, balloons, and *Copernicus* were superseded by IUE, which has observed hundreds of late-type stars in the 1200-3200 Å spectral region at both low and high resolution. These data in turn will be superseded by the more sensitive and versatile instruments on Space Telescope and the proposed *Columbus* mission. The capabilities of these instruments are summarized in Table 1.

It is important to recognize that instrumental spectral range and sensitivity limit the plasma temperature ranges that can be studied. For example, *Copernicus* was capable of observing only the Lyman alpha and Mg II lines in late-type stars (except for the very brightest stars like Capella), and thus could only observe plasma as hot as 10^4 K. IUE, on the other hand, can observe emission features of C IV

Table 1

Capabilities for Obtaining Ultraviolet Spectra of Cool Stars

Mission	Years Operational	Focal Plane Instrument	Spectral Range (Å)	Spectral Resolution ($\lambda/\Delta\lambda$)
Copernicus (OAO-C)	1972-78		Mg II, Lα lines	20,000 5,000
International Ultraviolet Explorer (IUE)	1978 →	Short Wavelength (SWP Camera)	1175-2000 "	10,000 250
		Long Wavelength (LWR,LWP Cameras)	2000-3000 "	10,000 400
Space Telescope (ST)	1987(?) →	High Resolution Spectrometer (HRS)	1175-3200 " 1175-1700	100,000 20,000 2,000
		Faint Object Spectrometer (FOS)	1150-8000 "	1,000 100
Columbus/Far Ultraviolet Spectroscopic Explorer	1991 →		900-1200 1200-2000 100-2000	30,000 10,000 1,000

at 1550 Å and N V at 1240 Å formed in plasma as hot as 150,000 K. The 912-1216 Å spectral range of Columbus (previously called FUSE) contains the strong resonance lines of O VI formed at 300,000 K, and the 100-912 Å spectral range, also observable by Columbus, contains lines formed in coronal plasma as hot as 3×10^7 K (Fe XXIV).

Even with the powerful spectrometers forthcoming on ST and Columbus, our ability to study nonradiative heating in stellar chromospheres and hotter layers will be limited. For one thing, use of ST will be severely constrained by intense competition for telescope time, so that it will be difficult to monitor specific stars for long periods of time or to observe a large sample of stars of a given type. Furthermore, it will be difficult to observe a large spectral interval at high resolution because the HRS can measure only a few Angstroms of spectrum at one time in the 100,000 resolution mode. Even for the nearest stars, Columbus observations in the Lyman continuum will be affected by interstellar absorption, but it should be able to observe a great many late-type stars at wavelengths below 300 Å.

Emission lines formed in the chromospheres and transition regions (TRs) of stars later than spectral type F0 are readily detected by IUE and the Ca II H and K lines are easily observed by telescopes on the ground. The important lines from plasmas at 4000-15,000 K are Ca II (3933, 3968 Å), Mg II (2796, 2803 Å), H I (1216 Å), C I (1657 Å), Si II (1808, 1817 Å), and O I (1305 Å mutliplet). At higher temperatures (2×10^4 - 2×10^5 K) the strongest available emission lines are of C II (1335 Å), Si III (1892 Å), C III (1909 Å), Si IV (1393, 1403 Å), C IV (1548, 1550 Å), and N V (1238, 1242 Å).

1.2. Previous Studies

The application of ultraviolet spectroscopy to the analysis of winds in red giants is a relatively new field. Some recent reviews that include aspects of this topic are by Brown (1983), Dupree (1981, 1982), Hartmann (1981, 1983), and Linsky (1981, 1982).

2. HOT PLASMA OR COOL WINDS: AN EITHER/OR SITUATION?

Before discussing the ultraviolet spectra of red giants in detail, it is important to determine where in the H-R diagram occur stars with chromospheres and hotter regions in their outer atmospheres. In particular, can we say on empirical grounds that stars either have cool winds or hot coronae but not both at the same time?

2.1. The X-ray "Dividing Line"

X-ray emission in the continuum and discrete lines can be produced by thermal processes (free-free, free-bound, and bound-bound) and, in principle, by nonthermal processes involving high energy particles in magnetic fields. These processes indicate a plasma at $\gtrsim 10^6$ K or nonthermal particles heated by nonradiative means. Prior to the launch of the Einstein X-ray Observatory, instrumental sensitivity limited the detection of possible hot coronae in the yellow and red giants to the X-ray luminous close binary systems of the RS CVn and Algol types.

A number of authors (e.g. Vaiana et al. 1981; Ayres et al. 1981; Helfand and Caillault 1982; Haisch and Simon 1982) have used the IPC and HRI instruments on Einstein to search for X-ray emission from giants and supergiants of spectral type G-M. Stars with detected hot coronae include such G supergiants as β Dra (G2 Ib-II), late G giants like β Cet (G9.5 III), early K giants like α Ser (K2 III), RS CVn binary systems, and binary stars with white dwarf components like 56 Peg (K0 II+WD). However, no giants cooler than about spectral type K2 III and no G-M supergiants were detected with very low upper limits. In the case of Arcturus (K2 III), for example, the upper limit corresponds to a surface flux 5×10^{-4} that of the mean quiet Sun (Ayres, Simon, and Linsky 1982).

The absence or small upper limits of X-ray emission in the upper right-hand portion of the H-R diagram led Ayres et al. (1981) and Haisch and Simon (1982) to propose a "dividing line" in the H-R diagram (see Fig. 1) separating stars with and without detected hot coronae. Needless to say the absence of detected X-rays in the cool stars is no proof of the absence of hot plasma in the outer atmospheres of these stars, only a determination of the upper limit to the emission measure of such hot plasma. The data so far, however, are highly suggestive that the M giants and supergiants, together with the K giants cooler than K2 III and the G-K supergiants do not have hot coronae. Gray (1981) has proposed that the disappearance of hot plasma as stars evolve across this "dividing line" is a consequence of the rapid decrease in rotational velocity as stars evolve to the right across this region of the H-R diagram.

In addition to the X-ray sensitivity problem just described, there are additional problems with this tentative conclusion that stars to the right of the X-ray "dividing line" do not have hot coronae: The first problem is that soft X-rays can be attenuated significantly by the cool gas in a wind, as is shown by the calculation for the 1/4 keV opacity of the wind in α Aqr (G2 Ib) by Ayres et al. (1981). Second, a number of the bright giants and supergiants to the right of the X-ray "dividing line," in particular, the hybrid stars (see below), exhibit He I 10830 Å absorption or emission (Zirin 1982; O'Brien 1980) which may be explained by recombination following ionization of He I by X-rays (e.g. Zirin 1975). Both problems require further study.

2.2. The Transition Region "Dividing Line"

In the initial survey of cool stars with the new IUE spacecraft, Linsky and Haisch (1979) found that essentially all stars later than

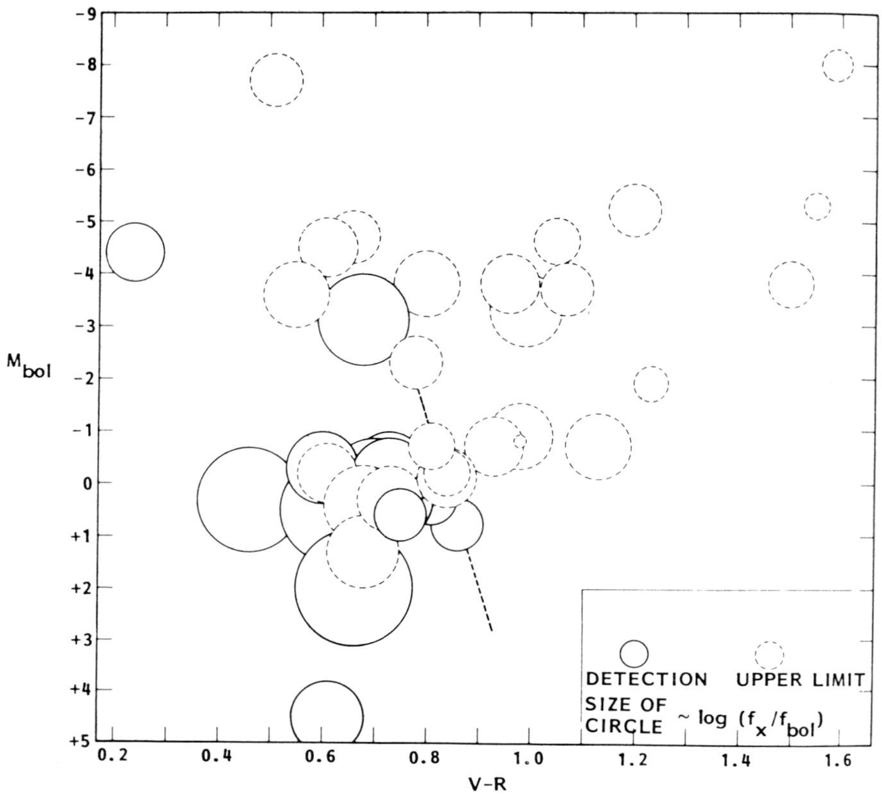

Fig. 1. An H-R diagram in which normalized soft X-ray fluxes (solid-line circles) or their upper limits (dashed-line circles) are denoted by circles whose sizes are proportional to f_x(0.2-4.0 keV)/f_{bol}. Also shown as a nearly vertical dashed line is the Linsky-Haisch (1979) transition region dividing line. (From Haisch and Simon 1982.)

early F spectral type show emission lines indicative of chromospheric plasma ($T < 10^4$ K), but that among the apparently single stars, only the F-M dwarfs and F-G giants clearly show evidence for transition region ($T \approx 10^5$ K) plasma. They therefore proposed a transition region "dividing line" in the H-R diagram running nearly vertical near spectral type K1 III separating stars with and without detected transition regions. Using a much larger data set, Simon, Linsky, and Stencel (1982) confirmed the existence of such a "dividing line" for apparently single stars (see Fig. 2) in the same location of the H-R diagram as that previously mentioned for X-rays. They showed that in general stars to the right of the line have C IV 1550 Å emission measure upper limits that are significantly below the emission measures generally inferred for stars to the left. In the case of Arcturus (K2 III) the

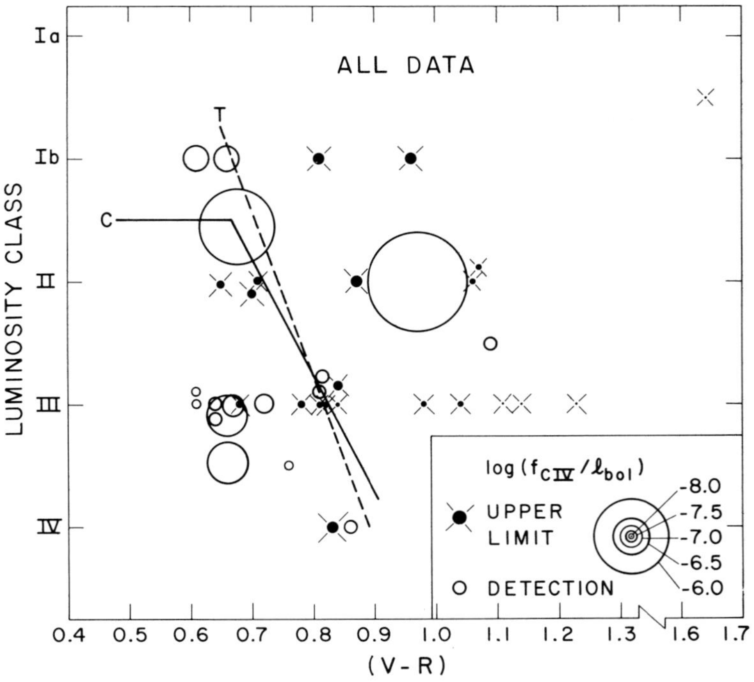

Fig. 2. An H-R diagram in which normalized C IV λ1549 fluxes or their upper limits are indicated for 39 stars by the size of the circles (see scale at lower right). The line marked C is that proposed by Ayres et al. (1981) to separate a region of the H-R diagram in which nearby stars generally have detected coronae (to the left) from a region in which no single stars have detected coronae (to the right and above). The line marked T is that originally proposed by Linsky and Haisch (1979) to separate a region of the H-R diagram in which nearby stars generally show evidence of 10^5 K plasma (to the left) from a region in which stars generally do not show evidence of any plasma hotter than 10^4 K. The smallest dots are upper limits in the log of ≲-8.5. (From Simon, Linsky, and Stencel 1982.)

emission measure upper limit for 10^5 K plasma is only 0.02 that of the quiet Sun. However, as was previously stated for the X-ray data, the absence of detected C IV emission in the cooler stars is no proof of the absence of 10^5 K plasma, only a determination of the upper limit of its emission measure. Nevertheless, I believe at present that there is no 10^5 K plasma in the outer atmospheres of such stars.

This picture that I have just presented is probably too simple. For example, the RS CVn-type close binary systems containing rapidly rotating K giants show strong C IV emission even though the giant star lies somewhat to the right of the dividing line indicated. A good example is σ Gem (K1 III+?). This suggests that stellar rotation may play a key role in determining the amount of hot plasma in the outer atmosphere of a star through the dynamo generation of magnetic fields and the large nonradiative heating rates due to MHD heating modes. Even though rotation rates for single stars decrease rapidly as stars evolve across this boundary (Gray 1981), the dividing line should be somewhat fuzzy due to a distribution of rotation rates at a given spectral type and more rapid rotation induced by the tidal forces of a companion star. It is noteworthy that the three stars with detected C IV emission to the right of the boundary in Figure 2 have all turned out to be binary systems -- 56 Peg has a white dwarf companion (Schindler et al. 1982), α UMa has an A dwarf companion, and α TrA appears to have an F dwarf companion (Ayres 1984). The roles that these companions may play in producing or enhancing the observed C IV emission is not known at this time. Also, stars located near the dividing line with very similar macroscopic parameters including rotation rates, like the Hyades K0 III giants (Baliunas, Hartmann, and Dupree 1983), can have very different C IV fluxes for no apparent reason. Thus the mechanisms responsible for the dividing line are not well understood.

The evidence to date, therefore, is that the M giants and supergiants do not have 10^5 or 10^6 K plasma in their outer atmospheres.

2.3. Evidence for the Onset of Cool Chromospheric Winds

Stars cooler and more luminous than these boundaries typically have large mass loss rates (Cassinelli 1979) as inferred from circumstellar absorption features and infrared emission from dust among the M supergiants or asymmetric Ca II and Mg II emission lines (Stencel 1978; Stencel and Mullan 1980) in the K stars. While the main thrust of this meeting is to study mass loss from the coolest luminous stars, it is important to recognize that mass loss does not begin abruptly among the M stars, but is important in the K giants and supergiants as well. This is illustrated in Figure 3. Permanent circumstellar absorption features indicative of large mass loss occur above and to the right of the line designated by the symbol CS (Reimers 1977). Thus the K supergiants have large mass loss rates. The less luminous K stars have insufficient mass loss to produce permanent blue-shifted circumstellar absorption features, but mass loss is indicated by asymmetries in self-reversed chromospheric resonance lines. The onset of such asymmetries as seen in the Ca II H and K lines is indicated by the line marked C (Stencel 1978), and in the Mg II h and k (2803, 2796 Å) lines by the line marked M (Stencel et al. 1980a) in Figure 3.

Asymmetric chromospheric resonance lines can be understood in the following way. In a static plane-parallel chromosphere characterized by increasing temperature and decreasing density with height, the line source function will initially increase with height above the base of the chromosphere if the line center optical depth exceeds the critical

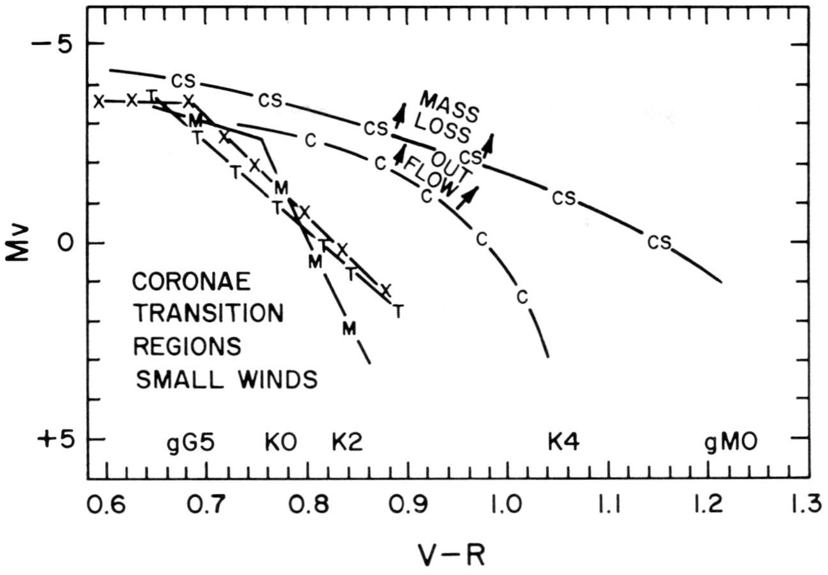

Fig. 3. Tentative dividing lines for different phenomena in the H-R diagram. The line designated by the symbol (×) separates the regions where hot coronae are observed (to the left) and not observed (above and to the right) in soft X-rays (Ayres et al. 1981). The line designated by the symbol (T) separates the regions where transitions regions are usually observed (to the left) and not observed (above and to the right) by emission in the C IV line (Linsky and Haisch 1979). The lines designated by the symbols (M) and (C) separate regions of small winds (to the left) and large winds (above and to the right) as determined by the asymmetry of the Mg II and Ca II resonance lines (Stencel et al. 1980a, Stencel 1978). The line designated by the symbol (CS) marks the onset of permanent circumstellar absorption features (above and to the right) proposed by Reimers (1977). (From Linsky 1981.)

value for thermalization. Higher in the chromosphere the source function decreases outward as radiation near line center can leak out after many scatterings. The net effect is a dark line core, equally bright emission features on either side, and rather dark line wings formed in the photosphere to either side of the emission features. The symmetry of the two emission features breaks down if we allow the chromosphere to expand such that mass flux is conserved. Then the expansion velocity increases outward as the density decreases correspondingly. As a result the most opaque part of the line (the dark core) is blue shifted the most (because it is formed highest in the chromosphere), and it covers up most of the emission feature on the blue side of the line. The result is an asymmetric line characterized by a blue-shifted core and brighter emission in the red peak, however, Athay (1976) has called attention to ambiguities in inferring flow velocities from asymmetric line profiles. This very simple explanation ignores the partial redistribution and extended geometry effects to be discussed below.

Line asymmetries in the Mg II h and k line are illustrated in the IUE spectra of Stencel et al. (1980b) shown in Figure 4. Here the two G stars (ε Leo and ζ Cyg) show relatively symmetric profiles, whereas the two K stars (α Boo and β UMi) show highly asymmetrical profiles. In interpreting these line asymmetries one should recognize that interstellar Mg II absorption features are typically present (Böhm-Vitense 1981) and can also produce asymmetries when the interstellar flow velocity differs substantially from the stellar radial velocity. Interstellar absorption cannot explain the asymmetry for α Boo, however, as it occurs at much smaller velocity than the main absorption features at -40 km s^{-1} (Ayres, Simon, and Linsky 1982).

The Ca II and Mg II curves in Figure 3 are significantly different, and between them lie stars with asymmetric Mg II and symmetric Ca II lines. Mullan and Stencel (1982) refer to these as "discrepant asymmetry" stars, and suggested that the discrepancy may be explained

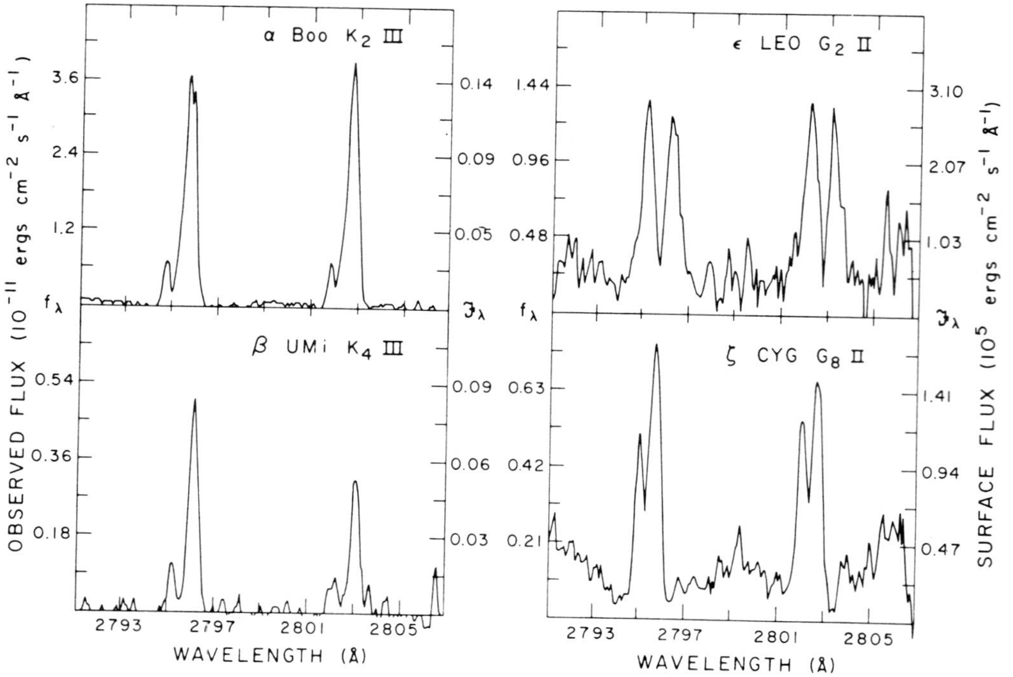

Fig. 4. IUE spectra in the region of the Mg II h(2803 Å) and k(2796 Å) lines in observed and surface flux units. (From Stencel et al. 1980b.)

by the smaller abundance and therefore smaller opacity of the Ca II lines which require larger mass loss rates to be observed. Mullan (1984a) has proposed that the discrepancy is instead a natural consequence of the Wilson-Bappu effect and the decrease in wind speed toward the upper right-hand corner of the H-R diagram. Since the Ca II lines are nearly a factor of two narrower than the Mg II lines, the blue emission peaks of the Ca II lines cannot be obliterated by wind absorption until the wind speed is sufficiently small. Thus the onset of Ca II asymmetry lies to the right and above the onset of Mg II asymmetry in the H-R diagram.

2.4. Hybrid Stars

The question of whether cool giants can have either hot plasma or cool winds but not both is complicated by the existence of the so-called hybrid stars. Hartmann, Dupree, and Raymond (1980, 1981) and Reimers (1982) showed that there is a class of stars that show evidence for C IV emission, indicative of 10^5 K plasma, and absorption features in the Mg II lines blueshifted by ~100 km s^{-1}, indicative of strong mass loss. These stars also have low velocity absorption features in the Mg II lines that are likely due to interstellar absorption (Drake and Linsky 1984). Some of these hybrid stars like α Aqr (G2 Ib) and β Aqr (G0 Ib) lie to the left of the transition region dividing line, while others like ι Aur (K3 II) and α TrA (K4 II) clearly lie to the right. These stars place this question into clear focus and have stimulated much thought, resulting in four proposed explanations.

(1) Hartmann et al. (1981) proposed that these stars have Alfvén-wave driven winds with a maximum temperature near 10^5 K, and the Mg II absorption features are formed by cool gas at the base of the flow (low velocity component) and in a cooling region beyond the region of maximum temperature (high velocity component).

(2) Linsky (1982) proposed that these stars have outer atmospheres with two components -- a few magnetically closed loops responsible for the 10^5 K emission, and large magnetically open regions (analogous to solar coronal holes) with cool outflowing gas. The presence of these loops may be due to dynamo processes as the hybrid stars are rapid rotators for their spectral classes (Brosius, Mullan, and Stencel 1985).

(3) Mullan (1984b) proposed that these stars have cool (~10^4 K) winds with 10^5 K shocked interfaces between high and low velocity components of the wind. These shocks would be analogous to corotating interaction regions (CIRs) on the Sun.

(4) Ayres (1984) found that α TrA (K4 II) has excess ultraviolet continuum emission that indicates the presence of a heretofore unknown young F dwarf companion. He argued that the F dwarf may be responsible for some or all of the 10^5 K emission lines and that other hybrid stars may also have F dwarf companions.

We do not know yet whether any of these proposed explanations is correct, but the data so far are consistent with the tentative conclusion that, except for the hybrids, all cool stars either have outer atmospheres with cool winds or hot plasma but not both.

3. PHYSICAL PROPERTIES OF CHROMOSPHERES IN RED GIANT STARS

3.1. Empirical Determination of Chromospheric Temperatures

As previously mentioned, we cannot yet determine whether the atmospheric region lying above the photosphere in the red giants and supergiants is an extended, nonradiatively heated chromosphere of relatively homogeneous properties or consists of isolated warm plasma lying behind shocks produced by outwardly propagating acoustic waves in an otherwise cool medium. In what follows I will assume the former, but one should not exclude the possibility of the latter. In any case it is important to know typical temperatures for the warm plasma indicated by the emission lines of Ca II, Mg II, C I, O I, S I, and Si II usually seen in the ultraviolet spectra of these stars.

Brown and Carpenter (1984) have shown that a good diagnostic of chromospheric temperatures of these stars is the flux ratio of the 2325 Å (^{4}P–^{2}P^o) multiplet to the 1335 Å (^{2}D–^{2}P^o) multiplet of C II. This ratio is primarily temperature dependent due to the different collisional excitation rates to the upper states, and only weakly density dependent through collisional de-excitation of the ^{4}P–^{2}P^o intersystem multiplet. They found (see Table 2) that for 11 K–M

Table 2

Electron Density, Chromospheric Temperature, and Geometric Extent of C II Emitting Regions

Star	Spectral Type	Coronal/Noncoronal	R_{chromo}/R_{photo}	Log N_e (cm^{-3})	T_e
α Cen B	K1 V	C	<1.001	10.9:	>11,800
β Cet	G9.5 III	C	<1.001	10.9:	>7,300
β Gem	K0 III	C	<1.001	10.3	>9,000
α Boo	K2 III	NC	2.0	8.5	8,100
α Tau	K5 III	NC	1.4	8.3	10,000:
β And	M0 III	NC	2.4	8.2	<8,300
γ Cru	M3.4 III	NC	1.4	8.2	8,200
α TrA	K2 II	hybrid	2.0	8.1	11.800
β Peg	M2.5 II-III	NC	3.2	7.9	8,700
β Gru	M3 II	NC	1.7	8.1	7,500
ε Gem	G8 Ib	NC	5.9	7.8	8,100
ε Peg	K2 Ib	NC	1.9	8.0	11,300:
λ Vel	K5 Ib	NC	2.1	8.3	7,800
σ CMa	M0 Iab	NC	4.8	8.0:	<6,900:
α Ori	M2 Iab	NC	2.1	7.5-8.1	8,900

giants and supergiants with no evidence for hot plasmas the chromospheric temperatures generally lie in the range 8400±1000 K. By contrast, the hybrid star α TrA (K4 II) appears to have a chromospheric temperature of 11,800 K and solar chromospheric models generally place the formation of the C II lines near 16,500 K (Lites, Shine, and Chipman 1978). The maximum chromospheric temperature in the plane-parallel model of α Ori (M2 Iab) computed by Basri, Linsky, and Eriksson (1981) is 7000 K, and in the spherically symmetric model of Hartmann and Avrett (1984) is 8050 K. Radio spectra place the chromospheric temperature of this star at ~10,000 K (Newell and Hjellming 1982) or 8600 K (Wischnewski and Wendker 1981). Thus the mean temperature of 8400±1000 K derived from the C II data seems to be a realistic value at least for the chromospheric layers of the red giants and supergiants where the C II lines are formed.

3.2. Empirical Determination of Chromospheric Electron Densities

There is at present only one direct method for determining chromospheric electron densities in the red giants without computing detailed models based on numerous assumptions. Stencel <u>et al</u>. (1981) showed that line flux ratios within the C II 2325 Å multiplet ($^4P-{}^2P^o$) are useful density diagnostics in the range $10^7 < n_e < 10^9$ cm^{-3}. The density sensitivity arises from the different collisional deexcitation rates for lines in the multiplet. Using this method, Brown and Carpenter (1984) found that for 11 K-M giants and supergiants the electron densities generally lie in the small range $7.8 < \log n_e < 8.3$ (see Table 2), whereas for three stars with coronae the chromospheric electron densities derived from Si III 1892 Å/C III 1909 Å ratios lie in the range $10 < \log n_e < 11$. The hybrid star α TrA lies in the latter range. They found no other apparent trends with spectral type or luminosity class. By comparison, chromospheric electron densities in the Hartmann-Avrett model of α Ori lie in the range $\log n_e = 6$–7.

3.3. Empirical Determination of the Geometrical Scales of Red Giant Chromospheres

Assuming spherical symmetry and constant electron density, Carpenter, Brown, and Stencel (1985) were also able to derive the outer radius of red giant chromospheres (R_{chromo}) from the C II electron densities and emission measures derived from the 2325 Å fluxes (cf. Stencel 1981). The R_{chromo}/R_{photo} ratios are also listed in Table 2. They found that the chromospheres of the three coronal stars are geometrically thin [$(R_{chromo}-R_{photo})/R_{photo} < 0.001$], whereas for the 11 noncoronal stars and α TrA the chromospheres are geometrically extended [$R_{chromo}/R_{photo} = 1.3$–$5.6$] with no apparent trend with spectral type or luminosity class. Relaxing the assumption of constant electron density leads to more extended chromospheres. These results provide important evidence that the geometrical scale of chromospheres changes drastically across the transition region dividing line.

The R_{chromo}/R_{photo} ratio derived for α Ori is 2.1 compared with 4 in the Hartmann-Avrett model and a similar value implied by the

Newell-Hjellming radio data. The 11 μm heterodyne interferometry of Sutton (1979) indicates that the inner radius of the α Ori dust shell is at least 5 R_{photo} and Low's (1979) infrared interferometry imply at least 10 R_{photo}. These data suggest that the constant n_e assumption of the C II method is probably unrealistic, but the basic result that red giant chromospheres are geometrically extended appears firm.

4. MASS LOSS RATE ESTIMATES DERIVED FROM ULTRAVIOLET SPECTRA

I now come to the main aspect of my talk -- the derivation of red giant mass loss rates from the analysis of ultraviolet spectra. The ultraviolet spectrum plays an important role in the determination of mass loss rates because the resonance lines of the dominant stages of ionization are located in the ultraviolet so that corrections for the uncertain degree of ionization are unimportant and the complex photospheric spectra are relatively weak.

4.1. Analysis of Asymmetric Emission Lines

As previously mentioned, outflow in the chromosphere can produce asymmetric emission profiles of optically thick resonance lines including Ca II H and K, Mg II h and k, the O I 1304 Å multiplet, and H I Lyman alpha (1216 Å). In addition, such subordinate lines as Hα and the Ca II infrared triplet lines (8498, 8542, 8002 Å) can have emission features or P Cygni type absorption-emission features near line center (e.g. Goldberg 1979).

Drake and Linsky (1983a) have developed a nonLTE computer code to infer the mass loss rate of cool stars by matching the observed asymmetric emission lines of Mg II and other species. This code assumes a spherically-symmetric extended chromosphere in which the expansion velocity increases monotonically outward. The transfer equation is solved in the comoving frame of the flow taking into account the nearly coherent scattering in the line wings. They made a complete partial redistribution (PRD) calculation rather than assuming the Sobolev approximation, because the expansion velocities in red giants are typically only a few times the maximum Doppler width and the computation must include the base of the chromosphere where the line wings are formed and the expansion velocity is very small. They modified a code written by P. Kunasz to include PRD using the comoving frame formulation developed by Mihalas, Kunasz, and Hummer (1975) and Mihalas _et al_. (1976).

In their calculations Drake and Linsky (1983a) explored the sensitivity of the Mg II h and k profiles to changes in the atmospheric parameters. They found as anticipated that the line profiles are very sensitive to the expansion velocities. This is shown in Figure 5, where mass flux conservative flow is assumed and the maximum expansion velocity (assumed to occur at twice the photospheric radius, 2 R_*) is set equal to 0.02, 1, and 3 times the Doppler width at R_{max} = 2 R_*.

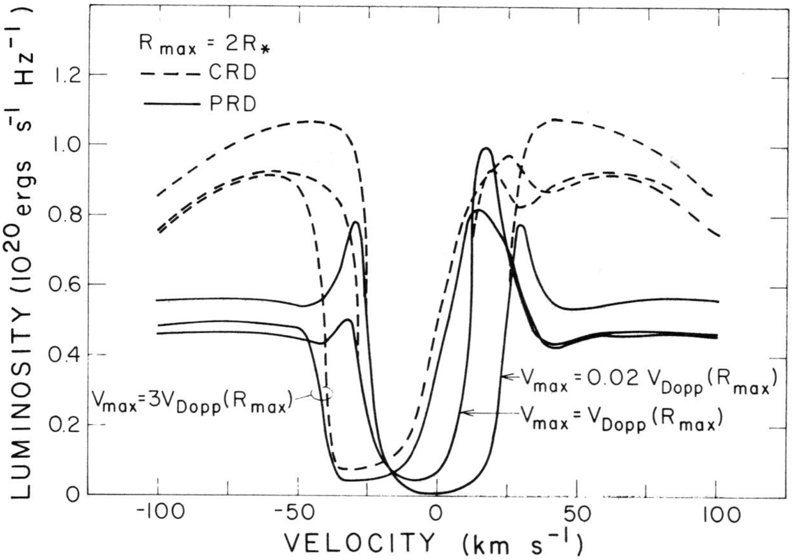

Fig. 5. Computed profiles of the Mg II k line for an early K giant star with $R_{max} = 2\ R_*$. CRD and PRD profiles are given for three different velocity laws. (From Drake and Linsky 1983a.)

The effect of increasing the maximum expansion velocity is to progressively obliterate the blue emission peak and to enhance the red emission peak. The PRD profiles agree with complete redistribution (CRD) near line center where Doppler redistribution dominates, but the PRD profiles lie significantly below the CRD profiles away from line center where the line opacity is nearly coherent scattering. They also found (see Fig. 6) a progressive enhancement of the emission peak on the red side with increasing chromospheric size (R_{max}) and chromospheric column density. These calculations indicate that asymmetric chromospheric profiles can be used to estimate chromospheric geometrical scales, expansion velocities, column densities, and thus mass loss rates. In a subsequent calculation (Drake and Linsky 1983b) they matched the Mg II k line of α Boo (K2 III) by an extended chromosphere model with $\dot{M} \approx 10^{-10}\ M_\odot\ yr^{-1}$.

Hartmann and Avrett (1984) computed a spherically-symmetric model for α Ori (M2 Iab) for which Alfvén waves drive an outflow of $\sim 10^{-6}\ M_\odot\ yr^{-1}$ and heat the extended chromosphere. This internally self-consistent one component model is in good agreement with the radio fluxes and various estimates of the mass loss rates. Their computed fluxes for lines of Ca II, Mg II, C II, and O I differ from observations, however, and profiles for the Mg II k and Hα lines are considerably different from the observed profile shapes. Additional calculations are needed to see what modifications to their model can match the ultraviolet emission line profile. Using the same computational

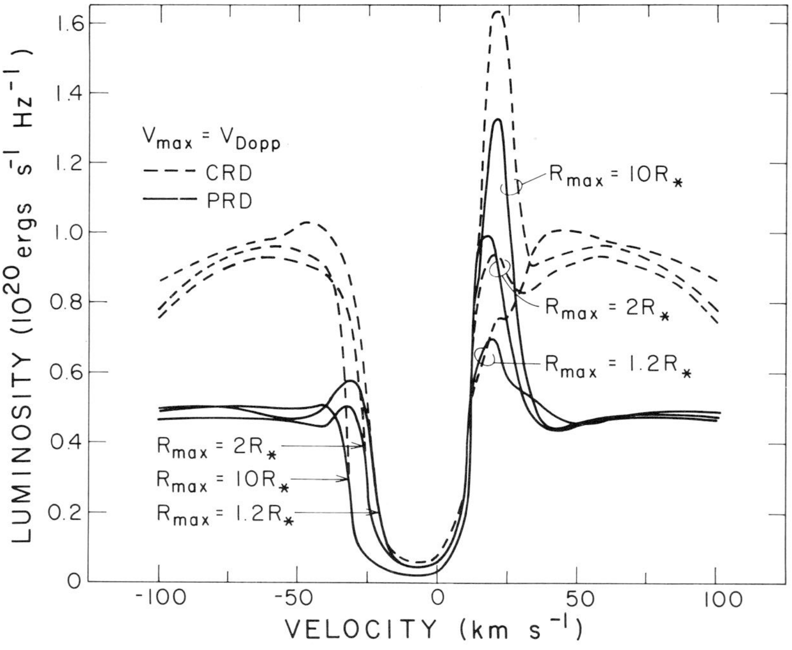

Fig. 6. Computed profiles of the Mg II k line for an early K giant star with $V_{max} = V_{Dopp}(R_{max})$ for radial extents of the chromosphere of 1.2 R_*, 2 R_*, and 10 R_*. (From Drake and Linsky 1983a.)

methods and Alfvén waves to drive and heat the expanding chromosphere, Dupree, Hartmann, and Avrett (1984) computed Hα line profiles for two Population II giants and found very small steady mass loss rates, $\dot{M} \ll 2 \times 10^{-9}\ M_{\odot}\ yr^{-1}$.

Present IUE instrumental sensitivity limits the available high resolution, high signal-to-noise ultraviolet line profiles to the Mg II lines in many red giants and a few additional lines of S I, O I, Fe II, and Si II in a few of the brightest stars (see Engvold et al. 1983). Judge (1983) has discussed mechanisms for pumping the 1820, 1826 Å multiplet of S I and Hempe and Reimers (1982) have discussed the radiative pumping of Fe II multiplet UV191 in the wind of 32 Cyg. These lines may turn out to be useful probes of the mass loss rates because they can be used to infer column densities of neutral species.

Carpenter (1984) analyzed the very rich IUE high dispersion 2300-3000 Å spectrum of Fe II consisting of 42 unmutilated, unblended lines of multiplets UV 1-3, 32-33, 35-36, and 60-64. He found a correlation of Fe II line asymmetry with intrinsic line strength that indicates a velocity field which is initially constant, then algebraically increases with radius to a maximum value and then decreases significantly before reaching an asymptotic flow speed far from the star. The mean velocity of the chromospheric regions emitting Fe II does not appear to differ substantially from the time-average of the photospheric velocity in this data set, but there is evidence that the two regions are not strongly coupled and thus that the chromosphere does not strictly follow the semi-periodic six year pulsations of the photosphere.

4.2. Analysis of Absorption Lines in the Antares (M1.5 Iab + B2.5 V) System

The presence of a hot main sequence star companion in the wind of a red giant or supergiant provides a powerful tool for determining the properties of this wind by absorption line spectroscopy. The resonance lines of most dominant stages of ionization in these winds are located in the ultraviolet, so that estimates of mass loss rates and wind temperatures and densities from absorption lines in the visible are severely compromised by highly uncertain corrections for ionization equilibria. The study of ultraviolet absorption lines formed in the winds of these stars when single is similarly hampered by their very weak ultraviolet continua. Both problems are solved when the red giant has a hot main sequence star companion to act as a bright continuum background against which absorption lines produced by the wind of the red giant can be easily measured. Furthermore, in those cases where the orbital period is not too many years, the hot star acts as a moving probe to study the red giant wind at different radial positions.

Antares is the brightest such system consisting of an M1.5 Iab star and a B2.5 V companion at a projected separation of 522 AU = 2"9, corresponding to 140 radii of the M supergiant. Van der Hucht, Bernat, and Kondo (1980) studied this system using high resolution 2000-2900 Å spectra obtained with the BUSS experiment, and Bernat (1982) studied this system with high resolution 1200-2700 Å spectra obtained with IUE. These two analyses are in basic agreement. They both place the B star somewhat in front of the M star and assume the M star wind to be spherically symmetric, homogeneous, time-independent, and to have reached its terminal velocity of 17 km s^{-1} before reaching the radial distance of the B star. Both studies also assume solar abundances, constant line broadening, and radiative ionization equilibrium determined from the radiation field of the M star with some contributions of the B star close to the B star.

Van der Hucht et al. used a curve-of-growth analysis of absorption lines of Cr II, Fe II, Ni II, and Zn II to derive column densities for each ion and a Doppler line broadening parameter of 8 km s^{-1}. Their final parameters are a mass loss rate for the M star of 7.1(±3.5) × 10^{-6} $M_\odot$ yr^{-1}, an inner radius of the M star expanding envelope of ~8.5 photospheric radii, and a total gas mass in the wind of ≈0.01 the mass of the M star. Bernat (1982) used a similar method of analysis but included the dominant stages of ionization of 11 elements. With this larger data base he determined a best fit mass loss rate of 6.4 × 10^{-6} $M_\odot$ yr^{-1}, in agreement with the analysis of the BUSS data. This agreement is likely fortuitous, however, as the derived mass loss rates differ by a factor of 40 depending on which ion is used. Bernat considered different possible sources for the large discrepancies, including interstellar absorption, but concluded that the source is most probably inadequacies in the simple assumptions made concerning the flow, in particular the complex interaction of the winds from the two stars.

Recently, Hagen (1984) has carefully reconsidered these Antares data. He argued that about 90% of the data, consisting of absorption lines from ground states, are most likely interstellar rather than circumstellar in origin. The remaining lines have P Cygni profiles and must be circumstellar in origin. His analysis using the techniques described in the next section yields consistent mass loss rates among the different ions in the range $4 \times 10^{-7} - 1 \times 10^{-6}$ $M_\odot$ yr^{-1}. This work points out some of the problems in deriving mass loss rate estimates from ultraviolet spectral lines, but it does not invalidate my belief that ultraviolet spectroscopy is a potentially powerful tool for probing the winds of red giant stars.

4.3. Analysis of P Cygni-Shape Line Profiles in ζ Aurigae/VV Cephei Systems

I next consider the thoughtful analysis of IUE spectra of ζ Aurigae/VV Cephei binary systems under way by astronomers at Hamburger Sternwarte in a series of papers by Hempe (1982, 1983), Che, Hempe, and Reimers (1983), Reimers and Schröder (1983), and Che and Reimers (1983). These systems are eclipsing binaries consisting of a K or M supergiant and a main sequence B star. Historically, two names have been used to describe these systems -- ζ Aur systems for those with K supergiant primaries, and VV Cep systems for those with M supergiant primaries. In fact, these systems are basically similar, and I will refer to systems of both types hereafter as ζ Aur systems. A summary of the properties of the five systems that these authors have investigated is given in Table 3.

In these systems the flow from the K supergiant is perturbed by the B star and must be treated as a three-dimensional problem. They compute the P Cygni-shape line profiles as a function of orbital phase that result from the resonant scattering of B star photons by ions in the expanding envelope of the K star. They compute line source functions using the Sobolev approximation in an escape probability formalism for a two-level atom, and then compute the emergent line profiles exactly by integrating over the three-dimensional geometry. The ionization equilibria in the wind is computed including ionization by photons from the B star, and for all these systems there is an H II region surrounding the B star. They assume Doppler line profile shapes and that the wind from the K star reaches terminal speed before it reaches the radial distance of the B star.

I have described their theoretical approach in some detail to give some feeling for the complexity of the problem. In practice,

Table 3

Summary of Data for the ζ Aur Systems

Parameter	ζ Aur		32 Cyg		31 Cyg		δ Sge		Boss 1985	
	B star	K star	B star	K star	B star	K star	B star	K star	B star	K star
Spectral Type	~B8	K4 Ib	~B8	K5 Iab	~B4	K4 Ib	~B9	M2 II	B2 V	M1 Iab
T_{eff} (K)	13,000	3,850	13,000	3,800	17,500	3,850	10,500	---	22,500	---
Mass ($M_\odot$)	5.6	8.3	4	8	6.2	9.2	2.8	8.5	---	---
Eclipse Radius ($R_\odot$)	5.1	130	---	---	4.0	135	2.7	---	---	---
log $L/L_\odot$	---	3.41	---	3.82	---	3.91	---	3.44	---	---
Mass Loss Rate (10^{-8} $M_\odot$ yr^{-1})	---	0.63	---	2.8	---	$\gtrsim 1$	---	2	---	10
Wind Speed (km s^{-1})	---	40	---	60	---	80	---	28	---	10
Line Broadening (km s^{-1})	---	30	---	25	---	20	---	20	---	20
Period (days)	972.2		1147.8		3784.3		3720		9752	
Semi-major axis (10^{13} cm)	6.83		7.65		17.64		3.67		20.4	
Inclination	90°		≈80°		90°		>70°		≈90°	

they first analyze the Fe II UV multiplet 9 lines at 1272-1277 Å because these are pure absorption lines that are simply modeled. A simple curve-of-growth analysis of these lines gives a good first approximation to the wind speed, line broadening parameter, and mass loss rate. With these initial parameters, they attempt to match the optically thick scattering lines of Fe II (1618, 2607 Å), Si II (1533 Å), and S II (1253 Å) which have P Cygni-type profiles. Examples of the fits to the observed line profiles of 32 Cyg at phase 0.78 are shown in Figure 7, and the fits to the Si II 1533 Å line in 31 Cyg at four phases are shown in Figure 8.

These authors are arriving at an interesting self-consistent picture for the winds in these ζ Aur systems. Typical mass loss rates are roughly 10^{-8} $M_\odot$ yr^{-1} with uncertainties of roughly a factor of 3, and the rates appear to be largest for the Iab supergiants. Wind speeds are 40-80 km s^{-1} for the K supergiants and smaller for the M supergiants. In all these stars the line broadening parameter is large (20-30 km s^{-1}), indicating large stochastic motions that could be produced by pulsations in the late-type stars. These parameters are likely more accurate than values derived from optical spectroscopy and infrared data.

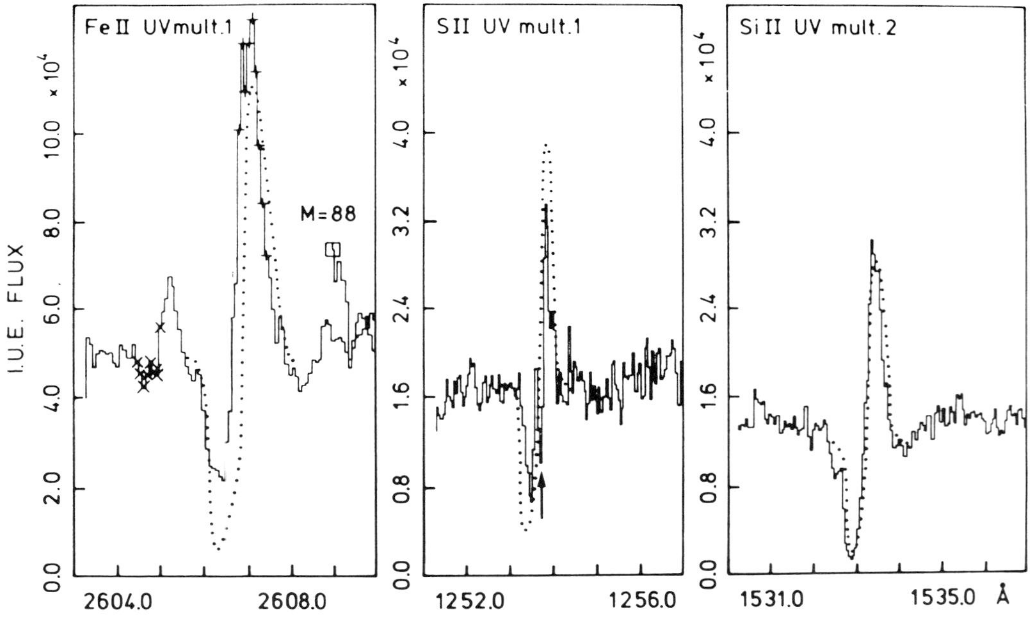

Fig. 7. Comparison of theoretical (••••) and observed Fe II multiplet 1, S II multiplet 1 (interstellar absorption indicated by arrow) and Si II multiplet 2 lines for 32 Cyg at phase ϕ = 0.78. (From Che, Hempe, and Reimers 1983.)

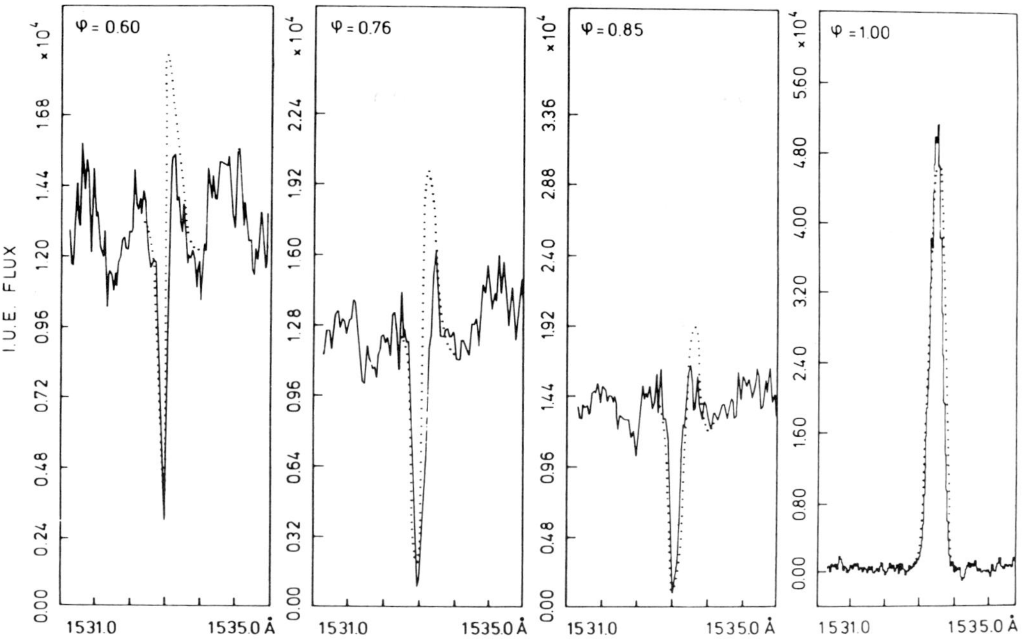

Fig. 8. Comparison of theoretical (····) with observed Si II multiplet 2 lines for 31 Cyg at various phases. $v_{wind} = 80$ km s^{-1}, $v_{sto} = 20$ km s^{-1}, $\dot{M} = 10^{-8}$ $M_{\odot}$ yr^{-1}. (From Che, Hempe, and Reimers 1983.)

Recently Schröder (1984) determined power law representations for the density and expansion velocity dependence on distance for the ζ Aur system. He places the location of the main energy input to the wind per unit mass between 0.5 and 5 photospheric radii. Also very recently, Che-Bohnenstengel (1984) estimated temperatures in the wind of 32 Cyg from the excitation of Fe II lines. The derived temperature in the wind beyond the B star is dependent on the assumed electron density, but is ~4800 K if $n_e/n_H \gtrsim 0.01$ at about 5 R_{photo} as suggested by the microwave observations of Drake and Linsky (1983a) and a comparison of direct determinations of n_e (Carpenter et al. 1985) with n_H (Schröder 1984). If, on the other hand, $n_e/n_H < 0.001$, then the wind far from the K star is at roughly 10,000 K. While it is tempting to argue for the cooler temperature, which would imply that the wind has cooled significantly at a distance of 5 R_{photo}, more realistic calculations of the Fe II excitation equilibrium are needed.

The analysis of the wind in ζ Aur itself by Che et al. (1983) is uncertain because none of the Fe II, S II, and Si II lines they modeled have emission components. Chapman (1981) studied the Mg II h and k lines that do have red-shifted emission components and blue-shifted circumstellar absorption components, but he inferred the maximum wind expansion velocity not from a detailed fit to the line profile but rather from measuring the longest wavelength on the emission line profile. His estimate of the wind speed ≈100 km s^{-1}, however, is likely an upper limit because the red edge of the emission peak indicates the line-of-sight velocity of atoms far behind the B star that are red-shifted by a combination of expansion and turbulent velocities. Thus his estimate of $\dot{M} = 2 \times 10^{-8}\ M_{\odot}\ yr^{-1}$ is probably also an upper limit.

Chapman (1981) called attention to the presence of strong C IV absorption features in ζ Aur, which is contrary to the general trend (see above) of no 10^5 K plasma in the atmospheres or winds of K supergiants. Since the C IV absorption is strongest at the time of secondary eclipse (B star in front), he argued that the absorbing hot plasma is located in an accretion column on the side of the B star away from the K supergiant. The phase dependence of the absorption features of C IV, Si IV, and Mg II and the emission lines during primary eclipse then led Chapman (1981), Stencel and Chapman (1981), and Ahmad, Chapman, and Kondo (1983) to propose that any realistic analysis of the circumstellar envelopes of ζ Aur systems should include the winds of both stars, an accretion bow shock around the B star where the two winds interact, and a hot column of shocked gas behind the B star and accreting onto it. Thus the physical processes occurring in the circumstellar environments of these systems as indicated by their ultraviolet spectra are likely quite complex and interesting.

5. CONCLUSIONS

Although red giants emit only a tiny fraction of their radiative output in the ultraviolet, the study of ultraviolet emission and absorption line profiles provides powerful and in some cases unique methods for inferring the properties of the extended, expanding outer atmospheres of these stars and their mass loss rates. In particular, binary systems containing a red giant and a hot star are invaluable in ascertaining accurate mass loss rates. Instrumental sensitivity has limited the stars and spectral lines that could be studied until now, but the high sensitivity and spectral resolution of the spectrographs on Space Telescope and the proposed Columbus/Far Ultraviolet Spectrograph Explorer missions should stimulate rapid advances in this field.

This work was supported in part by NASA grants NAG5-82 and NGL-06-003-057 through the University of Colorado. I would like to thank my colleagues Drs. T. R. Ayres, A. Brown, K. G. Carpenter, S. Drake, D. M. Gibson, and F. M. Walter for stimulating discussions and for permission to describe unpublished work. I would also like to thank Drs. D. Reimers, R. E. Stencel, L. A. Willson, and other participants at the Third "Trieste" Workshop at the National Solar Observatory (Sunspot) for discussions on this topic.

REFERENCES

Ahmad, I. A., Chapman, R. D. and Kondo, Y. 1983, Astr. Ap., 126, L5.

Athay, R. G. 1976, The Solar Chromosphere and Corona: Quiet Sun (Dordrecht: Reidel).

Ayres, T. R. 1984, submitted to Ap. J. (Letters).

Ayres, T. R., Linsky, J. L., Vaiana, G. S., Golub, L. and Rosner, R. 1981, Ap. J., 250, 293.

Ayres, T. R., Simon, T. and Linsky, J. L. 1982, Ap. J., 263, 791.

Baliunas, S. L., Hartmann, L. and Dupree, A. K. 1983, Ap. J., 271, 672.

Basri, G. S., Linsky, J. L. and Eriksson, K. 1981, Ap. J., 251, 162.

Bernat, A. P. 1982, Ap. J., 252, 644.

Böhm-Vitense, E. 1981, Ap. J., 244, 504.

Brosius, J. W., Mullan, D. J. and Stencel, R. E. 1985, Ap. J., in press.

Brown, A. 1983, in Cool Stars, Stellar Systems, and the Sun, eds. S. L. Baliunas and L. Hartmann (Berlin: Springer-Verlag), p. 282.

Brown, A. and Carpenter, K. G. 1984, Ap. J. (Letters), in press.

Carpenter, K. G. 1984, Ap. J., in press.

Carpenter, K. G., Brown, A. and Stencel, R. E. 1985, Ap. J., in press.

Cassinelli, J. P. 1979, Ann. Rev. Astr. Ap., 17, 275.

Chapman, R. D. 1981, Ap. J., 248, 1043.

Che, A., Hempe, K. and Reimers, D. 1983, Astr. Ap., 126, 225.

Che, A. and Reimers, D. 1983, Astr. Ap., 127, 227.

Che-Bohnenstengel, A. 1984, submitted to Astr. Ap.

Drake, S. A. and Linsky, J. L. 1983a, Ap. J. (Letters), 274, L77.

_____. 1983b, in Cool Stars, Stellar Systems, and the Sun, eds. S. L. Baliunas and L. Hartmann (Berlin: Springer-Verlag), p. 350.

_____. 1984, Ap. J., in press.

Dupree, A. K. 1981, in Effects of Mass Loss on Stellar Evolutions (Dordrecht: Reidel), p. 87.

_____. 1982, Advances in Ultraviolet Astronomy: Four Years of IUE Research, NASA Conference Publication 2238, p. 3.

Dupree, A. K., Hartmann, L. and Avrett, E. H. 1984, Ap. J. (Letters), 281, L37.

Engvold, O., Kjeldseth Moe, O., Jensen, E., Jordan, C., Stencel, R. and Linsky, J. 1983, in Cool Stars, Stellar Systems, and the Sun, eds. S. L. Baliunas and L. Hartmann (Berlin: Springer-Verlag), p. 359.

Goldberg, L. 1979, Q.J.R.A.S., 20, 361.

Gray, D. F. 1981, Ap. J., 251, 155.

Hagen, H. J. 1984, Masters thesis, University of Hamburg, Germany.

Haisch, B. M. and Simon, T. 1982, Ap. J., 263, 252.

Hartmann, L. 1981, in Solar Phenomena in Stars and Stellar Systems, eds. R. M. Bonnet and A. K. Dupree (Dordrecht: Reidel), p. 331.

_____. 1983, in Solar and Stellar Magnetic Fields: Origins and Coronal Effects, IAU Symp. 102, ed. J. O. Stenflo (Dordrecht: Reidel), p. 419.

Hartmann, L. and Avrett, E. H. 1984, Ap. J., in press.

Hartmann, L., Dupree, A. K. and Raymond, J. C. 1980, Ap. J. (Letters), 236, L143.

_____. 1981, Ap. J., 246, 193.

Helfand, D. J. and Caillault, J. P. 1982, Ap. J., 253, 760.

Hempe, K. 1982, Astr. Ap., 115, 133.
_____. 1983, Astr. Ap. Suppl. Ser., 53, 339.
Hempe, K. and Reimers, D. 1982, Astr. Ap., 109, 36.
van der Hucht, K. A., Bernat, A. P. and Kondo, Y. 1980, Astr. Ap., 82, 14.
Judge, P. G. 1983, in Cool Stars, Stellar Systems, and the Sun, eds. S. L. Baliunas and L. Hartmann (Berlin: Springer-Verlag), p. 353.
Linsky, J. L. 1977, in The Solar Output and Its Variations, ed. O. R. White (Boulder: Colorado Associated University Press), p. 477.
_____. 1980, Ann. Rev. Astr. Ap., 18, 439.
_____. 1981, in Physical Processes in Red Giants, eds. I. Iben, Jr. and A. Renzini (Dordrecht: Reidel), p. 247.
_____. 1982, in Advances in Ultraviolet Astronomy: Four Years of IUE Research, NASA Conference Publication 2238, p.17.
Linsky, J. L. and Avrett, E. H. 1970, P.A.S.P., 82, 169.
Linsky, J. L. and Haisch, B. M. 1979, Ap. J. (Letters), 229, L27.
Lites, B. W., Shine, R. A. and Chipman, E. G. 1978, Ap. J., 222, 333.
Low, F. J. 1979, in High Angular Resolution Stellar Interferometry, eds. J. Davis and W. J. Tango (Sydney: Univ. Sydney), p. 15.
Mihalas, D., Kunasz, P. B. and Hummer, D. G. 1975, Ap. J., 202, 465.
Mihalas, D., Shine, R. A., Kunasz, P. B. and Hummer, D. G. 1976, Ap. J., 205, 492.
Mullan, D. J. 1984a, Ap. J., 283, 303.
_____. 1984b, Ap.J., 284, 769.
Mullan, D. J. and Stencel, R. E. 1982, Ap. J., 253, 716.
Newell, R. T. and Hjellming, R. M. 1982, Ap. J. (Letters), 263, L87.
O'Brien, G. 1980, Ph.D. Thesis, University of Texas.
Reimers, D. 1977, Astr. Ap., 57, 395.
_____. 1982, Astr. Ap., 107, 292.
Reimers, D. and Schröder, K. P. 1983, Astr. Ap., 124, 241.
Schindler, M., Stencel, R. E., Linsky, J. L., Basri, G. S. and Helfand, D. J. 1982, Ap. J., 263, 269.
Schröder, K. P. 1984, Astr. Ap., in press.
Simon, T., Linsky, J. L. and Stencel, R. E. 1982, Ap. J., 257, 225.
Stencel, R. E. 1978, Ap. J. (Letters), 223, L37.
_____. 1981, SAO Special Report No. 392, p. 137.
Stencel, R. E. and Chapman, R. D. 1981, Ap. J., 251, 597.
Stencel, R. E., Linsky, J. L., Brown, A., Jordan, C., Carpenter, K. G., Wing, R. F. and Czyzak, S. 1981, M.N.R.A.S., 196, 47P.
Stencel, R. E. and Mullan, D. J. 1980, Ap. J., 238, 221.
Stencel, R. E., Mullan, D. J., Basri, G. S. and Linsky, J. L. 1980a, in The Universe at Ultraviolet Wavelengths: The First Two Years of IUE, ed. R. D. Chapman, NASA Conf. Publ. 2171, p. 317.
Stencel, R. E., Mullan, D. J., Linsky, J. S., Basri, G. S. and Worden, S. P. 1980b, Ap. J. Suppl. Ser., 44, 383.
Sutton, E. C. 1979, Ph.D. Thesis, University of California, Berkeley.
Vaiana, G. S. et al. 1981, Ap. J., 245, 163.
Wischnewski, E. and Wendker, H. J. 1981, Astr. Ap., 96, 102.
Zirin, H. 1975, Ap. J. (Letters), 199, L63.
_____. 1982, Ap. J., 260, 655.

DENSITIES, TEMPERATURES AND GEOMETRIC EXTENTS OF C II EMITTING REGIONS IN THE WINDS OF LUMINOUS, LATE-TYPE STARS

Kenneth G. Carpenter, Alexander Brown
Joint Institute for Laboratory Astrophysics, University of Colorado and National Bureau of Standards, Boulder, CO 80309
and
Robert E. Stencel
NASA Headquarters, Astrophysics Div., Washington, DC 20546

We summarize techniques for using IUE observations of the C II UV 0.01 and UV 1 emission line multiplets at 2325 Å and 1335 Å to estimate the electron density (N_e) and temperature (T_e) in, and the geometric extent of, the chromospheres of late-type stars. The results of applying these techniques to a sample of 15 stars observed with the IUE satellite are discussed.

Stencel et al. (1981) and Stencel and Carpenter (1982) show that the emission line ratios within the UV 0.01 multiplet ($2s^2 2p\ ^2P - 2s2p^2\ 4p$) are useful diagnostics of N_e in these chromospheres. Relative line fluxes measured in IUE high resolution LWR spectra are used with a theoretical calibration of line ratios versus N_e (Stencel et al. 1981) to obtain estimates of N_e for all of the stars in our sample, except the three earliest-type stars whose densities are determined from the Si III λ1892/C III λ1909 ratio.

Estimates of the electron temperatures in the chromospheres are obtained by comparing the observed ratio of the C II UV 0.01 and UV 1 multiplet fluxes with the theoretical calibration of the ratio versus N_e and T_e by Hayes and Nussbaumer (1984). The UV 0.01 fluxes are taken from the same spectra used to determine N_e, while the UV 1 multiplet fluxes are measured in low resolution SWP spectra taken with IUE. Brown and Carpenter (1984) present the details of this T_e-estimation and compute the values of N_e and T_e listed in Table I.

The radial geometric extent of the C II emitting regions around these stars are estimated from the total C II UV 0.01 multiplet flux, the N_e and T_e derived above, and estimates of the photospheric angular diameter and radius and chromospheric C II abundance. Carpenter, Brown and Stencel (1985) present derivations of the relevant equations and describe an iterative technique for their solution. They also give the ancillary data for the 15 stars in our sample and compute the values of R/R_* and $R/R_\odot$ given in Table I.

M. Morris and B. Zuckerman (eds.), Mass Loss from Red Giants, 55–56.

TABLE I

Characteristics of Late-Type Stellar Chromospheres

Star	Spectral Type	Coronal/ Noncoronal	log N_e	T_e	R/R_*	$R/R_\odot$
α Cen B	K1 V	C	10.9	>11,800	<1.001	0.9
β Cet	G9.5 III	C	10.9	>7,300	<1.001	11
β Gem	K0 III	C	10.3	>9,000	<1.001	11
α Boo	K2 III	NC	8.5	8,100	2.0	54
α Tau	K5 III	NC	8.3	10,000	1.4	66
β And	M0 III	NC	8.2	<8,300	2.4	72
γ Cru	M3.4 III	NC	8.2	8,200	1.4	170
α TrA	K2 II	hybrid	8.1	11,800	2.0	180
β Peg	M2.5 II-III	NC	7.9	8,700	3.2	320
β Gru	M3 II	NC	8.1	7,500	1.7	360
ε Gem	G8 Ib	NC	7.8	8,100	5.9	770
ε Peg	K2 Ib	NC	8.0	11,300	1.9	310
λ Vel	K5 Ib	NC	8.3	7,800	2.1	420
σ CMa	M0 Iab	NC	8.0	<6,900:	4.8	2200
α Ori	M2 Iab	NC	7.9	8,900	2.1	1100

R = stellar chromospheric radius; R_* = stellar photospheric radius; $R_\odot$ = solar photospheric radius.

The physical characteristics of the C II emitting regions (i.e., chromospheres) around a sample of 15 late-type stars as derived by the techniques outlined above are summarized in Table I. The results indicate that coronal stars (those with hot plasmas in their outer atmospheres, as indicated by C IV and N V emission in 1200-2000 Å IUE spectra) have very thin ($r \ll R_*$), relatively high density chromospheres, while the non-coronal stars (no C IV, N V emission) have geometrically thick, relatively low-density (log $N_e \sim 8$) chromospheres extending on average out to 2.5 photospheric radii.

REFERENCES

Brown, A., Carpenter, K.G.: 1984, Astrophys. J. (Lett.), in press.
Carpenter, K.G., Brown, A., Stencel, R.E.: 1985, Astrophys. J., in press.
Hayes, M.A., Nussbaumer, H.: 1984, Astron. Astrophys., 134, 193.
Stencel, R.E., Carpenter, K.G.: 1982, in Advances in Ultraviolet Astronomy: Four Years of IUE Research, eds. Y. Kondo, J.M. Meade and R.D. Chapman, NASA CP 2238, p. 243.
Stencel, R.E., Linsky, J.L., Brown, A., Jordan, C., Carpenter, K.G., Wing, R.F., Czyzak, S.: 1981, Monthly Notices Roy. Astron. Soc., 196, 470.

RESULTS FROM OPTICAL INTERFEROMETRY

Jacques M. Beckers
Advanced Development Program
National Optical Astronomy Observatory *
Tucson, Arizona 85726

ABSTRACT. The techniques of optical speckle and Michelson interferometry have both contributed significant information on the diameter, the atmospheres, and the envelopes of red supergiant stars. In this paper, I summarize the observations obtained to date and preview future plans.

1. INTRODUCTION

Next to the sun, the largest angular diameter stars are red supergiant and Mira variable stars. Their diameters are as much as 50 milliseconds of arc which is well below the angular resolution of even the space telescope but which is within the resolution capability of interferometric techniques using large telescopes or multielement interferometers. The techniques of speckle and Michelson interferometry have been described elsewhere (e.g. Labeyrie 1970, 1978). In this summary, I will report on the results of observations using these techniques. Observations are generally limited to the larger and brighter objects (α ORI, α SCO, α HER, α TAU, α BOO, o CET, etc.) and mostly refer to either diameter observations of the star as a function of wavelength and time or attempts at observations of the 2D image structure. Since the techniques are very data intensive and complex the number of research groups involved is small and the results are still very limited.

2. STELLAR DIAMETERS

Stellar diameters are derived from the variation of fringe visibility of the object as a function of angular frequency using either two element Michelson interferometers or speckle images. Instrumental and atmospheric fringe visibility effects are removed by calibration

*Operated by the Association of Universities for Research in Astronomy, Inc., under contract with the National Science Foundation.

M. Morris and B. Zuckerman (eds.), Mass Loss from Red Giants, 57–61.

against unresolved stars. The fringe visibility spectrum is interpreted in terms of the object diameter and its center to limb variation in intensity. Both measurement and interpretation include uncertainties. These are probably responsible for much of the α ORI diameter scatter shown in figure 1 taken from a recent paper by Balega et al. (1982). Diameters range from 30 to 77 millisecond of arc. Some of these variations are related to spectral features. All observers agree on an increase of diameter of 10-20% in the TiO molecular bands and perhaps

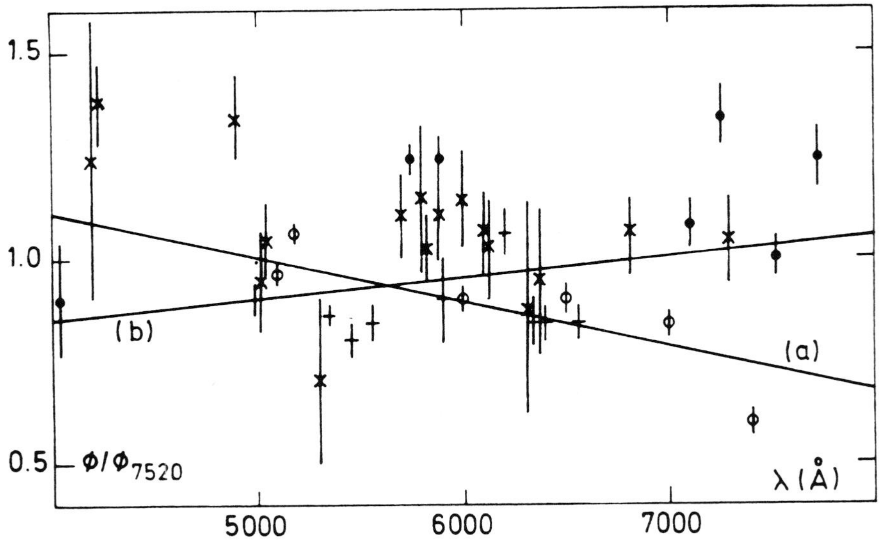

Figure 1: Uniform disc angular diameter of Betelgeuse normalized with respect to the diameter at λ 7520 Å ($\emptyset$ 7520 = 50 msa) versus wavelength. Symbols refer to: • this paper, 0 Lynds et al. (1976) and Welter and Worden (1980), + Roddier et al. (1981), and x other measurements reported by White (1980). The straight lines are linear regressions: **a** $\emptyset/\emptyset_{7520} = -10^{-4}$ ($\pm 3\ 10^{-6}$) λ (Å) + 1.49 (±0.02) with a correlation coefficient r = - 0.48. Excluding data from Lynds et al. and from Welter and Worden would lead to: **b** $\emptyset/\emptyset_{7520} = 5 \times 10^{-5}$ ($\pm 6\ 10^{-6}$) λ (Å) + 0.65 (±0.03) with r=0.23 (from Balega et al. 1982).

more in spectral lines like Hα. These increases are believed to be due to the extended atmosphere of this type of star and will in the future lead to improved stellar atmospheric models of these stars. It has been suggested that some of the other variations in the diameter of α ORI are caused by temporal changes in the diameter of the star.

3. STELLAR IMAGES

A number of techniques have been tried with limited success to give images of supergiant stars and their envelopes. The well known image of α ORI by Lynds et al. (1976) was derived by means of the so-called shift-and-add algorithm of speckle images. It uses data from the 3.8 meter telescope at Kitt Peak and is therefore limited in resolution to the radius of its Airy disk (≈ 33 millisecond of arc at 600 nm) which means less than 1.5 resolution elements per stellar diameter. To really adequately resolve the stellar disks of stars like α ORI, α SCO and α HER, one needs a substantially larger telescope. The phased MMT with its 686 cm baseline has a resolution of 18 msa at 600 nm which is substantially better and an improvement in this type of imaging may therefore be expected in the future.

The University of Arizona and Nice groups have been observing two-dimensional fringe visibilities of α ORI. Figure 2 from Roddier et al. (1983) is a two-dimensional map of the fringe visibility of α ORI. This type of observational data cannot distinguish between stellar

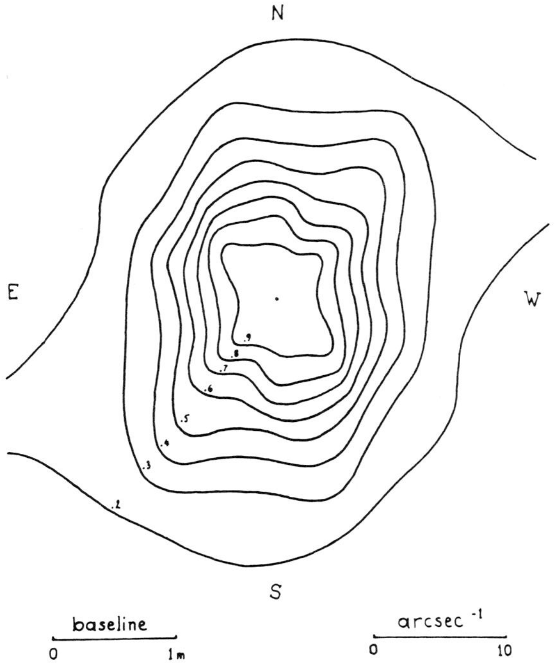

Figure 2: Two-dimensional map of the visibility of the fringes produced by Betelgeuse at λ = 5348 A (Δλ = 90 A) on 1980 November 30 (from Roddier and Roddier 1983).

image structure 180° in position angle apart. The data in Figure 2, therefore, demonstrates anisotropic image structure with an extension to either the SW or NE direction, but it cannot distinguish between structure in these two directions. Computational techniques have been proposed which derive image information from two dimensional visibility functions as shown in figure 2. They depend on the fact that the scene to be imaged is positive everywhere. So far, those techniques have been used on stellar data only with limited success.

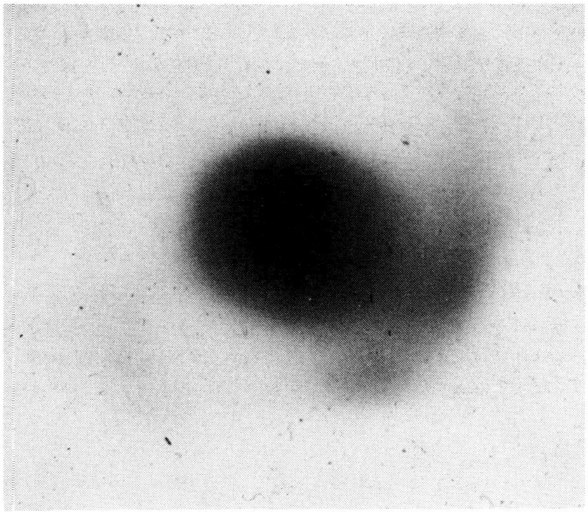

Figure 3: Image of α ORI obtained from interferograms in the pupil plane (courtesy C. and F. Roddier).

The Roddiers are presently experimenting with pupil interferometry observations of α ORI and other stars. They use a pupil rotating shearing interferometer (Roddier et al. 1978) to obtain an array of both fringe visibility and relative phase at different 2D Fourier frequencies. Figure 3 is an image of α ORI obtained from pupil interferograms. It, for the first time, shows a structure in the continuum outside the stellar disk at 2.5 R_* with an intensity of about 10% of the stellar disk. The Roddiers interpret this as evidence for the existence of a highly anisotropic dust layer around the star with a temperature of about 1750 K. We (Beckers et al. 1982, 1983) using the technique of Differential Speckle Interferometry obtained images of the Hα envelope around the star extending at the 1% level to about 2 R_* and at the 0.1% level to about 3 R_*. We find, however, that the results are very sensitive to proper detector calibration. New observations using a phased MMT and good calibrations are now being analysed. Because the MMT has an altazimuth mount, the image rotates across the field of view of the speckle camera so that it will be possible to separate instrumentally caused anisotropies from object anisotropies.

4. ANTICIPATED FUTURE DEVELOPEMENTS

It appears that the field of red supergiant imaging in optical wavelengths is ready for major advances. Pupil interferometry will result in unambiguous images of supergiants in continuum wavelengths. Differential speckle interferometry is developing rapidly and will result in images in lines. Images in opposite wings of emission lines will result in Doppler shift observations. With the phased MMT those observations can be done at resolutions of 15-20 msa. With the NNTT in the future, 5 msa resolution observations are a possibility. We also anticipate that observations of the supergiants α SCO and α HER will compete and maybe overtake the α ORI observations. These two stars are accompanied by bluer, small angular size, relatively bright companions. The angular distance of these companions from the red supergiant (2.9 and 4.9 arc sec respectively), is within the so-called isoplanatic patch so that the bluer companion can be used to determine by deconvolution or crosscorrelation the supergiant image.

REFERENCES

Balega, Y., Blazit, A., Bonneau, D., Koechlin, L., Foy, R., and Labeyrie, A. 1982, *Astron. and Astrophys.* **115**, 253.

Beckers, J. M., Hege, E. K., Murphy, H.P., and Burnette, F. 1982, *Bull. Am. Astron. Soc.* **14**, 918.

Beckers, J. M., and Hege, E. K. 1983, SPIE Proceedings, **445**, "Instrumentation in Astronomy V," p. 462.

Labeyrie, A., 1970 *Astron. and Astrophys.* **6**, 85.

Labeyrie, A., 1978 Ann. Rev. *Astron. and Astrophys.*, **16**, 77.

Lynds, C. R., Worden, S. P., and Harvey, J. W. 1976, *Astrophys. J.* **207**, 174.

Roddier, F., Roddier, C., and Demarcq, J. 1978, *J. Optics* (Paris) **9**, 145

Roddier, C., Roddier, F.., and Vernin, J. 1981, Proc. ESO Conference on "Scientific Importance of High Angular Resolution at Infrared and Optical Wavelengths," p. 165.

Roddier, C., and Roddier, F. 1983, *Astrophys. J.* **270**, L 23.

Welter, G. L. and Worden, S. P. 1980, *Astrophys. J.* **242**, 673.

White, N.M. 1980, *Astrophys.* J. **242.** 646.

Comment added by the Author: This paper was not intended to be a full review of the application of interferometric techniques to the imaging of red giants and supergiants. I therefore have not discussed all published observations. My intent was to summarize my knowledge of the sizing and imaging of these objects.

HIGH ANGULAR RESOLUTION INTERFEROMETRIC OBSERVATIONS OF BETELGEUSE IN THE VISIBLE

F. Roddier
C. Roddier
M. Karovska
University of Nice, France

A new high angular resolution interferometric technique, based on pupil plane observation and rotation shearing of the wavefront, has been developed in our laboratory and compares favorably with speckle interferometry. A two-dimensional map of the visibility of the fringes produced by Betelgeuse has been obtained from data recorded at the C.F.H. telescope on November 30, 1980, in the continuum at λ = 5348Å. The map (C. Roddier and F. Roddier, 1983) shows evidence for departure from circular symmetry. It has been interpreted as produced by a (possibly non-uniform) stellar disk surrounded with an irregular envelope.

We present here an attempt to reconstruct an image from these data. The most satisfactory convergence has been obtained with the maximum entropy algorithm described by Gull and Daniell (1978). A perspective representation of the two-dimensional irradiance distribution in the reconstructed image is given on figure 1. Although this solution is by no means unique, it is entirely consistent with the data. It shows a highly limb darkened stellar disk partially surrounded with a semi-circular shell or halo.

The infrared emission of dust particles has been detected at several tens of stellar radii from Betelgeuse, (Bloemhof, Townes and Vanderwyck, 1984). Our image suggests that dust is also present at a distance of the order of 2.5 stellar radii from the stellar center and scatters light in the visible as suggested by Tsuji (1978). The question arises whether dust can survive so close to the stellar disk. A detailed study of the dust formation process by Draine (1981) shows that, in the case of Betelgeuse, clean silicate grains would start to condense at 1.8 stellar radii, in spite of the hot surrounding gas, strongly supporting our interpretation.

A new set of interferometric data was obtained in February 1982 using the McMath solar telescope at Kitt Peak. A new map of the fringe visibility has been derived within the 1.5 m maximum baseline allowed by the telescope. The new map shows that appreciable changes have occurred in 14 months. The visibility curves are closer to circular symmetry and the visibility decreases more slowly as the length of the baseline increases, implying that at least part of the envelope has been washed

M. Morris and B. Zuckerman (eds.), Mass Loss from Red Giants, 63–64.

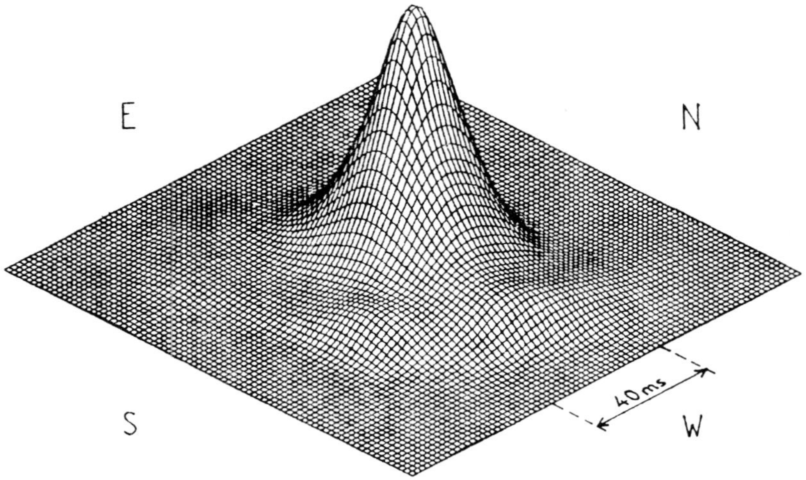

Figure 1: Irradiance distribution in the reconstructed image of Betelgeuse. Orientation is given modulo 180°

out. This result suggests an intermittent mechanism for the ejection of material since a large amount of dust was apparently condensed or condensing in 1980 in possible relationship with the rapid magnitude and polarization changes occurring at that time (Guinan, 1983). The infrared dust envelope may well have been similarly ejected in the past and blown away by radiative pressure (Draine, 1981).

The new map also displays a sinusoïdal modulation which may be produced by a stellar companion at a radial distance r = 0.45" ± 0.05" and a position angle PA = $85^{\circ} \pm 5^{\circ}$ (mod. 180°). The magnitude difference between Betelgeuse and this companion is estimated to be around 3.5. Since the ultra-violet spectrum of Alpha Orionis does not show evidence for a hot companion, it can be a cool giant star. Such a companion could play a role in the intermittent ejection of material, if it happens to come close enough to the main star to produce a tidal wave as suggested by the interferometric observations made by Wilson (1941).

REFERENCES

Bloemhof, E.E., et al.: 1984, Astroph. J., 276, L-21.
Draine, B.T.: 1981, in Physical Processes in Red Giants (I. Iben Jr. and A. Renzini, edit., D. Reidel, pub.) p. 317.
Guinan, E.F.: 1983, in: Cool Stars, Stellar Systems and the Sun, Proceedings of the Cambridge Workshop, (S.L. Baliunas and Lee Hartmann edit.).
Gull, S.F., Daniell, G.T.: 1978, Nature, 272, p. 686.
Roddier, C., Roddier, F.: 1983, Astrophys. J., 270, L-23.
Tsuji, T.: 1978, Publ. Astron. Soc. Japan, 30, p. 435.
Wilson, R.H., Jr.: 1941, pub. of the Univ. of Pennsylvania, Astron. series, vol. VI, p.1

MASS LOSS FROM RED GIANTS: INFRARED SPECTROSCOPY

P. G. Wannier
Jet Propulsion Laboratory, California Institute of Technology
4800 Oak Grove Drive, Pasadena, California 91109

ABSTRACT. A discussion is presented of IR spectroscopy, particularly high-resolution spectroscopy in the ~1-20 micron band, as it impacts the study of circumstellar envelopes. The molecular bands within this region contain an enormous amount of information, especially when observed with sufficient resolution to obtain kinematic information. In a single spectrum, it is possible to resolve lines from up to 50 different rotational/vibrational levels of a given molecule and to detect several different isotopic variants. When high resolution techniques are combined with mapping techniques and/or time sequence observations of variable stars, the resulting information can paint a very detailed picture of the mass-loss phenomenon.

To date, near-IR observations have been made of 20 molecular species. CO is the most widely observed molecule and useful information has been gleaned from the observed rotational excitation, kinematics, time variability and spatial structure of the its lines. Examples of different observing techniques are discussed in the following sections.

1. INTRODUCTION

The near infrared is ideal for observations of circumstellar material, containing spectral lines of many molecular species. Some of these species are present in the photospheres of cool stars while others form in the outflowing circumstellar material. The region of the spectrum from 1-20 microns contains several windows available to earth-bound observers (figure 1). Within these windows, astronomical observations can be made of species not abundant in the Earth's atmosphere. By making use of the doppler shifts of astronomical objects, observations can even be made of telluric species of modest abundance, such as CO and HCl.

In order to derive kinematic information from IR observations of circumstellar envelopes (CSE's), spectral resolution is a must. The expansion velocities of CSE's fall typically in the range of 3-30 km/s implying a need for a resolution of 1-10 km/s or a fractional resolution of 3×10^4 to 3×10^5. Such resolutions have been achieved with

M. Morris and B. Zuckerman (eds.), Mass Loss from Red Giants, 65–79.

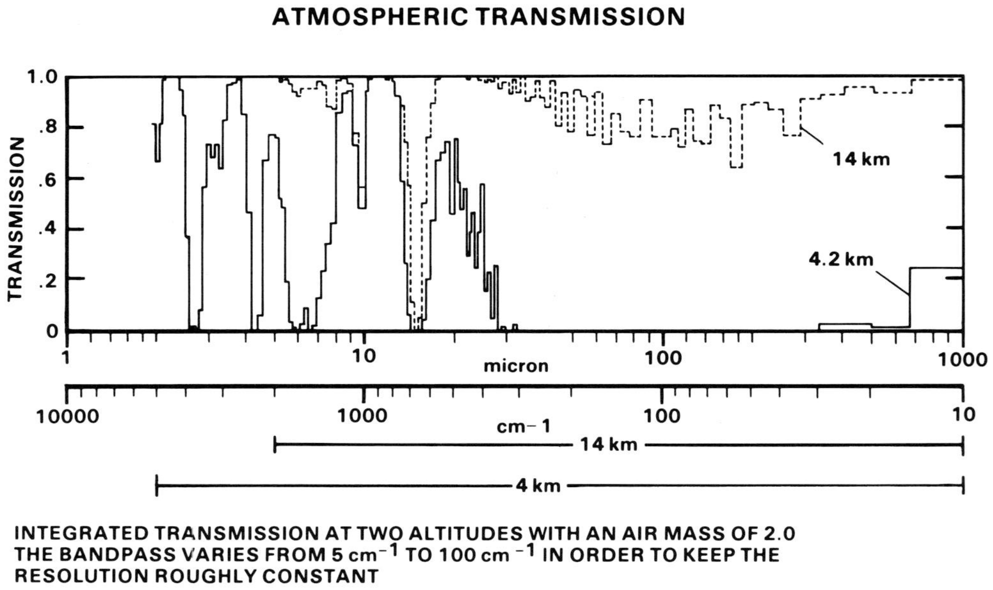

Figure 1. The atmospheric transmission is shown as a function of wavelength from altitudes of 4.2 km (Mauna Kea) and of 14 km (Kuiper Airborne Observatory). The near-IR observing windows are considerably improved by observing at airplane altitudes.

long-travel fourier transform spectrometers and even greater resolutions are available with heterodyne mixers. The most productive instrument to date has been the long-travel fourier transform spectrometer installed on the Mayall 4-m telescope on Kitt Peak having a total available travel of ~1m.

There have been several comprehensive reviews of IR spectroscopy in stars (Spinrad and Wing, 1969; Merrill and Ridgway, 1979 and references therein) and material in the earlier reviews will not be repeated here. The technology of IR spectroscopy has been discussed by several authors, for example Soifer and Pipher (1978) and Storey (1985).

Since 1979, great progress has been made in exploiting the very high spectral resolution available with modern instruments, especially in conjunction with time-sequence observations and mapping techniques. The cool, extended photospheres and the warm, dusty envelopes of late-type giants give rise to strong near-IR continuum radiation, against which the cooler molecular species in the outer envelope are readily observed in absorption. Most observations of late-type stars make use of absorption spectra, yielding line-of-sight information outside the region where the IR continuum is formed. Studies of molecular abundances, molecular excitation and line-of-sight kinematics are discussed in section 2 below.

In the case of regular Miras, the IR continuum forms at the photosphere and absorption-line spectra yield information about the entire envelope, including the critical region where material is being accelerated. Time-sequence observations sampling all phases of the photospheric pulsation have been used by Hinkle and co-workers to study the relation of the "Mira" shock to the initiation of mass-loss. These observations are discussed in section 3.

When a vibration/rotation photon is absorbed in the envelope, the absorbing molecule is generally de-excited by radiating another IR photon, giving rise to a process known as resonant scattering and creating, around the central IR continuum source, a halo of near-IR emission lines. The emission lines can be observed using mapping techniques and can thus indicate the spatial distribution of material newly accelerated away from the central star. Two such studies are discussed in section 4.

2. NEAR-IR SPECTRAL LINES

2.1 The Molecular Spectra

Even with low-resolution spectra, it is possible to gain abundance information by contrasting line strengths of different molecular species. It is also possible to follow the chemical processing which occurs in the expanding, cooling envelope by contrasting molecular abundances at different distances from the central star. In carbon-rich envelopes the transfer of outward momentum from the radiation field to the expanding envelope is thought to proceed by means of carbon-bearing grains: perhaps silicon carbide close to the photosphere and amorphous carbon grains at several stellar radii. Because the composition of the grains is not directly observable, the best avenue for understanding their formation is through related molecular species.

Prior to the review of Merrill and Ridgway, 15 molecular species had been observed in stars (table I). Of these, thirteen were observed by means of vibrational transitions: stretching and bending modes which give rise to bands containing a multitude of rotational and vibrational lines. The hydrogen bond stretch is the most important of the vibrational modes, accounting for the transitions of eight of the thirteen observed molecular species and giving rise to the astronomical importance of the 3000-4000cm^{-1} region of the spectrum. Only two of the 15 molecules, C_2 and CN, have important electronic spectra in the 1-15 micron spectral region.

Since 1979, five additional molecular species have been detected (table I). Methane has been observed by Hall and Ridgway (1978) at 3.3 microns and C_2H was detected by means of its electronic spectrum (Keady 1984). Ammonia (NH_3) and C_2H_4 have been observed using a heterodyne spectrometer with a velocity resolution of 0.2 km/s (Betz et al., 1979; Betz, 1981). Finally, and most recently, the silicon analog of methane, SiH_4 (silane), has been detected by Goldhaber and Betz (1984) and also observed by Ridgway and Keady (1985), the result

Table I: Molecular Infrared Bands in Stars

	Molecule	Bands	Wavenumber range (cm^{-1})
	C_2	$b^1\Pi_u - x^1\Sigma_g^+$ (Phillips)	$\sigma \gtrsim 4000$
		$A'^3\Sigma_g^- - X'^3\Pi_u$ (Ballik-Ramsey)	
	C_3	ν_3 $\nu_1+\nu_3$	1850 3200
	C_2H_2	ν_1 (C-H stretch)	3200-3400
	CH	$\Delta\nu=1$	2500-3000
*	CN	$A^2\Pi - X^2\Sigma$ (red)	$\sigma \gtrsim 2500$
*	CO	$\Delta\nu=1$ $\Delta\nu=2$ $\Delta\nu=3$	2000 4000 6000
*	CS	$\Delta\nu=2$	2500-2600
	HCl	$\Delta\nu=1$, $\Delta\nu=2$	2500-4500
*	HCN	ν_3 (C-H stretch)	3200-3400
	HF	$\Delta\nu=1$	4000-4400
	H_2	$\Delta\nu=1$ (quadrupole)	4000-6000
*	H_2O	$\nu_3, 3\nu_2$, $\nu_2+\nu_3-\nu_2$, $\nu_2+\nu_3$, $\nu_1+\nu_2$	2800-6200
	NH	$\Delta\nu=1$	2500-3400
*	OH	$\Delta\nu=1$ $\Delta\nu=2$	2500-3200 5000-6400
*	SiO	$\Delta\nu=1$ $\Delta\nu=2$	1200-1250 2400-2500
	CH_4	ν_3	3020
*	C_2H	$A^2\Pi - X^2\Sigma$	4000,4100
*	NH_3	ν_2	950
	C_2H_4	ν_7	949
	SiH_4	ν_4	910

Information is given about the 20 molecular species observed to date in red giant stars. Original references for the first 15 species are listed by Merrill and Ridgway (1979). The last 5 species in the table are discussed in Section 2.1 and references are listed there. The 20 molecules contain 7 different elements and provide information about abundances in the stellar envelopes. They can also be used to probe the developing chemistry in the outflowing wind. In addition, each molecule has isotopic variants which can yield accurate isotope abundance information relatively free from chemical effects. An asterisk in the first column indicates those molecules also observed at radio wavelengths in red giant envelopes.

of a spectral scan of the 8-14 μ atmospheric window. Silane is especially interesting because it is a silicon-bearing molecule. In oxygen-rich stars, silicate grains are thought to mediate the momentum transfer so that the chemistry of silicon-bearing molecules is of special interest. (Here, and throughout this paper, the term "oxygen-rich" is used only as a way to distinguish those "normal" stars with [O]>[C], from those stars which are carbon-rich, with [C]>[O].) The continued detection of molecular species provides important new probes of the chemistry in CSE's, but the results are limited without the ability to spectrally resolve the individual rotational lines.

2.2 Rotational Structure

The first useful step in high spectral resolution comes at a resolution ($\lambda/\Delta\lambda$) of about 3×10^3, when it is possible to resolve the rotational transitions within a band. At the same time, it becomes possible to unscramble overlapping vibrational bands resulting from isotopic substitution and/or different initial vibrational states. As an example, consider the fundamental vibrational transition of CO, $\Delta v=1$. A hard selection rule is $\Delta J=+-1$, giving rise to the R and P branches respectively. In each branch, the rotational excitation is easily derived from the run of line strengths. Because the stretching motion is slightly anharmonic, there is a shift between transitions from v=0-1 (ground-state CO) and v=1-2 (vibrationally excited CO). Finally, the slight shift in vibrational energies of the different isotopic variants give rise to an interleaving of the lines of CO, ^{13}CO, $C^{18}O$ and $C^{17}O$ (figure 2). The ability to distinguish between vibrational levels within a band therefore makes available information about vibrational excitation, rotational excitation and relative isotopic abundances. This information becomes especially powerful for studies of CSE's when it is possible to locate material within the CSE by means of its doppler line shift.

2.3 Velocity Profiles

The power of Doppler line profiles to diagnose the action in extended circumstellar envelopes is seen most dramatically by looking at an actual spectrum. Profiles of 4.6 μ CO in V Hya from Wannier and Sahai (1985) are shown in figure 3, with absorption features not only at the asymptotic expansion velocity of the envelope, but at several other velocities as well. It is obvious that the absorption features from the v=0-1 transition of ^{12}CO are quite different from those of the less abundant ^{13}CO molecule and that both species produce lines which differ from the v=1-2 transition of ^{12}CO.

For line-of-sight observations, velocity resolution sometimes makes it possible to place material along the line-of-sight. If there are variations in the expansion velocity, then the rotational spectrum at each velocity yields an excitation temperature which is used together with an assumed radius-velocity relation. For material near the star, it is thereby possible to study the material as it is accelerated outwards. Within the stellar photosphere, it is possible to follow the

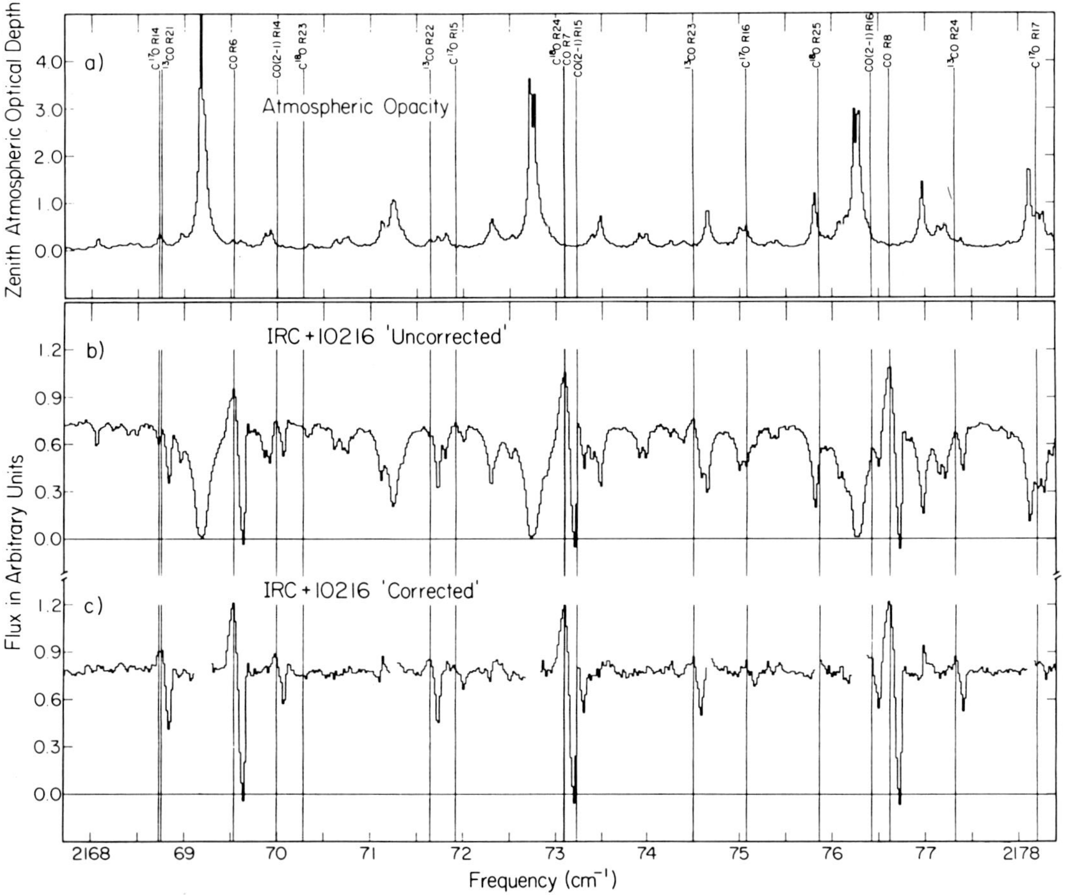

Figure 2. Data from Sahai and Wannier (1985) showing a small portion of a raw 5 micron spectrum. Shown are a plot of the earth's atmospheric opacity, an uncorrected spectrum and a spectrum corrected for atmospheric effects. The data are taken with a small annular observing aperture (see text) and contain lines of CO, $C^{18}O$, $C^{17}O$, ^{13}CO, and lines from the second excited vibrational level of CO. Other lines, although present in principal, are too weak to show up on this spectrum.

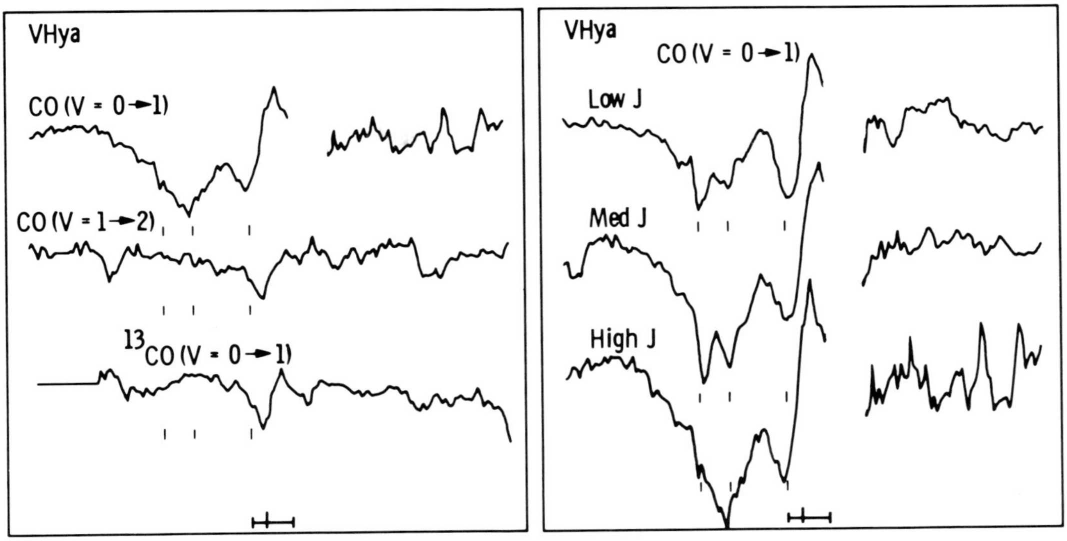

Figure 3. Data from Wannier and Sahai (1985) showing CO fundamental band line profiles of V Hya. The symbol at the bottom of each figure gives the velocity information. The center, tallest vertical hash-mark is the systemic velocity, the lefthand short mark is the (blue-shifted) expansion velocity based on radio observations and the right-hand short mark is displaced from the systemic velocity by 20 km/s. Prominent absorption features are seen at large negative velocities which vary in character according to rotational level (lower figure) indicating the different temperatures/densities of the gas at different velocities. The high-velocity gas is almost not seen for ^{13}CO or for vibrationally excited CO (top figure). An emission feature is seen at the systemic velocity corresponding to resonantly scattered radiation. To the red (right on the figure) the ground-state lines of CO are truncated by telluric absorption.

action of the Mira pulsation (section 3.). Surprisingly, observations of the cooler gas in the outer envelopes have also shown interesting velocity structure, indicating either a secular variability in the expansion velocity or, in a few cases, a staged acceleration of the envelope. Finally, velocity resolution allows for studies of material not in the direct line of sight and having a different Doppler shift. Where the telescope spatial resolution is poorer than the size of the continuum source, this material is seen as giving rise to P-Cygni type profiles with an emission feature red-shifted relative to the main absorption dip (figure 3; figure 8).

The first survey of CO line profiles in giants was made by Bernat (1981) who analyzed nine giant stars. He found evidence for absorption features at several different velocities in 3-4 of the nine objects observed. Interestingly, he found no clear association of either increasing or decreasing expansion velocity with radius, indicating that the expulsion and acceleration of material is considerably more complex than had been generally assumed. The separation of velocity components as deduced from these observations is typically of the order of hundreds of years, much greater than the acoustic period (~1 year) and much less than the thermal time-scales in these objects.

It is clear that the wealth of information available from up to a hundred line profiles originating from material at different densities and temperatures along the line of sight is difficult to interpret without resorting to models of considerable sophistication. To do an adequate job, such models must handle all observed transitions in a self-consistent way, dealing with acceleration of the material and taking account of 1) the IR continuum radiation from the central star and from the dust and 2) of radiative transfer in all appropriate lines, yielding the molecular excitation as a function of radius.

The most complete such models to date have been constructed for the carbon-rich object IRC+10216 assuming steady flows in the envelope, (Keady 1985, Sahai and Wannier, 1985), matching line profiles of the entire CO vibration-rotation spectrum. In Keady's model, the velocity was assumed to be steady, but was allowed to vary (increase) with radius. The resulting velocity steps are associated by Keady with staged acceleration caused by successive episodes of grain condensation. The model of Sahai and Wannier uses more analytic techniques and is engineered to deal with the particular aperture shapes used by them to map the resonantly scattered component of the CO fundamental band. To date, there has been no substantive effort to deal with the non-steady state flows evident from the observations of Bernat (1981), though considerable progress has been made on time-scales appropriate for regular mira-type acoustic pulsations.

3. TIME VARIABILITY

Late-type giant and supergiant envelopes are unstable to a variety of acoustic pulsations, giving rise to their observed variability. It would be surprising indeed if these phenomena, involving supersonic motions in the outer photosphere in addition to changes in the effective radius and temperature, were not intimately related to the copious observed mass-loss rates. High-resolution IR spectroscopy is unchallenged in its ability to probe the action in the photospheres of variable stars.

Particularly interesting studies are the time sequence observations of R Leo (Hinkle, 1978) and Chi Cyg (Hinkle et al., 1982), showing how the extended envelopes behave at different phases of the Mira pulsation. Both studies rely on the CO overtone bands ($\Delta v=2$ and $\Delta v=3$) to probe into the outer regions of the stellar photosphere. Because the overtone bands result from the small anharmonic part of the CO stretching vibration, their line strengths are respectively ~100 and 10,000 times smaller than the CO fundamental band at 4.6 μ. These observations, in combination with fundamental band observations offer the possibility of seeing precisely how the dissipating Mira shocks interact with the inner part of the circumstellar mass outflow.

Some interesting results from the study of Chi Cyg are presented in Figures 4-6, showing the observed radial velocities and excitation temperatures of the second overtone CO observations as a function of optical phase. It is clear that the second overtone band of CO provides a direct view of the hot (2000-4000 K) outer photospheric component traversed by the outwardly propagating shock. The complete interpretation of such data is clearly a significant chore, with velocity profiles for many different rotational levels, isotopic variants, vibrational levels and overtone bands, each as a function of stellar phase. A most intriguing result which comes from contrasting observations in the different overtone bands is the appearance, above the shocked regions in both stars, of a second, clearly distinguishable component of the gas: apparently a layer of levitated, quiescent gas at a radius corresponding to a CO excitation temperature of ~1000 K. Such a temperature is close to that for the condensation of silicate grains and may offer a clue as to how mass-loss is originated (from the Mira shocks) and accelerated (by subsequent acceleration by radiation pressure on grains). This layer is prominent in the first overtone band, especially in the lower rotational levels. A third distinguishable component of gas associated with the star is a component of infalling material at ~1500-2000K, seen primarily at the intermediate rotational levels of the first overtone band and representing, possibly, material partially expelled by passage of a previous Mira shock, but not accelerated sufficiently to leave the star.

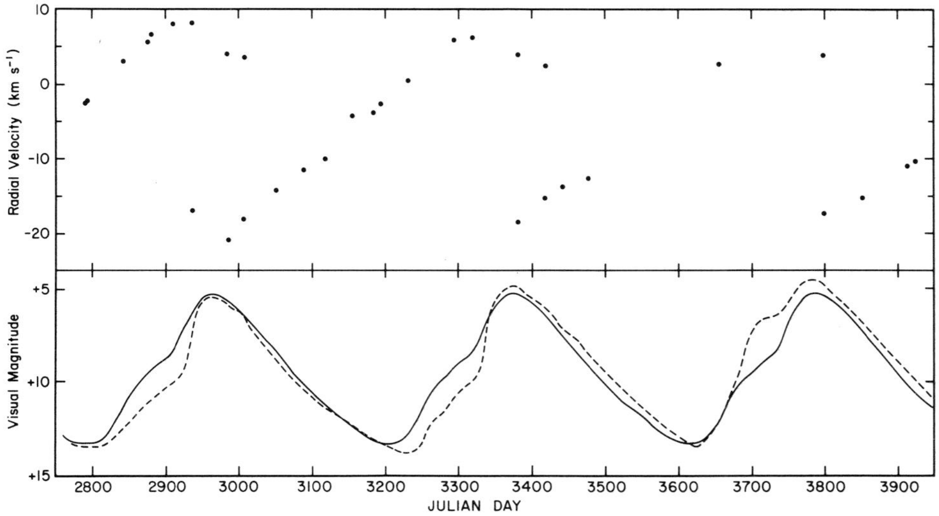

Figure 4. Data from Hinkle et al. (1982) using the second overtone band of CO toward Chi Cyg in a time-sequence study. In this figure, the observed radial velocity is shown along with the observed visual magnitude for the same period. It is apparent that the observations are sensitive to the photospheric material actually accelerated in the 'Mira' shock which accounts for the observed variability.

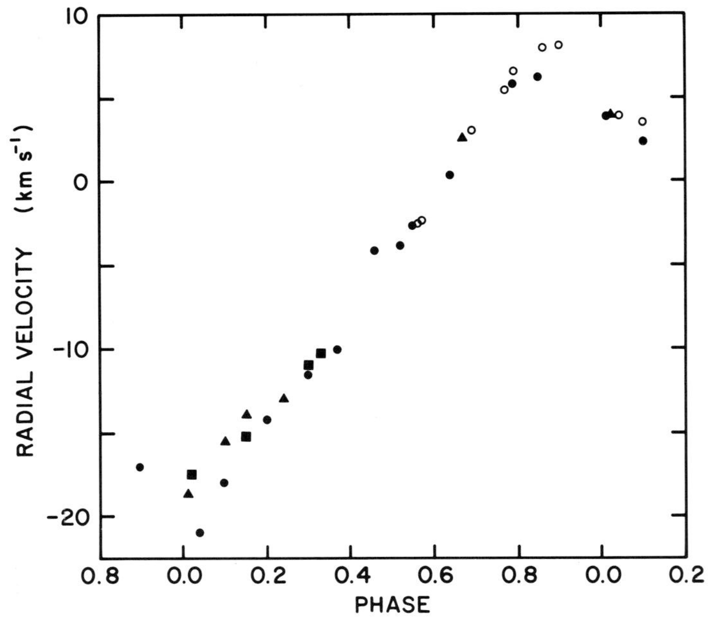

Figure 5. Similar to figure 4, except the data have been folded with the known period of the star. Between phases .1 and .8, the material is in almost constant deceleration. Between phases .9 and 1.2, material from two successive shocks are seen simultaneously. The different plotting symbols correspond to different cycles, identified by their starting Julian Day number in figure 6.

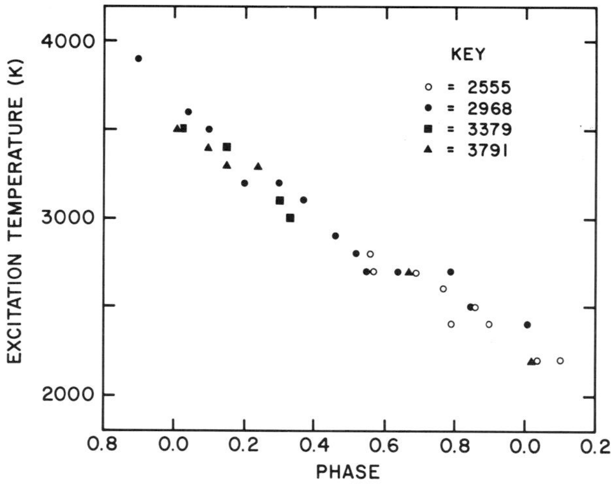

Figure 6. Similar to figures 4 and 5 except that excitation temperature is shown as a function of phase. It is apparent that the temperature decreases as the shock propagates outward.

A fourth discrete component of the gas, not treated by Hinkle et al., is the cool CSE which dominates the CO fundamental band. The qualitative results from these two early studies have recently been corroborated by Hinkle et al. (1984) from observations of five additional Miras, R And, R Aql, R Cas, T Cep and o Ceti (the prototype "Mira") and comprise the first close look at the action of Mira-type variables. A corresponding study of carbon-rich stars is rendered more difficult because of the higher opacity of their envelopes, but observations are planned.

4. SPATIAL MAPPING

All of the results described above make use of absorption lines formed by using the stellar photosphere or the opaque, inner dusty envelope as a continuum source and depend on an assumed temperature-radius relation in order to correctly place the material along the line-of-sight. Direct two-dimensional observations are needed to measure the actual temperature-radius relation. In addition, line-of-sight observations cannot answer questions about the symmetry of the mass loss process, yet it is known that related stellar phenomena are often strongly asymmetric (cf, Wannier, 1983). For example, many planetary nebulae and/or their halos have a bipolar symmetry indicating that an axis, defined either from stellar rotation or binary orbits, must affect their formation (Calvet and Peimbert, 1983; Cohen, 1985). Azimuthal structure in CSE's may well indicate the presence of either polar or equatorial winds and may point to a role for surface stellar fields or the importance of a close companion for driving winds in a binary system.

Spectral line mapping techniques make use of a simple physical characteristic of CSE's, namely the low collision rates that apply throughout the extended envelopes. When a vibration/rotation photon is absorbed, spontaneous re-emission is the preferred route for vibrational de-excitation, giving rise to another vibration/rotation photon in a process known as resonant scattering (see figure 7). Corresponding to every line of sight absorption feature, this creates a halo of IR line emission around the central continuum source and provides the opportunity for spatial mapping of the CSE. When the star is observed with low spatial resolution, the resonant scattering gives rise to an emission feature at the systemic velocity and red-shifted relative to the prominent blue absorption dips corresponding to the expanding envelope seen in absorption (figure 8; figure 3).

Two different groups have observed 4.6 μ CO radiation resonantly scattered in extended CSE's. These studies have detected extended structure on scales both large and small compared to a typical seeing limit of ~2 arcsec. The larger-scale observations were made by Sahai and Wannier (1985) with a large area detector and an FTS spectrometer using annular-shaped observing apertures of different sizes, masking the bright central star with its deep blue-shifted, absorption lines while passing emission at the stellar rest velocity. Prominent emission lines were observed in 7 out of 14 objects observed (cf, fig 8)

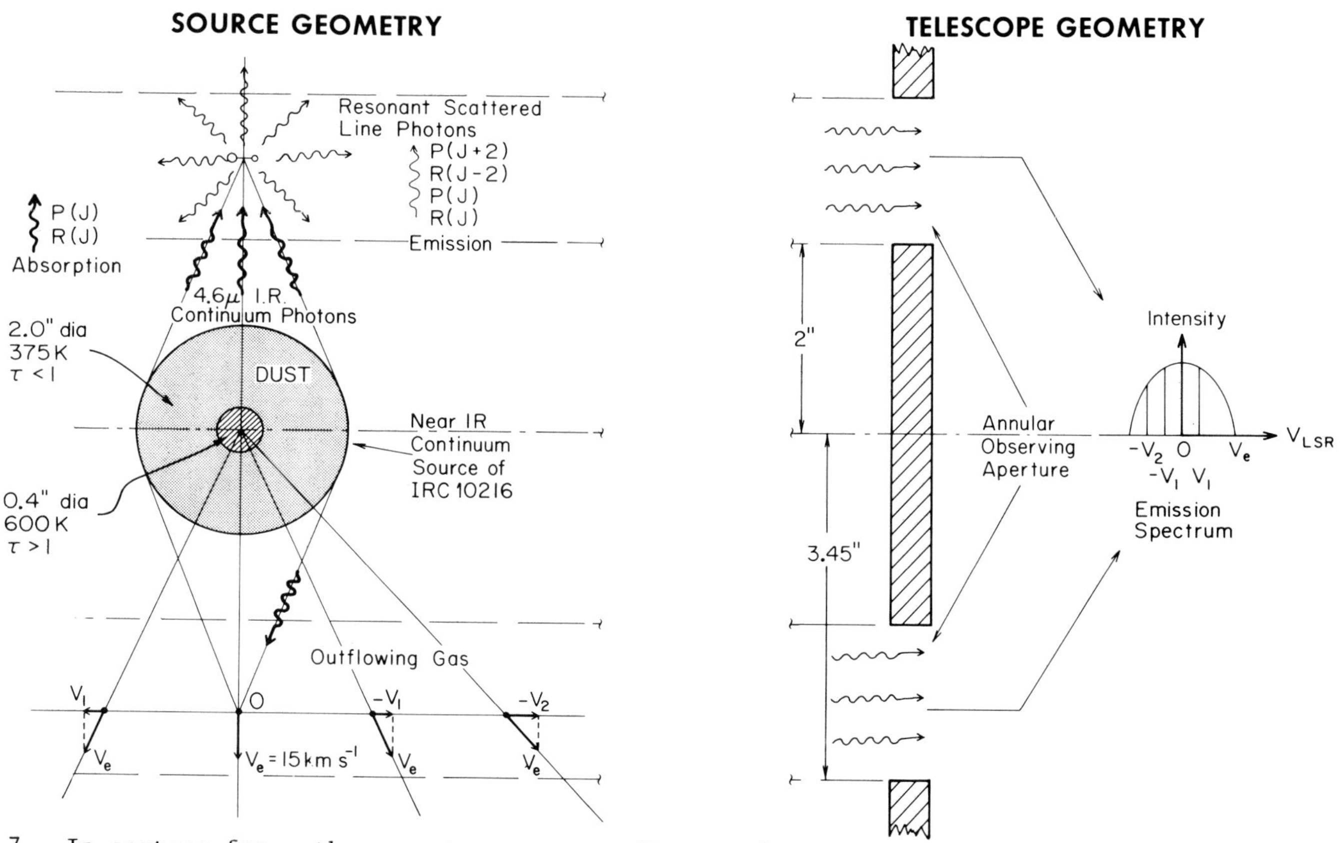

Figure 7. In cartoon form, the geometry corresponding to observations of resonantly scattered radiation is depicted. Continuum emission at 4.6 microns is incident on a CO molecule (top left) initially in rotational state J and an R(J) or P(J) photon is absorbed. Because of the low collision rate, de-excitation is by re-emission of one of the four transitions listed. An annular observing aperture is sensitive primarily to the resonant emission with the velocities on the Doppler profile corresponding to different radial distances as indicated in the figure at the left.

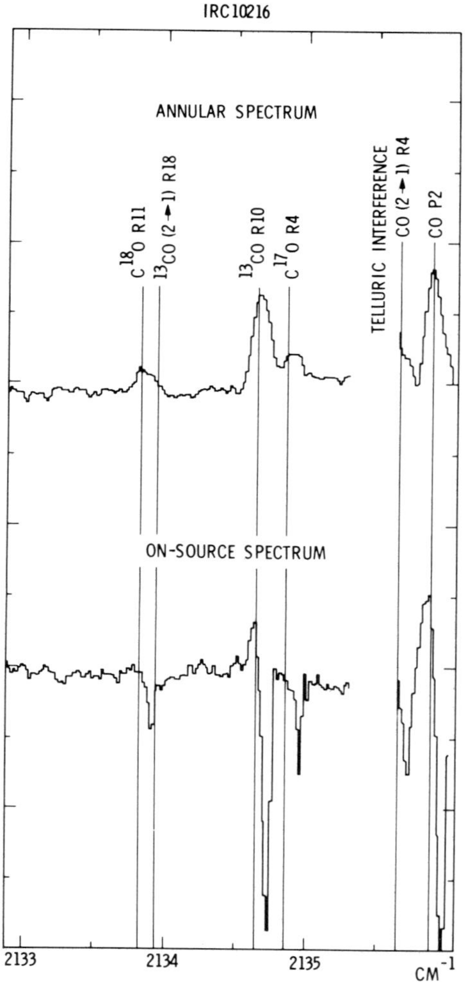

Figure 8. Data from Sahai and Wannier (1985) show a few of the many transitions in the fundamental CO band. Emission lines, seen through an annular observing aperture (top) are contrasted to the on-source absorption-line spectra (bottom). It is apparent that the absorption-line spectra contain cancelling absorption and emission features which make interpretation difficult and, indeed, have led to misinterpretations in the literature. The annular spectrum is apparently free from such effects and unambiguous rotational excitation temperatures can be derived.

and radial distributions of CO emission and excitation temperature have been obtained. In IRC+10216, the observed rotational excitation temperatures were almost twice those previously determined from modelling. In addition, there is apparently a high velocity shell of expanding material in the inner envelope indicating some sort of episodic expulsion of material. There are plans to extend the observing technique to include full two-dimensional maps by use of offset circular observing apertures.

Smaller-scale structure of CO in CSE's has been measured by a method combining the technique of speckle interferometry with high spectral resolution. Thus far, this technique has been applied only to IRC+10216 (Dyck et al., 1983) and Mira (Beckwith et al., 1985). The initial results of this technique indicate that all observed lines are spatially resolved and larger than the continuum (dust) radiation. Larger radii are measured for the lower-excitation lines, in keeping with expectations, and it should be possible to obtain a run of excitation temperature with radius down to ~0.1-0.2" resolution.

5. CONCLUSIONS

Near-IR spectroscopy offers a richness of data which is now placing pressure on models and on theoretical interpretation. Twenty molecules have been observed in the near-IR, offering information about chemical abundances and about chemical processing in the expanding material. The vibrational and electronic band structures of molecules in the near-IR offer the spectroscopist a huge array of tools, all available simultaneously in a single spectrum. Different lines originate from different vibrational and rotational levels as well as from species of different isotopic composition, providing information about radiation levels, kinetic temperatures, densities and nuclear enrichment. Doppler profiles yield kinematic information and the different vibrational overtone bands offer a range of line strengths which probe different regions in the photospheres and outer envelopes. Time sequence observations provide details about the interaction of acoustic pulsations with the observed outflows and resonantly scattered radiation provides a means to map the distribution of material in two dimensions. Speaking from experience, this author can attest to the difficulty of adequately handling the information available in even a few spectra, each containing up to 50-100 line profiles.

Even with existing observations, the pressure is clearly on astronomers to adequately handle the wealth of information by suitable interpretation and by adequate modelling of the envelopes. Self-consistent solutions must be sought which deal with molecular excitation in the presence of collisions and radiative transfer in an expanding envelope. The condensation of dust in the envelope both affects the radiation field and, through radiation pressure, can affect the kinematics of the material. In the next several years, there will surely be a dramatic increase in our knowledge of mass-loss from supergiant stars and from stars on the asymptotic giant branch as a result of near-IR spectroscopy.

REFERENCES

Beckwith, S., Dyck, H. M. and Zuckerman, B., 1985, Ap. J., (submitted).
Bernat, A. P., 1981, Ap. J., 246, 184.

Betz, A. L., McLaren, R. A. and Spears, D. L., 1979, Ap. J. (letters), 229, L97.
Betz, A. L., 1981, Ap. J. (letters), 244, L103.
Calvet, N. and Peimbert, M., 1983, Rev. Mexicana Astron. Astrof., 5, 319.
Cohen, M., 1985, (these proceedings).
Dyck, H. M., Beckwith, S. and Zuckerman, B., 1983, Ap. J. (letters), 271, L79.
Goldhaber, D. M. and Betz, A. L., 1984, Ap. J. (letters), 279, L55.
Hall, D. N. B. and Ridgway, S. T., 1978, Nature, 273, 281.
Hinkle, K. H., 1978, Ap. J., 220, 210.
Hinkle, K. H., Scharlach, W. W. G., and Hall, D. N. B., 1984, Ap. J., (suppl.), 56, 1.
Hinkle, K. H., Hall, D. N. B. and Ridgway, S. T., 1982, Ap. J., 252, 697.
Keady, J., 1984, (private communication).
Keady, J., Hall, D. N. B. and Ridgway, S. T., 1985, Ap. J., (in press).
Merrill, K. M. and Ridgway, S. T., 1979, Ann. Rev. Astron. Astrophys., 17, 9.
Ridgway, S.T. and Keady, J., 1985, (to be published).
Sahai, R. and Wannier, P. G., 1985, Ap. J., (submitted).
Soifer, B. T. and Pipher, J. L., 1978, Ann. Rev. Astron. Astrophys., 16, 335.
Spinrad, H. and Wing, R. F., 1969, Ann. Rev. Astron. Astrophys., 7, 249.
Storey, 1985, Proceedings of the 1984 U.R.S.I. International Symposium on Millimeter and Sub-Millimeter Wavelength Radio Astronomy, J. Gomez-Gonzales, ed.
Wannier, P. G., 1983, Cool Stars, Stellar Systems, and the Sun, Springer-Verlag, New York, N.Y., p. 296.
Wannier, P. G. and Sahai, R., 1985, (in preparation).

CO VIBRATION ROTATION LINES FROM CIRCUMSTELLAR SHELLS

R. Sahai
Owen's Valley Radio Observatory, Caltech
P.G. Wannier
Jet Propulsion Laboratory, Caltech

We have conducted an investigation of the inner part ($\sim$ 2-3") of red giant star envelopes through observations of emission in the 4.6 μm vibration rotation lines of CO produced by resonance scattering. A novel experimental technique, employing an annular aperture to mask the central infrared continuum source, was used to obtain line profiles at a mean angular offset of 2.7$\pm$0.7" from the center. The results for the carbon-rich Mira variable IRC+10216 are presented. The KPNO Fourier Transform Spectrometer at the 4m telescope, was used to measure the IRC+10216 spectrum at $\lambda\lambda$ = 2100-2200 cm^{-1}, with a resolution of 3.5 km s^{-1}, including CO lines from P(9) to R(14). In addition to the expected emission up to the expansion velocity of the envelope, both CO and ^{13}CO show evidence of high velocity features in the blue wing of the line, beyond the 15 km s^{-1} expansion velocity (Fig. 1).

We develop an analytical model to calculate the excitation of these lines due to thermal emission from dust, including a) the trapping of radiation in the optically thick lines, and b) 'seeing' effects. The model emission intensity is found to vary in a simple manner with distance from center, velocity, J value of the lower state, and the kinetic temperature. Our model shows the V=1 rotational level populations (except J=0) to be in equilibrium with the kinetic temperature despite being excited by infrared radiation. By fitting the model to the data (Fig. 2) we derive, for the IRC+10216 inner envelope (where we assume $T_{kin} \sim r^{-0.7}$), i) the kinetic temperature at radius r = 2" to be $\sim$ 244 K, roughly twice the extrapolation of the Kwan and Linke (1982) thermodynamic model, forcing revision of analyses of absorption lines in which the radial information was derived assuming a temperature profile (known a priori), and ii) the gas density in the inner regions of the envelope to be characterised by a mass loss rate larger than 6 10^{-5} $M_{\odot}yr^{-1}$ for a $[CO]/[H_2]$ ratio of 3 10^{-4}.

REFERENCE

Kwan, J., and Linke, R. 1982, Ap. J., 254, 587.

M. Morris and B. Zuckerman (eds.), Mass Loss from Red Giants, 81–82.

Fig.1 (right). The average infrared 'emission' line profile from IRC+10216, in CO and ^{13}CO, shown with the CO J=2-1, V=0 line. Both infrared line profiles show evidence of unusual high velocity emission in their blue wings beyond the expansion velocity (15 km s^{-1}) of the extended IRC+10216 envelope.

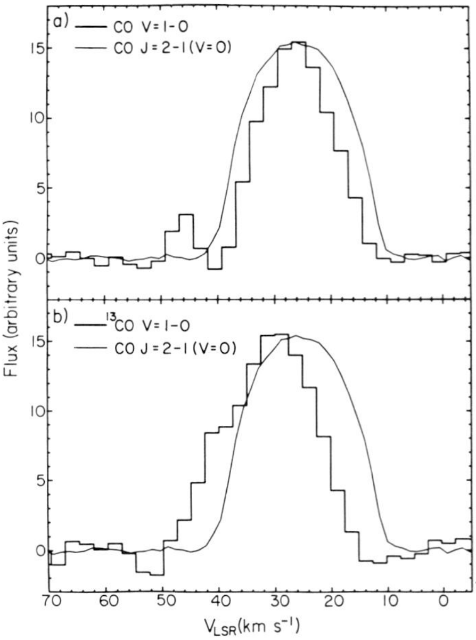

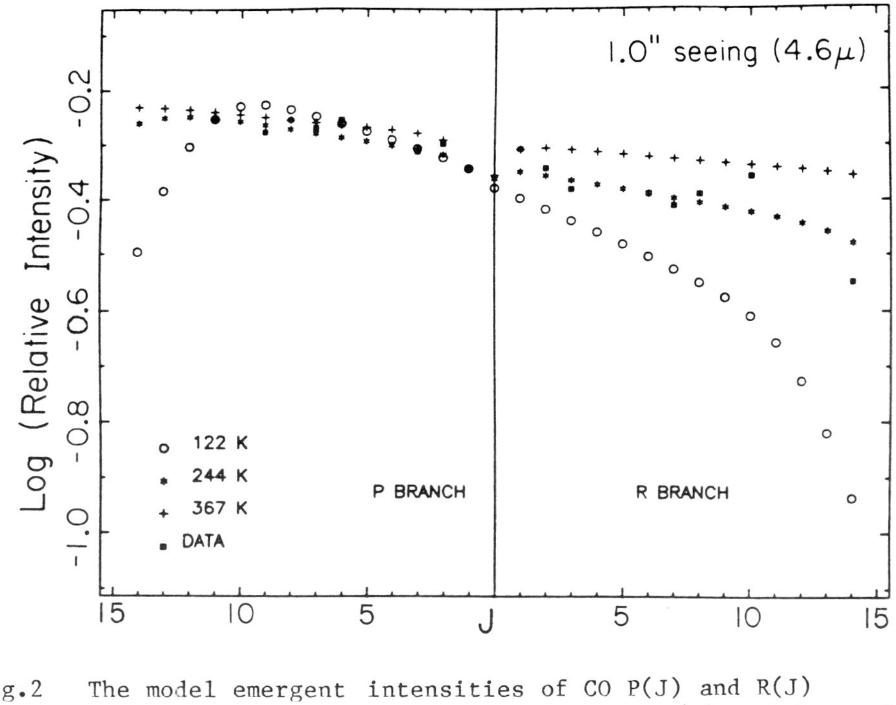

Fig.2 The model emergent intensities of CO P(J) and R(J) lines for three different normalisations of T_{kin} (r) - indicated by T_{kin} (2") - shown with the data.
The models have been scaled to make the predicted P(1) intensity equal to the observed. Data points are P(9), P(8), P(7), P(6), P(3), P(2), P(1), R(1), R(2), R(3), R(6), R(7), R(8), R(10), and R(14).

CIRCUMSTELLAR DYNAMICS OBSERVED BY INFRARED HETERODYNE SPECTROSCOPY

A.L. Betz and D.M. Goldhaber
Space Sciences Laboratory
University of California, Berkeley

The photospheres of Mira variables are known to exhibit complex episodic motions, and one may ask whether the circumstellar regions also show such behavior. The answer to this question may help us understand how mass-loss is initiated in stars prior to the formation of dust-grain condensates. Observations of CO absorption lines in the infrared by Hall (1980) indicate that in o-Ceti the circumstellar gas is grouped into several velocity components. Jura (1984) has interpreted these data as evidence for both expanding and contracting motions of the circumstellar material.

In 1981 we searched for lines from NH_3 at wavelengths near 10μm and were successful in detecting vibrational absorption from 5 different rotational levels. The heterodyne spectrometer used in these observations of o-Ceti is described elsewhere (Betz, 1981). The ammonia lines show multiple velocity components similar to CO but not all the same velocities. In addition, the NH_3 lines vary in velocity on a month-to-month basis. This behavior is illustrated in Figure 1, where separate observations of the aQ(2,2) line are shown. In July, 1981, o-Ceti was just before visual maximum (ϕ ~0.8) and the ammonia seen in absorption is blue-shifted 8 km/s with respect to the stellar velocity of 46 km/s (LSR). In October, however, the gas is expanding more slowly at only 4 km/s relative to the star. Also an indication of absorption redward of the stellar velocity begins to appear at 54 km/s (LSR). The reality of this red-shifted component is confirmed by more complete observations of other ammonia lines. In addition, the other data reveal a highly blue-shifted absorption component centered around +24 km/s (LSR), which implies a second component of gas expanding with a velocity of 22 km/s relative to the star. All the velocity components for ammonia are quoted approximately because they vary somewhat from line-to-line (depending on excitation) and from time-to-time (depending on stellar phase). The accuracy of each of the separate velocity measurements is actually better than 0.1 km/s. The components at 40 km/s (6 km/s outward expansion) and at 54 km/s (8 km/s infalling motion) are also seen in the infrared CO observations (Hall, 1980); but the "shell"

M. Morris and B. Zuckerman (eds.), Mass Loss from Red Giants, 83–84.

at 24 km/s (LSR) was not reported for CO, perhaps because of interference by terrestrial CO absorption. In September, 1982, additional observations of 7 NH_3 lines at a stellar phase ~0.1 showed weak blue-shifted absorption but strong red-shifted absorption from infalling ammonia at 54 km/s (LSR). In addition, sharp P-Cygni type emission was detected in several ammonia lines near 78 km/s (LSR), which at present bears no obvious relation to the gas observed near the stellar velocity of 46 km/s.

The abundance of ammonia observed in o-Ceti is very high, and suggests that nitrogen is enhanced in the circumstellar material. The NH_3 lines appear saturated at ~0.82 of the continuum intensity, and their relative intensities are not in accord with what is expected from optically-thin gas. Consequently a rotational temperature cannot be reliably estimated. A lower limit for the total NH_3 column density is ~10^{16} cm^{-2}, which is comparable to that observed in heavily obscured supergiants such as IRC+10216, VY CMa, and NML Cygnus. The mass loss rate in o-Ceti is much lower than that of the super-giants, however. Even though the peak ejection rate of o-Ceti may be as high as 10^{-5} $M_\odot$/yr, the gravitational return of most of this material back to the star limits the net mass-loss rate to be an order of magnitude less at ~10^{-6} $M_\odot$/yr. These conclusions are supported by more complete data presented in a manuscript now being prepared for submission to Ap. J. This work was supported in part by the NSF under grant AST-8204589.

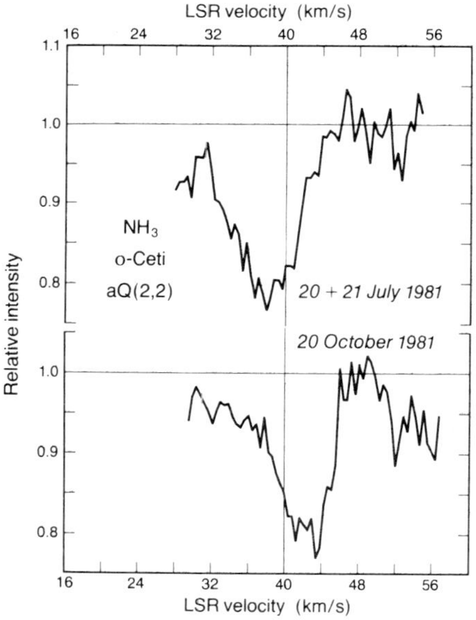

Figure 1.

Ammonia in o-Ceti

REFERENCES

Betz, A.L.: 1981, in Laser Spectroscopy V, Springer-Verlag, pp. 81-88.
Hall, D.N.B.: 1980, in Interstellar Molecules, D. Reidel, pp. 515-524.
Jura, M.: 1984, Ap. J., 282, pp. 200-205.

THE NEAR CIRCUMSTELLAR ENVIRONMENT OF MIRAS

Kenneth H. Hinkle
Kitt Peak National Observatory
National Optical Astronomy Observatories[1]
P. O. Box 26732
Tucson, Arizona 85726

Time series spectra were used to analyze the CO $\Delta v=2$ spectrum (2.3-2.5 μm) of 5 Miras and 1 SRa variable. A similar analysis previously was undertaken for the S-type Mira χ Cyg (Hinkle, Hall and Ridgway 1982). The $\Delta v=2$ spectrum of long period variables is complex with high excitation lines originating entirely in the (IR) photosphere and the lower excitation lines being formed through the photosphere and one or two extended atmospheric regions. The photospheric lines are characterized by large amplitude velocity and temperature changes with a mean excitation temperature of $\sim$3000K (Hinkle, Scharlach and Hall 1984). In contrast, the extended atmospheric regions exhibit much smaller velocity changes plus lower and near stable excitation temperature.

The first extended atmospheric region has an excitation temperature of $\lesssim$2000K. This region is most easily probed in the 3-1 lines R32 through R35. Velocities and temperatures show a continuity with the infalling IR photospheric material at the last phases it is detectable in the $\Delta v=3$ and high excitation $\Delta v=2$ lines. Apparently, the 2000K gas left the photosphere as dense, hot material in the previous pulsation(s) and is observed falling back cooler and less dense. The infalling material is sufficiently less dense than the expanding material that it rains back onto the photosphere for more than a light cycle. Infalling material of similar velocity and temperature contributes the majority of the absorption lines seen in the blue-violet spectrum.

The second extended region (not seen in the SRa variable) is $\sim$1000K gas that is probably a reservoir for the mass loss process. This region of the stellar atmosphere contributes broad and deep cores to the low excitation 2-0 lines. While the cores of these lines change velocity by a few kilometers per second over time scales of thousands of days, these changes are tied to changes in the line depth. We feel that the velocity changes are not actual velocity changes but result from the blended line profile changing in response to changes in the column densities of the various contributing lines.

---

[1]Operated by the Association of Universities for Research in Astronomy, Inc., under contract with the National Science Foundation.

M. Morris and B. Zuckerman (eds.), Mass Loss from Red Giants, 85–86.

Velocities of the stationary components measured in CO $\Delta v=2$ for the program Miras and χ Cyg are given in the table. Note the strong correlation between the CO velocities and the systemic and blue-violet absorption line velocities. The systemic velocity has been determined from thermal microwave lines when available and otherwise from the CO $\Delta v=3$ velocity curve (Hinkle, Scharlach and Hall 1984). Note that o Ceti has the worst agreement between the systemic and blue-violet absorption line velocities and the $\Delta v=2$ CO velocities. This may indicate an asymmetry in the extended portions of the stellar atmosphere caused by the hot companion.

Table I: Heliocentric Velocity Comparisons

Star	Systemic Velocity	Blue-violet Absorption Line Velocity	Average CO $\Delta v=2$ Velocities Component "$\sim$1000K"	"$\lesssim$2000K"
R And	-22:	-11	-23	-12
R Aql	+31:	+34	+33	+35
R Cas	+16.4	+23	+17	+23
T Cep	-15:	-12	-15:	-11
o Cet	+54	+60	+57	+63
χ Cyg	-8.5	0	-7	0
X Oph	-75:	-71	--	-71

REFERENCES

Hinkle, K. H., Hall, D. N. B., and Ridgway, S. T. 1982, Ap.J. **252**, 697.
Hinkle, K. H., Scharlach, W. W. G., and Hall, D. N. B. 1984, Ap. J. Suppl., in press.

THE INFRARED CHARACTERISTICS OF CIRCUMSTELLAR SILICATE GRAINS

W. Schutte, Rijksuniversiteit Leiden, The Netherlands

A.G.G.M. Tielens, NASA-Ames Research Center
Moffett Field, CA 94035

ABSTRACT. We have made a theoretical study of the infrared emission from circumstellar shells around late-type giants, with the aim of deriving the infrared characteristics of the silicate grains condensing in these shells. A large grid of models has been compared with observations of optically visible Miras, IRC sources and OH/IR stars. From fitting the observed relation between the color temperature and the strength of the 10μm feature, it is concluded that the ratio of the 3.5 to 10μm absorption efficiencies of the dust is about 0.25, a factor of 2 less than a previous determination. Detailed modeling of the 2 to 13μm spectrum of OH26.5+0.6, IRC+10011 and R Cas yielded a similar ratio. These detailed models also show that the shape of the 10μm feature, particularly around 8 and 13μm, varies from source to source. The derived 10μm feature is narrower for larger dust column densities. These observed differences in the intrinsic shape of the 10μm feature are not due to differences in size of the condensing particles. Probably they are related to structural or compositional differences in the condensing silicates.

I INTRODUCTION

To a large extent the structure of circumstellar shells around late-type giants losing mass is determined by the properties of the dust condensing in these envelopes (Kwok 1975, Tielens 1983). In that respect the infrared characteristics of the dust are of particular importance, because the mass-loss is ultimately driven by radiation pressure on dust grains and most of the stellar flux is radiated in the infrared. Fortunately, the infrared spectrum of circumstellar dust shells is very sensitive to the infrared characteristics of the dust. Theoretical model fits to observed infrared spectra can therefore be used to determine these properties. This paper presents the results of such a study.

M. Morris and B. Zuckerman (eds.), Mass Loss from Red Giants, 87–93.

II METHOD

We have developed a numerical code for radiative transfer in dusty circumstellar envelopes, following the quasi-diffusion method outlined by Leung (1975). The free parameters in the calculations are the dust density distribution, the stellar flux distribution incident on the inner boundary, the radii of the inner and outer boundary of the dust shell, and the dust absorption and scattering properties. A large grid of models has been calculated to study the influence of these free parameters on the emergent spectra. Details of these calculations will be presented elsewhere (Schutte and Tielens 1985). Here we will merely analyze existing infrared observations in order to determine the infrared characteristics of the circumstellar dust.

In all of the calculations the dust density distribution, $n_d(r)$, is given by a simple analytical formula. Far from the inner radius of the dust shell n_d follows an r^{-2} dependence, appropriate for a constant velocity outflow. Close to the star n_d steepens due to radiative acceleration of the dust. The density distribution as a function of the dimensionless radius x ($=r/r_*$, where r_* is the stellar radius) is completely described by the total dust column density and a parameter describing the steepness near the inner boundary.

The stellar flux distribution is assumed to be given by a black body at an effective temperature, T_{eff}. Note that the stellar radius, or equivalently the stellar luminosity, is not a free parameter in the model grid calculations. Essentially this is because the emergent spectrum is only dependent on the density and temperature distribution as a function of the dimensionless radius x and we choose to vary x.

Three models for the dust absorption efficiencies, Q_λ, are shown in figure 1. These dust models differ only in the shape of the 10μm feature, particularly around 8 and 13μm. Dust model 1 has a seemingly sharper 10μm feature than dust model 3. It should be emphasized, however, that the full width at half maximum does not change very much from one model to another. In the near-infrared ($\lambda \leqq 8\mu m$) it is assumed that the absorption efficiencies follow a λ^{-1} law, appropriate for small particles with a constant absorptivity. The ratio of the near-infrared to 10μm efficiencies is a free parameter in the calculations. Note that for a given optical depth the infrared spectrum is only sensitive to this ratio and not to the absolute values. This is because the dust temperature is a balance between emission and absorption of radiation.

III THE NEAR-INFRARED ABSORPTION CROSS SECTION

The ratio of the near-infrared to 10μm cross section of the dust can be determined by comparing the near-infrared flux with the strength of the 10μm feature in emission and absorption (Jones and Merrill 1976, Bedijn 1977). We have compiled near and mid infrared data of Miras and OH/IR stars from the literature (Merrill and Stein 1976a,b, Evans and Beckwith 1977, Werner et al. 1980, Herman 1983). For each source the colour temperature, T_c, between 3.5 and 12.5μm and the strength of the 10μm feature, A(9.7μm), has been determined. The latter is measured logarithmically by assuming a linear continuum from 8.7 to 12.5μm, e.g.

$$A(9.7\mu m) = \log_{10}(F/F_c)$$

where F and F_c are the observed and continuum flux at 9.7μm. Note that in this notation A(9.7μm) is positive for emission features. These data are plotted in figure 2. There clearly is a relation between A(9.7μm) and T_c. The sources with the lowest color temperature also have the largest 10μm absorption feature. This is not too surprising, since the derived color temperature is very sensitive to the near infrared optical depth through the circumstellar shell. For a larger optical depth we look less deep into the envelope. As a consequence the temperature of the radiating dust will be lower. Essentially, the A(9.7μm)-T_c relation measures, therefore, the ratio of the 3.5 to 10μm absorption cross section. This is illustrated by curves 1 and 2 in figure 2, which differ by a factor 2 in this ratio. Each curve is constructed from a set of models which differ only in the total dust column density. Curve 1 assumes dust model 1 (figure 1). Curve 2 assumes the same dust model, except for a lower ratio of the 3.5 to 10μm absorption efficiencies, which is half that assumed for dust model 1. Obviously curve 1 provides a much better fit to the observations than 2. Actually, the observations suggest that this ratio may even be somewhat larger than assumed in model 1. However, the A(9.7μm)-T_c relation also depends to some extent on the exact shape of the 10μm feature. This is because the continuum, used to measure A(9.7μm), is determined from the flux in the wings of the 10μm feature. This is illustrated by curve 1 and 3 in figure 2, which differ in the shape of the 10μm feature (curve 1 versus 2 in figure 1). Other parameters, notably the effective temperature of the star and the inner radius of the dust shell, only influence the A(9.7μm)-T_c relation for the more optically thin cases, in particular, the Miras. This makes it more difficult to determine the absorption efficiency ratio for these sources. It probably also explains the somewhat larger spread of the observational data for these sources.

IV THE 10μm FEATURE

The shape of the 10μm feature in the emitted spectrum of a circumstellar dust shell is very sensitive to the assumed shape of the 10μm dust feature. Using a trial and error method the inherent 10μm dust absorption properties can thus be determined from an observed flux distribution. In figure 3 two model calculations are compared to the observations of OH26.5+0.6, a typical OH/IR Mira variable. The models differ in the assumed shape of the 10μm dust feature. One uses the best fit 10μm feature determined for this source (curve 1 in figure 1). The other assumes the 10μm shape derived from observations of Orion (Gillett et al. 1975), which is similar to curve 3 in figure 1. For both models the other free parameters have been varied to produce the best possible fit to the observed spectrum from 1.65 to 13μm. From this comparison it is obvious that for OH26.5+0.6 the absorption efficiencies at 8 and 13μm relative to that at 10μm are much less than derived for the dust in Orion. In contrast, a similar analysis for R Cas, an optically visible Mira, yields a much better fit with a broad 10μm absorption feature, similar to that of Orion (fig. 4). For IRC 10011, which has an optical depth through the circumstellar shell in between OH26.5+0.6 and R Cas, the 8 and 13μm absorption efficiencies relative to that at 10μm are also in between those derived for these two sources (e.g., curve 2 in figure 1, Schutte and Tielens 1985). Finally, it should be noted that the detailed fit of the 2-13μm spectrum of each of these sources yields a similar ratio for the near-infrared to 10μm efficiencies. The derived ratio is in accordance with the analysis of the A(9.7μm)-T_c relation (section III).

The derived differences in the shape of the 10μm feature are not due to differences in the size of the condensing particles. First, the derived differences at 13μm are larger than those at 8μm, opposite to what is expected for size-related changes. Second, the derived ratio of the near-infrared to 10μm efficiency does not vary between the three sources, again indicating that a change in size is not the dominant factor. Instead these differences in the shape of the 10μm feature are probably related to variations in the structure or composition of the condensing silicate in these sources. One possible interpretation is that this is the consequence of the increased temperature in the condensation zone of the silicates when the optical depth increases. In this picture the radius of this zone (5-10 R_*) is set by the pulsations of the star, rather than the condensation temperature of the silicate material. The temperature of the condensing silicates will then be much less than the equilibrium condensation temperature and a highly amorphous material will result (Tielens 1983). A circumstellar dust shell with a larger optical depth will have a higher temperature in the condensation zone, because of the back heating of the shell on the condensation zone, resulting in a more crystalline silicate. Laboratory experiments have shown that the higher the temperature of the grains, the less amorphous the material and the broader the 10μm feature (Day 1974).

The shape of the 10μm feature derived for the dust in Orion (Gillett et al. 1975) is very similar to that derived for R Cas. This might suggest, perhaps, that a large fraction of the interstellar silicates are produced in the envelopes of optically thin Miras. One should, however, be cautious drawing such a conclusion. In an HII region, like Orion, a considerable contribution to the flux in the wings of the 10μm feature can come from graphite particles. This tends to produce a seemingly broader 10μm feature (Tielens and de Jong 1979). It is also possible that silicates in the interstellar medium are further processed. For example, laboratory experiments have shown that bombardment with cosmic rays produces a more amorphous material and therefore a broader 10μm feature (Kratschmer and Huffman 1979).

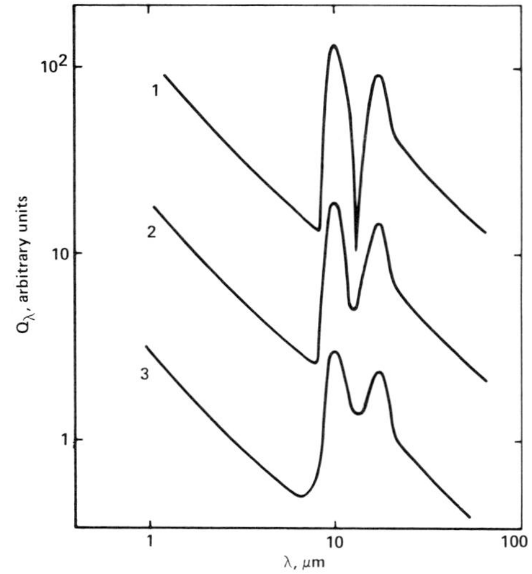

Figure 1: The three different dust models assumed in these calculations. They differ only in the shape of the 10μm feature, particularly around 8 and 13μm. Curves 1,2 and 3 give the best fit to the observed 2 to 13μm spectrum of OH26.5+0.6, IRC+10011 and R Cas, respectively. The shape of the 10μm feature derived for the dust in Orion (Gillett et al. 1975) is quite similar to that of curve 3.

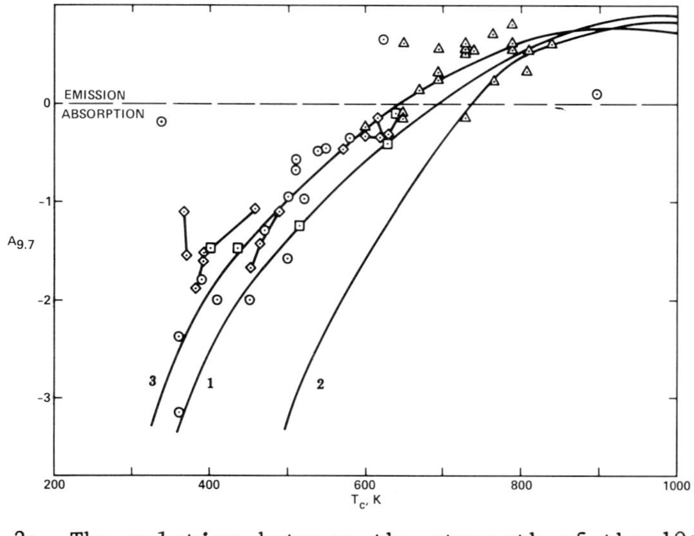

Figure 2: The relation between the strength of the 10μm feature, A(9.7μm), and the color temperature, T_c. Data are taken from Merrill and Stein 1976ab (Δ), Evans and Beckwith 1977 ( ), Werner et al. 1980 ( ) and Herman 1983 (o). The curves are model calculations for different dust models. Curve 1 assumes dust model 1, shown in figure 1. Curve 2 assumes the same dust model except for the near infrared to 10μm ratio, which is lower by a factor of two. Curve 3 assumes dust model 2 in figure 1.

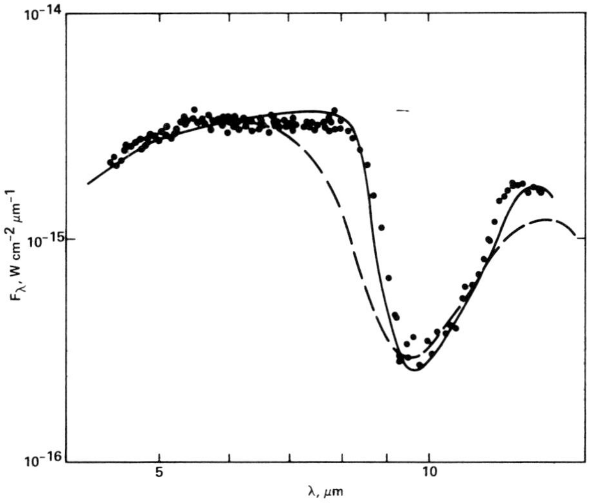

Figure 3: A comparison of the model calculations with the observations of OH26.5+0.6 (Forrest et al. 1978). The solid line gives the best fit model, assuming dust model 1 from figure 1. The dashed line gives the best fit model, using the Orion 10μm feature.

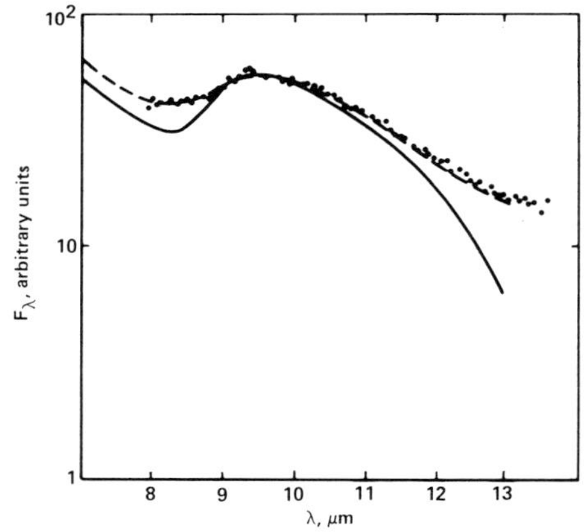

Figure 4: A comparison of the model calculations with the observations towards R Cas (Forrest et al. 1974). The solid line gives the best fit model using dust model 1 from figure 1. The dashed line assumes dust model 3 from figure 1.

REFERENCES

Bedijn, P. 1977, Ph.D. Thesis Leiden.
Day, K.L. 1974, Ap.J., 192, L15.
Evans, N.J. and Beckwith, S. 1977, Ap.J., 217, 729.
Forrest, W.J., Gillett, F.C. and Stein, W.A. 1974, Ap.J., 195, 423.
Forrest, W.J. et al. 1978, Ap.J., 219, 114.
Gillett, F.C. et al. 1975, Ap.J., 200, 609.
Herman, J. 1983, Ph.D. Thesis Leiden.
Jones, T.W. and Merrill, K.M. 1976, Ap.J., 209, 509.
Kratschmer, W. and Huffman, D.R. 1979, Astro. Space Sci., 61, 195.
Kwok, S. 1975, Ap.J., 198, 583.
Leung, C.M. 1975, J.Q.S.R.T., 16, 559.
Merrill, K.M. and Stein W.A. 1976a, PASP, 88, 285.
Merrill, K.M. and Stein, W.A. 1976b, PASP, 88, 294.
Schutte, W. and Tielens, A.G.G.M. 1985, in preparation.
Tielens, A.G.G.M. 1983, Ap.J., 271, 702.
Tielens, A.G.G.M. and de Jong, T. 1979, A.A., 75, 326.
Werner, M.W. et al. 1980, Ap.J., 239, 540.

INFRARED SPATIAL INTERFEROMETRY

Steven Beckwith
Astronomy Department
Cornell University
Ithaca, NY 14853-0352

ABSTRACT. Infrared spatial interferometry of red giant stars is a powerful way to study the distribution of gas and dust in the circumstellar envelopes. The circumstellar dust around many giant stars is easily resolved at infrared wavelengths. The dust distributions sometimes deviate markedly from spherical and even reflection symmetry in the objects most easily resolved. The light intensities suggest dust formation radii larger than expected on the basis of pure theory, and may mean that mass loss is not driven by radiation pressure on dust grains, although there are different interpretations. Radial density and temperature distributions expected for uniformly outflowing envelopes heated solely by the central stars are consistent with the observed emission sizes versus wavelength, but are not strongly constrained by the data.

1. INTRODUCTION

Most red giant stars are powerful sources of infrared radiation; in fact, some are the brightest infrared objects in the sky outside of the solar system. The infrared light often comes from thermally radiating dust in circumstellar envelopes produced by mass loss from the stars, envelopes which are substantially larger than the stellar photospheres, extending to at least several thousand stellar radii. For nearby stars, the angular sizes of the infrared continuum radiation is larger than the diffraction-limited resolution of the largest telescopes, so it is possible, at least in principle, to spatially resolve the dust. Circumstellar gas shows up through infrared absorption and emission lines, providing an additional probe of circumstellar structure. Images of the dust and gas emission give us the opportunity to directly study the distribution of matter in the envelopes. This distribution should, in turn, trace the mass loss history of the stars.

Spatial interferometry refers to any method for achieving diffraction-limited resolution by interfering light from separate telescopes or separate parts of one telescope, including both filled and synthesized apertures. At infrared wavelengths, interferometry is

M. Morris and B. Zuckerman (eds.), Mass Loss from Red Giants, 95–113.

at a very rudimentary stage. It is difficult to synthesize entire apertures, and it has thus far been impossible to construct two-dimensional images with diffraction-limited resolution; image reconstruction may not be far away, however. Nonetheless, the circumstellar shells of red giant stars are routinely resolved at infrared wavelengths, both in the continuum and in lines. It has become relatively easy to obtain interferometric results toward a wide variety of stars.

If the interferometric techniques are rudimentary, then interpretations of the results are even more so. As we shall see, there are several ways of resolving circumstellar shells, each of which has its idiosyncrasies. For example, large contrasts in dust or gas density show up distinctly in existing interferograms, but small changes generally cannot be distinguished from the (often unknown) systematic uncertainties in the data. The likely success of any scientific investigation carried out to date will depend on whether the sought-after effects are striking or subtle. This situation might change when image reconstruction is possible; however, in this review we will concentrate on the restrictions imposed by the techniques, and their effect on the scientific interpretation.

2. INFRARED INTERFEROMETERS

Three common types of interferometers are in use for the study of red giant stars: heterodyne interferometers, Michelson interferometers, and speckle interferometers. The range of angular resolution, depending on wavelength and telescope, is similar for all three techniques, between a few tenths and a few arcseconds, corresponding to spatial resolutions between about five and a few hundred stellar radii for nearby giant stars. In general, speckle interferometers provide the greatest sensitivity, but sensitivity is usually not an important factor in giant star observations; there are plenty of bright stars.

Heterodyne interferometry is, as the name implies, almost identical to twin telescope radio interferometry. Infrared light at the foci of two telescopes is mixed with light from an infrared laser on a solid state detector. Mixing modulates the light at the sum and difference of the frequencies of the laser light and the incoming radiation from the sky, the lower frequency being in the radio band. Standard radio frequency instruments detect and amplify the lower frequency component, and the signals from the two telescopes ultimately combine in a radio frequency mixer to give a real-time interference signal (Johnson, Betz, and Townes 1974).

Observations have thus far been restricted to 11 μm by the CO_2 laser used as a local oscillator. The McMath auxiliary telescopes at the Kitt Peak National Observatory provide a fixed east-west baseline of 5.5 m, thus limiting the spatial frequency coverage. Good spatial resolution occurs mainly in right ascension. The relative phase between the two telescopes is recorded but not the absolute phase (at least phase information is not published), so the complex visibilities are obtained without absolute positional information.

A Michelson interferometer operates on one telescope in a similar fashion as the early optical interferometer described by Michelson and Pease (1921). The primary mirror is imaged onto a mask with two holes. The masked light is re-imaged to place the focal plane on a spatial filter. The two holes mimick two small telescopes, and the interference pattern modulates the focal plane image. A servo-controlled delay introduced into the light path through one of the holes scans the fringe pattern across the spatial filter, modulating the detector output by the fringe contrast at the servo frequency. The spatial filter restricts the spatial frequency to that set by the spacing of the holes in the mask (McCarthy and Low 1975).

By varying the spacing and orientation of the holes in the mask (and, of course, the spatial filter), one obtains the fringe visibility at different spatial frequencies and position angles. Phase information is lost, and only the modulus of the visibility remains. Although the concept is quite simple, the actual observing procedure is cumbersome, if complete visibilities are desired. The Michelson interferometer has therefore largely been abandoned to the speckle interferometer for use with a single telescope.

Speckle interferometry makes use of the fact that atmospheric turbulence smears an image by creating a large number of diffraction-limited images at different positions in the focal plane. The individual images move around rapidly but can be seen if the exposure time for the blurred image is short compared with the changes in the turbulence pattern. The individual images are called speckles, since that is exactly what they look like in photographs, and basically contain the diffraction-limited information sought by the observer. Rapid exposures give images which are filtered in the Fourier transform plane to take out the blurring caused by the presence of many speckles. Use of many exposures allows an increase of signal-to-noise ratio essentially as the square root of the number of exposures, at least in the absence of systematic fluctuations in the seeing or the recording apparatus.

Infrared speckle interferometers are currently restricted by the lack of panoramic detectors, so two-dimensional exposures cannot be made. Use of individual photodetectors makes it relatively easy to produce one-dimensional images by scanning the two-dimensional image rapidly across a narrow slit in the focal plane, and focusing the light from the slit onto the infrared detector. The slit width is usually comparable to the size of a diffraction-limited point source. At infrared wavelengths, each image normally consists of a small number of speckles, so that very few speckles are in the slit at any one time (on average). The low-spatial frequency blurring caused by the many speckles is removed by ratioing the Fourier transforms of the stars to be studied to standard calibration stars whose images are smaller than the diffraction limit.

The use of a slit restricts the information to one dimension on the sky, just as in heterodyne and Michelson interferometry. Rotation of the slit yields information at different position angles. The huge advantage provided by speckle interferometry is that the entire telescope aperture is used at once, and all spatial frequencies are sampled simultaneously. Sibille, Chelli, and Lena (1979); Selby,

Wade, and Sanchez-Magro (1979); Howell, McCarthy, and Low (1981); and Dyck and Howell (1982) describe the instrumentation in some detail.

The high-spatial resolution provided by the infrared interferometers described above has thus far been obtained at the expense of real imaging capability. Interferometers give the spatial Fourier transform of the intensity distribution from the sky, commonly called a visibility function. In general, visibilities are complex functions of two inverse spatial dimensions. All three techniques yield only one-dimensional visibility functions at present, and often only the moduli of the complex visibilities are recorded. Inferences about the intensity distribution result from comparisons between the one-dimensional visibility curves and the spatial Fourier transforms of model intensity distributions. Future improvements to the instruments and data reduction techniques could give sufficiently complete visibilities for image reconstruction, and it is this direction which will probably have one of the greatest impacts on red giant research.

It is perhaps worth mentioning one more method recently presented by Bloemhoff, Townes, and Vanderwyck (1984). At sites with particularly good seeing, Mauna Kea in Hawaii being one of the best examples, the images are often relatively undistorted by seeing for several seconds to several minutes at wavelengths longer than about 5 μm. During such times, direct images can be made with carefully controlled scans across linear arrays. Bloemhoff, Townes, and Vanderwyck obtained diffraction-limited images at 10 μm of the stars α Ori and α Sco clearly showing the dust shell emission. We will examine these results in some detail in the next section.

3. CIRCUMSTELLAR DUST

When the wavelength resolution of the interferometer is low, or when the passband does not contain spectral features of particular interest, the observations mainly sample thermally emitted radiation from dust grains. (Although at shorter wavelengths, scattering can be important, this discussion will be confined to thermal emission.) The distribution of the radiation depends on the temperature and density distribution of the dust. In the absence of images, it is useful to characterize this distribution in terms of the fewest parameters possible. In this way, it might be possible to answer several restricted questions:

1. Where does the dust form in the envelopes?
2. What is the average radial dust density and temperature distribution?
3. Is the mass distribution spherically symmetric?
4. Is the dust smoothly distributed, or does it occur in clumps?
5. To what extent does dust affect mass loss from the stars?

The latter question, while it may not necessarily follow directly from the observations, is of sufficient general interest to be included.

Interferometry in its present state can be a very good technique to address question (1) and, to some extent, questions (3) and (5).

Question (4) is difficult to attack working with one-dimensional visibilities, but it should be a prime candidate for study with two-dimensional images. Question (2) will be difficult to address, even in principle. In practice, the visibilities are not very sensitive to the form of the radial temperature and density distribution of the dust, at least those close to the best model predictions, but it is possible to check for consistency with model calculations for the light distribution.

The number of red giant stars observed interferometrically is now more than 20. For most of the stars, the amount of interferometric information is limited and can be found in the references which follow. Three stars deserve special mention. The first, α Ori, is one of the best studied of all the stars. The dust is optically thin at all wavelengths (McCarthy, Low and Howell 1977), and both visual and infrared interferometric observations exist (Sutton et al. 1977; Howell, McCarthy, and Low 1981) as well as a direct image taken by Bloemhoff, Townes, and Vanderwyck (1984). It is an excellent candidate for model fits to the expected intensity distribution. The second is IRC+10216, an extraordinarily strong middle-infrared source (5 to 10 μm) with a thick circumstellar shell. Even at 11 μm, the shell is not optically thin (Crabtree and Martin 1979; Sutton et al. 1979). IRC+10216 has been observed by lunar occultation at two wavelengths (Toombs et al. 1972). It is quite asymmetric, both in visual (Becklin et al. 1969) and infrared light, with the amount of asymmetry varying with wavelength and radius (McCarthy, Howell, and Low 1980; Mariotti et al. 1983; Dyck et al. 1984).

The third star is Mira. Mira is the progenitor of the long-period variable stars with a period slightly less than one year. The 11-μm shell emission, and possibly the size, varies by about 30% with phase (Sutton et al. 1978; McCarthy, Howell and Low 1978), the effect caused almost entirely by the variation in the luminosity of the star which heats the dust as opposed to sudden condensation of dust when the temperatures fall. The absolute sizes constrain the dust distribution to a narrow range and imply that very little dust can be close to the stellar photosphere, differing with the earlier theoretical expectations about dust formation radii. This result holds for α Ori and IRC+10216 also.

Two characteristics of the dust distribution are common to all the stars thus far observed. The dust appears to form at a reasonably well-defined inner radius of order ten stellar radii seen at wavelengths where the shells are optically thin (Sutton et al. 1977, 1978, 1979). This radius is well outside the stellar photosphere, even taking into account the uncertainties in the location of the photosphere (Labeyrie et al. 1977), and well inside the extended shell seen through millimeter wave molecular emission and OH maser emission (Dyck et al. 1984). Unfortunately, it is almost always much larger than the theoretical dust condensation radius (Salpeter 1974, 1977), which is close to the photosphere. The formation radius will be discussed in more detail below. Also, the apparent size of dust emission varies slowly as a function of observing wavelength, with the longer wavelength emission appearing more extended (Foy et al. 1979;

McCarthy, Howell, and Low 1980; Mariotti et al. 1983; Dyck et al. 1984). Although generally consistent with the notion that shorter wavelength emission arises from hotter dust which is preferentially closer to the stars, the observations do not yet provide quantitative information about the dust distribution in the shells.

3.1. Simple Interpretation of the Visibility Curves

Ideally, the visibility curves can be inverted via Fourier transforms to give the light distributions. The lack of phase information vitiates this approach to existing data. Instead, visibility curves are calculated from model light distributions and compared with the observations to determine the matter distributions. Several excellent models for infrared light distribution from circumstellar shells around red giants appear in the literature (for example, Jones and Merrill 1976; Rowan-Robinson and Harris 1983; and references therein), where the radiative transfer equation is solved self-consistently.

To see how the visibility curves bear on the questions posed at the beginning of this section, it is useful to examine the theoretical visibility curves for some simple light distributions. For our purposes, very crude (and slightly unrealistic) models should suffice for comparison with the observations, since the uncertainties in the latter are often greater than the error in the former. Thus, the following discussion describes those features of the matter distribution in circumstellar shells that are most prominent in the observed visibilities. All the calculations are for optically thin dust emission at 10 μm, where the dust emission is near its peak intensity.

Figure 1 shows plots of the emission expected from two model dust distributions. In figure 1a, the emission intensity as a function of distance from the central star is plotted as it would appear in a linear cut across the diameter of the circular emission pattern on the sky. We assume the dust forms suddenly at radius r_0, where r_0 is either 20 $R_\star$ or 60 $R_\star$, $R_\star$ being the radius of the stellar photosphere, the dust density subsequently falls off with radius as r^{-2} after the formation radius, and the temperature falls off as $r^{-0.4}$. The strong spikes in the emission profile come at the inner edge of the dust shell as a result of limb-brightening.

In figure 1b, the emission profile is shown as it would appear in a slit scan of the circular image, using an infinitely narrow slit. This is the distribution sensed in typical infrared speckle interferometry, and it is similar to that seen with the other techniques, since they effectively average over one dimension. The really sharp features in figure 1a disappear, and the averaged emission looks smooth. Apparently, the dust formation radius dominates the size and peak intensity of the averaged profile. If the density is at all close to that expected from uniform mass outflow at a constant velocity, $\rho \sim r^{-2}$, most of the emission comes from dust very near the formation radius.

Figure 1c shows the visibility curves (normalized to unity at f=0) obtained from Fourier transforms of the averaged emission profiles of figure 1b. The visibilities fall off from unity to a

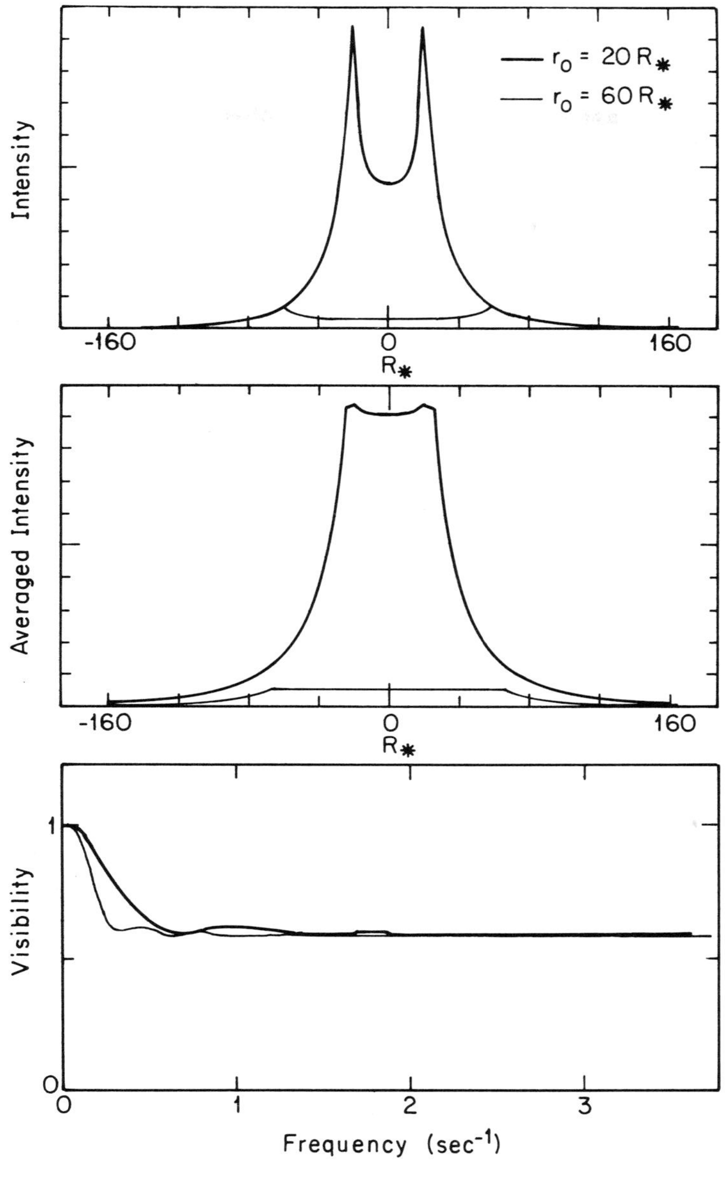

Figure 1 - Top (1a) shows calculated emission profiles at 11 μm for dust forming at two different radii around a red giant. Middle (1b) shows the emission averaged along one dimension as it would appear in a slit scan. Bottom (1c) displays the different visibility curves including the unresolved stellar emission, assuming a constant fraction of the visibility comes from the unresolved source. The integrated emission would not be the same in the two cases.

constant value at high spatial frequencies. I purposely included emission from the (unresolved) star in these visibilites for comparison with actual data, and it is the stellar emission which gives rise to the constant visibility at high-spatial frequencies. The dust emission causes the drop at low spatial frequencies; the ripples result from the sharp edges in the inner distribution.

It should be immediately clear that the visibilites are mainly sensitive to the dust formation radius, r_0, in this simple example. The spatial frequency at which the visibility drops to one-half its final value is inversely proportional to r_0. For the purposes of comparison, the ratio of dust emission to stellar emission has been held constant. In a real model, there would be much less dust emission from the model with $r_0 = 60\ R_\star$, so the visibility at high spatial frequencies would be much closer to unity. It should be apparent, however, that the visibility curves yield an accurate measure of the ratio of dust emission to stellar emission.

The visibility data obtained so far are generally consistent with models where the dust density falls off as $r^{-\alpha}$ and temperature as $r^{-\beta}$ for $\alpha \sim -2$ and $\beta \sim -0.4$ (Sutton et al. 1978, 1979; Dyck et al. 1984). Careful calculations show the temperature distribution coefficient β is slightly different from -0.4 (e.g., Rowan-Robinson and Harris 1983), but not by a large amount. There is no obvious disagreement between the theory and observations concerning the radial dependences. It should be noted, however, that there is good evidence for departures from spherical symmetry in at least IRC+10216, so the models are certainly oversimplified.

It is difficult to distinguish the effects of different power law dependences for the dust temperature and density in the visibility curves, at least for "reasonable" modifications. This is due in part to the loss of detail caused by averaging over one dimension and in part to the fact that changes in the slope of the "wings" of the emission profiles show up only at very low spatial frequencies which are normally undersampled by the interferometers. Sutton et al. (1978, 1979) suggest that density distributions flatter than $\rho \sim r^{-1.5}$ are inconsistent with their data for Mira and IRC+10216, but it is not clear how they arrive at this conclusion given figure 2 which shows plots of two model distributions with differing density laws: $\rho \sim r^{-2}$ and $\rho \sim {}^{-1.5}$. The three parts of the figure are the same as in figure 1.

There is very little difference between the emission distributions seen by the interferometers and even less in the final visibilities. With very good signal-to-noise ratios, it might be possible to distinguish between these two cases from the visibility curves only. In practice, the signal-to-noise ratios are never good enough to differentiate the subtleties shown here. The same statement holds for the temperature distributions as well. For smoothly varying temperatures where the dependence differs from $r^{-0.4}$, the visibility curves are relatively insensitive to the form of the temperature distribution. Different wavelengths exhibit much different sensitivities to the form of the emission "wings," and it may be possible to place much stronger constraints on the form of the density and temperature distribution. No one has yet derived α or β (or different dependences) from visibility data.

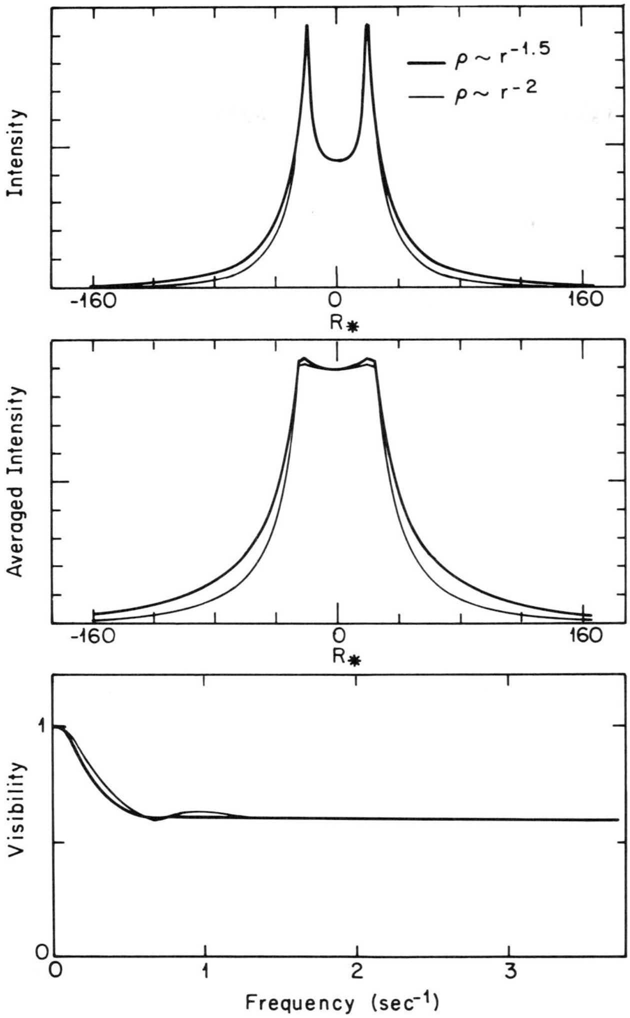

Figure 2 - Top (2a) shows calculated emission profiles at 11 μm for two different dust distributions. In both cases, no dust exists within an inner radius (20 R_*) and has a power law dependence outside this radius. The middle (2b) and bottom (2c) parts of the figure are analogous to figure 1.

3.2. Comparison of Visibility Data with an Image of α Ori

For the star α Ori, both a visibility curve and an image exist at the same wavelength. Howell, McCarthy, and Low (1981, henceforth HML) present a 10 μm visibility curve for position angle zero (north-south) at low spatial frequencies combined with high spatial frequency data for position angle 90° from Sutton (1979); the two data sets are nicely consistent, inasmuch as the extended dust emission shows up only in the low frequency data set. Bloemhoff, Townes, and Vanderwyck (1984, henceforth BTV) obtained a direct 10-μm image from an east-west scan under diffraction-limited seeing conditions. Although the image is only in one dimension, it is a true slice of the actual two-dimensional image, and there is no slit which averages the light distribution in the orthogonal direction.

From the visibility curve, shown in figure 3a, HLM fitted model intensity distributions and derived an inner radius for the dust shell of 60 $R_\star$ (~ 2"), $R_\star$ being the photospheric radius assumed to be ~ 0.03". BTV, on the other hand, see a distinct peak in the dust emission at about 35 $R_\star$ and see emission to within 0.5" or 20 $R_\star$ from the star; figure 3b shows their one-dimensional scan. The actual image shows a marked asymmetry in the dust emission from one side to the other, but the form of the distribution, in particular the radius of the peak emission, is very similar on both sides apart from the overall normalization. The image shape shows that the Gaussian distribution used by HML to fit the visibility curve is a quite reasonable approximation to use. The asymmetry seen in the image should not strongly affect the derived inner radius determined from the visibility data.

One way to resolve this discrepancy is to assume the inner radius is elliptical when projected on the sky, and the north-south and east-west radii are different. Since another star, IRC+10216, appears elliptical at shorter wavelengths, this is at least a plausible explanation and emphasizes the need for complete position angle coverage. BTV do have north-south data from their multiple detector scan. Although they state it is poorly calibrated, their data show the north-south dust emission peaks about 1" on either side of the star, consistent with their east-west data and in contradiction to the derived size from the HML visibility curve.

There is at least one alternative which reconciles the two data sets and points out the real uncertainties in inner radii derived from visibility curves. Figure 4 presents a plot of the shell emission only with the stellar emission subtracted from the actual data. There is some inaccuracy in this plot, since I measured the data from the article, and BTV are not responsible for the following conclusions. The symmetrized and smoothed data appear in figure 4b. These were averaged along the orthogonal direction, assuming a circularly symmetric distribution, and Fourier-transformed to give the visibility curve shown in figure 4c. The visibility data from HML also appear in figure 4c.

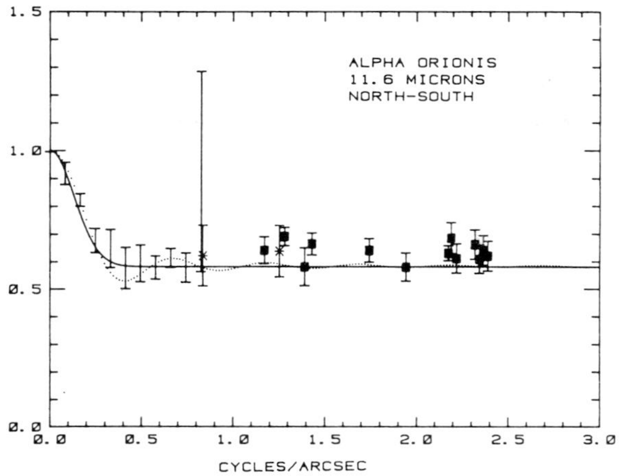

Figure 3a: Visibility data from Howell, McCarthy, and Low (1981) at low frequencies, and from Sutton (1979) at high spatial frequencies for a different position angle. The solid line represents a fit to the data for a gaussian light distribution with a 60 stellar diameter FWHM.

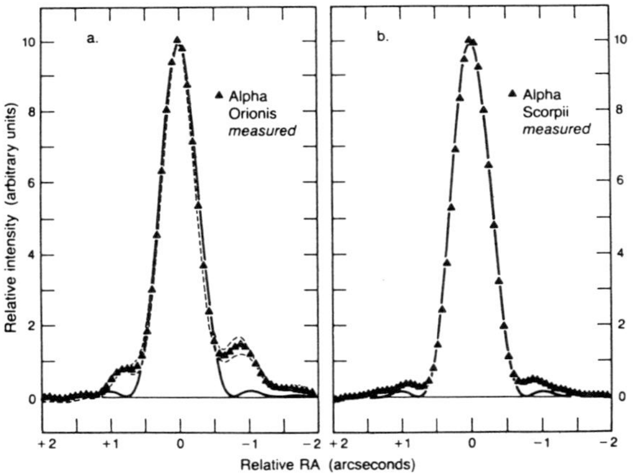

Figure 3b: Direct scans of the stars Alpha Orionis and Alpha Scorpii at 10 microns from Bloemhoff, Townes, and Vanderwyck (1984). The solid lines are diffraction-limited instrumental profiles for response to unresolved sources.

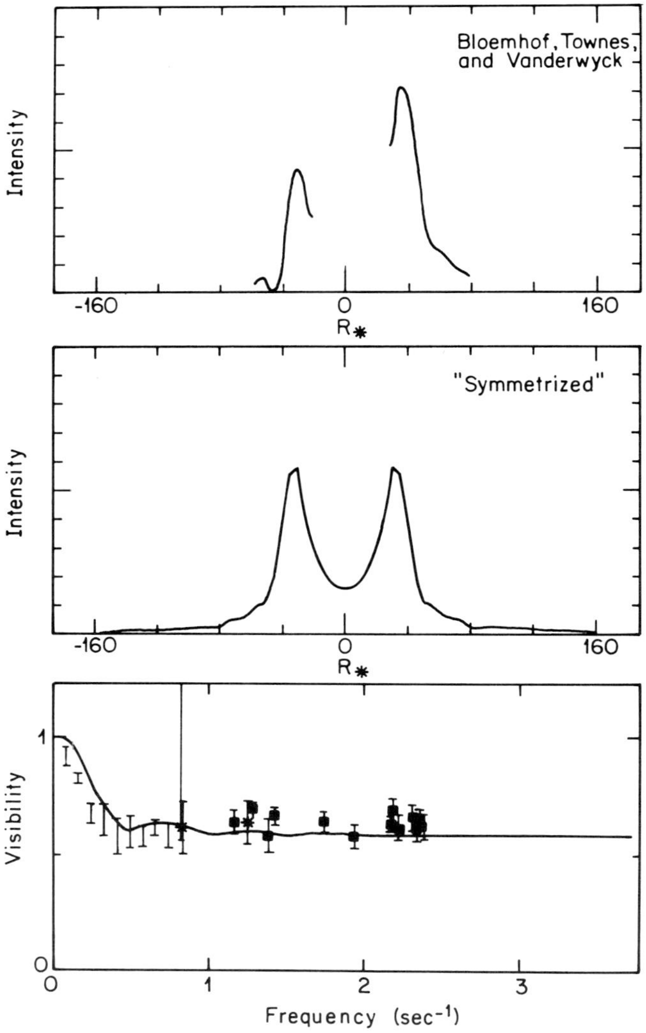

Figure 4 - Top (4a) displays the emission from Bloemhoff, Townes, and Vanderwyck in excess of the diffraction-limited stellar image. Middle (4b) is a symmetrized version of (4a) computed for convenience to produce the (slit averaged) visibility curve at the bottom (4c). The visibility data presented by Howell, McCarthy, and Low (1981) appear in the bottom figure for comparison.

This exercise points out two aspects of the light distribution as it relates to visibility curves. First, the inner radius corresponds to the peaks of the dust emission. Although BTV see emission as close as 20 $R_\star$ from the star, this may simply result from projection of emission at larger radii in front of and behind the star. This interpretation is supported by the (admittedly uncertain) rapid falloff of shell emission within the 35 $R_\star$ peaks. The inner radius derived from the visibility curve produced by transforming the image data is ~ 35 $R_\star$. Second, the visibility data of HML actually do provide a pretty good fit to the image data, if we assume the errors are correlated from point to point. HML used a one-dimensional speckle interferometer to obtain the visibility data. Normally, the uncertainties are determined from a comparison of many visibility curves which vary owing to fluctuations in the seeing. The point-to-point variations in these curves are correlated, and the uncertainties in model fits to the data are larger than those normally determined from least-squares fitting. With this proviso, the two data sets are quite consistent. It is also possible that a rather extended region of low-surface brightness can affect the interferometry, because of slit averaging, but be almost transparent to the direct imaging, because of low signal-to-noise ratios.

The comparison presented here emphasizes two important limitations to the present interferometric techniques which should be overcome when phase information is retained. The dust emission is not necessarily circularly symmetric, nor symmetric in reflection about any line through the stars. It is therefore important to get good position angle coverage whenever possible. Furthermore, the uncertainties in the speckle data can be correlated, so very good signal-to-noise ratios should be sought. Recent results with speckle interferometers show good signal-to-noise ratios can result in most cases (e.g., McCarthy 1982; Dyck et al. 1984). It is encouraging that in almost all cases where visibility data from different groups using different techniques exist for the same stars, the data agree nicely to within the combined uncertainties.

3.3. Aspects of the Observed Visibilities

One might hope that more radical departures from the smoothly varying distributions will show up in the visibilities. For example, many of these stars vary with periods of approximately a few years, and it may be that mass is ejected sporadically in separate shells (e.g., Hinkle 1978; Hinkle, Hall, and Ridgway 1982, and references therein). If the period is also the time between separate shell ejections, the separation between different shells will be of order the outflow velocity times the period. For the rather extreme case of a five-year period with a 50 km s^{-1} outflow speed, the separation between shells will be only ~ 10^{15} cm, subtending an angle of 0.2" at 200 pc. This separation would just barely be resolvable with existing telescopes, and the averaging effect of one-dimensional data would make the shells difficult to detect.

Dust and gas might be ejected in many dense clumps close to the stars (e.g., Alcock, this volume), a quite different picture of the mass loss than assumed for the calculations presented above. There is already the suggestion of clumping in α Ori from the asymmetry of the image shown by BTV. Such clumps will be difficult to detect unless there is a small number of relatively large condensations. Since the angular resolution of infrared interferometers is still not sufficiently good to resolve stellar photospheres, it seems unlikely that we can see the small clumps which might form from instabilities near the stellar surfaces (Salpeter 1974).

There is not yet any good evidence for clumping or separate shells in these stars. In IRC+10216, the light distribution is very smooth in at least the east-west direction. The systematic uncertainties in the visibilities probably make it impossible to identify clumps or shells anyway, although it is conceivable that some of the asymmetries seen in stars such as IRC+10216 result from huge clumps (or small individual clouds) close to the stars. It is also possible that departures from spherical symmetry result from disk or bipolar geometries seen in several very evolved objects (Ney et al. 1975, Westbrook et al. 1975).

It may be possible to see the effects of mass loss which varies on much longer time scales than the pulsation periods, time scales of about tens to hundreds of years, say. Hinkle, Hall, and Ridgway (1982) suggest that such long-period mass loss might occur in χ Cyg and other stars, based on their spectroscopic study of long-period variables. No one has discovered such effects in the spatial interferometric data, but low contrast will make them difficult to see. In addition to the averaging effect discussed above, the power in the visibility curves tends to be dominated by the dust closest to the red star. The spatial distribution of infrared light from the near infrared should be very useful to find long-term variations. The cold and more extended dust tends to have the greatest effect on the spatial distribution of the long wavelength light. Systematic studies of both the long and short wavelengths will be important for the study of long-period mass loss variations.

As mentioned above, the interferometric data suggest inner radii for the dust envelopes which are rather large, typically more than ten stellar radii. More importantly, the equilibrium temperatures of the dust grains are 1000 K or less in the carbon star IRC+10216 and 500 to 700 K in the oxygen-rich star α Ori at the inner dust shell boundaries. These temperatures are about a factor of 1.6 smaller than the equilibrium temperatures for grains to condense, assuming graphite will dominate in IRC+10216 and silicates will dominate in α Ori (Salpeter 1974), suggesting a possible problem with the theory. Spectroscopic evidence presented by Hinkle (1978); Bernat et al. (1979); and Hinkle, Hall, and Ridgway (1982), for example, shows that the region between the photosphere and inner boundary of the circumstellar shell is quite complicated and time-variable, with shock waves traversing the region every stellar period dissociating CO and other molecules, and possibly inhibiting the formation of dust grains. There appears to be a quasi-stationary layer with a temperature 800 K many stellar radii from the photosphere. To really test the nucleation theory, it will be necessary to consider this highly time-variable region. Another possibility is discussed below.

Can the dust drive the mass loss as has been supposed for many years? The bulk of the interferometry data suggests that the dust forms predominantly very far from the stellar photospheres, much too far to drive the mass loss close to the surfaces. Of course, shock waves generated in the lower photosphere might levitate matter to the region where dust forms, and subsequent mass loss occurs via radiation pressure on dust grains (e.g., Zuckerman 1980). It is also possible that we are being fooled by the infrared light. For the dust to be effective in driving the mass loss, it is necessary only to have an optical depth larger than unity at visual wavelengths, where most of the stellar luminosity appears (see Goldreich's discussion later in this volume). This optical depth provides very little infrared opacity (e.g., Johnson 1968), and correspondingly little emission which would in any case be difficult to separate from photospheric emission. Should refractory cores form close to the stars, the dust might be effective for the mass loss. Subsequent growth of volatile mantles at large distances from the star could result in a huge increase in infrared opacity, which would dominate the visibility curves.

This possibility is at least suggested by Roddier's remarkable image of dust near the photosphere of α Ori (also in this volume), a result which seems to suggest the dust forms twice -- once near the star and again at the inner edge of the infrared shell. A definitive answer will most likely require angular resolution sufficient to resolve the stellar photospheres, and perhaps even two-dimensional images as well, a prospect best suited to a next-generation telescope or long baseline interferometer.

4. CIRCUMSTELLAR GAS

The gas in the circumstellar envelopes absorbs and reradiates infrared light in many molecular lines which are seen in high resolution spectra (Merrill and Ridgway 1979). By isolating these lines and using one of the spatial interferometry techniques described in section II, one can study the spatial distribution of the reradiated light, and hence of the gas. This has been done for CO lines from two red giants, IRC+10216 (Dyck, Beckwith, and Zuckerman 1983) and Mira (Beckwith, Dyck, and Zuckerman 1985). By using a grating spectrometer with a small entrance slit, they isolated individual vibration-rotation lines of CO near 4.6 μm and applied the one-dimensional speckle interferometry technique described in section II. Although the velocity resolution is poor and entire lines are seen at once, in both cases the light scattered by the CO molecules subtends large enough angles to be easily resolved with large ground-based telescopes.

In IRC+10216, lines from several different rotational levels of both ^{12}CO and ^{13}CO have approximately the same extent on the sky, the full widths at half maximum being ~1.3" compared with the dust emission FWHM ~ 0.5". The different lines span a wide range of radial optical depths measured at the line center over a path from radius $\sim 10^{16}$ cm to ∞. For example, the optical depth of the ^{12}CO P(14) line is approximately unity over this radial path, whereas the ^{12}CO P(3) line has an optical depth greater than one hundred over the same distance. Since the two lines have similar angular extents, photons evidently scatter very close to their point of origin (in this case, the dust emission radius) and become radiatively decoupled from the gas in the outer envelope. There is no obvious correlation of angular size with line parameters for the 4.6 μm CO lines around IRC+10216.

In Mira, the angular extents of several lines are different with lines from high rotational levels (and correspondingly low radial optical depths) having smaller extents than lines from lower energy levels. The scattering process in Mira must somehow be different than in IRC+10216. Mira has a smaller mass loss rate than IRC+10216 by two to three orders of magnitude, and the outflow velocity is also smaller.

The differences probably can be explained by detailed scattering calculations (Natta, Beckwith, and Salpeter 1985), but the basic results follow from the Sobolev escape probability formalism. If the outflow reaches a constant velocity much higher than the thermal or "microturbulent" gas velocities and the radial optical depths are large, photons tend to interact only in small regions within the envelope and can escape along non-radial lines, because of the relative Doppler shifts of the photons and molecules in different parts of the envelope. In IRC+10216, the outflow velocity $v_\infty \sim 15$ km s^{-1} is much larger than the velocity dispersion $\Delta v \sim 1$ km s^{-1}, and the mass loss rate is sufficiently large, $dM/dt \sim 10^{-4}$ $M_\odot$ yr^{-1}, that the angular sizes result mainly from the form of the velocity field and depend little on the line parameters. Mira, on the other hand, has a much smaller velocity and mass loss rate, $v_\infty \sim 4$ km s^{-1} and $dM/dt \sim 10^{-7}$ $M_\odot$ yr^{-1}, so a much larger part of the envelope interacts with each photon. Then the velocity field and the line parameters combine to give the observed sizes.

Spatial interferometry of molecular lines has some obvious advantages and disadvantages over broad band interferometry for the study of the envelopes. Perhaps the most important advantage is that a different component, the gas, is observed; and since the dust and gas may not be in strict thermodynamic equilibrium (e.g., Kwan and Linke 1982), there might be differences in the gas and dust densities (or relative abundances) and temperatures. There are often a huge number of different lines available for study from several molecules. Whereas both the density and temperature of the dust determine the dust emission at any radius, the excitation temperature of molecular levels can be independent of both the volume and column densities, so it is in principle possible to separate the effects of temperature and density through observation. It should be particularly interesting to see if the scattered light from the gas in stars like IRC+10216 is as asymmetric as the dust emission, for example.

Lines give the added advantage of having different velocity components which may come from different projected radii. Should there be some acceleration of the gas in the outer envelopes, perhaps owing to radiation pressure on dust grains, it may be possible to see the effects of this acceleration as size variations at different line velocities. Should the observed sizes be relatively smooth functions of velocity, comparison of different lines might yield observations of the excitation temperature at well-defined radii for comparison with models.

Unfortunately, very high-spectral resolution is required to isolate the lines or parts of the lines, and the corresponding signal-to-noise in the visibility curves tend to be small. At present, only gas from the brightest stars can be observed with very high angular resolution, although all three of the stars mentioned in section (3) fall into this category. Since these three stars provide the bulk of our knowledge of red giant envelopes determined from infrared interferometry, the combination of very high spatial with very high spectral resolution, when applied to many different lines, might prove to be a very useful probe of the physical conditions around the stars.

5. FUTURE DIRECTIONS

Several tantalizing results emerge from the spatial interferometry discussed in this review. The dust formation radius appears to be larger than expected from pure theory by a modest factor, one which is important to understand for theories of the grain formation and the envelope dynamics. The marked asymmetry in some of the stars suggests that dust formation may be a sporadic process governed by as yet undiscovered physical conditions near the stars, or that the overall symmetry for dust envelopes is disk-like as opposed to spherical.

High spatial resolution observations can help to unravel these phenomena, but to make significant progress we probably require imaging capability, and perhaps much higher spatial resolution. There are ways of reconstructing images from single-dimensional speckle data even when the absolute phase is unknown, and application to the study of red giant stars may come within the next few years. Two-dimensional images are very important. Since optical observers have had access to panoramic arrays such as CCD's for many years, the techniques of reconstructing two-dimensional images from array data should be well known by the time two-dimensional infrared detectors are available for astronomy. Two-dimensional pictures should go a long way toward revealing the actual dust distribution around red giants.

Higher spatial resolution might be provided by a very large new telescope, such as the proposed 15m telescope, but it is not yet clear whether the increase in resolution will be sufficient to make a qualitative difference in red giant studies. For example, if the scattered light seen by Roddier toward α Ori is to be studied at infrared wavelengths (to test for thermal emission, say), the spatial resolution at mid-infrared wavelengths will need to be comparable to present optical resolution. This argument suggests that an increase of a factor of ten is important, and if dust forms very close to the photospheres of most giant stars, a factor of ten may not be enough. There is therefore a need for long baseline interferometers to probe the innermost regions of the circumstellar envelopes.

The possibility of very good velocity resolution of the gas through speckle spectroscopy is very appealing. Should it be possible to obtain velocity versus position information for the envelopes of red giants in the same way millimeter wave CO lines provide such information about molecular clouds, we might obtain a wealth of information about the dynamics of the circumstellar matter. This area can benefit greatly from new infrared technology, better detectors and better spectral filters; and it seems likely that the future holds a great deal of promise.

* * *

I would like to thank the Aspen Center for Physics for their hospitality during June 1984 where most of this work was carried out, and the Alfred P. Sloan Foundation for support which has allowed me to carry out research in this field.

REFERENCES

Becklin, E. E., Frogel, J. A., Hyland, A. R., Kristian, J., and Neugebauer, G. 1969, Ap. J. (Letters), **158**, L133.

Beckwith, S., Dyck, H. M., and Zuckerman, B. 1985, in preparation.

Bernat, A. P., Hall, D. N. B., Hinkle, K. H., and Ridgway, S. T. 1979, Ap. J. (Letters), **233**, L135.

Bloemhoff, E. E., Townes, C. H., and Vanderwyck, A. H. B. 1984, Ap. J. (Letters), **276**, L21.

Crabtree, D. R. and Martin, P. G. 1979, Ap. J., **227**, 900.

Dyck, H. M., Beckwith, S., and Zuckerman, B. 1983, Ap. J. (Letters), **271**, L79.

Dyck, H. M., Zuckerman, B., Leinert, C., and Beckwith, S. 1984, Ap. J., **287**, 801.

Dyck, H. M. and Howell, R. 1982, Astron. J., **87**, 400.

Foy, R., Chelli, A., Sibille, F., and Lena, P. 1979, Astron. Ap., **79**, L5.

Hinkle, K. H. 1978, Ap. J., **220**, 210.

Hinkle, K. H., Hall, D. N. B., and Ridgway, S. T. 1982, Ap. J., **252**, 697.

Howell, R. R., McCarthy, D. W., and Low, F. J. 1981, Ap. J. (Letters), **251**, L21.

Johnson, H. L. 1968, in Nebulae and Interstellar Matter, ed. B. M. Middlehurst and L. H. Aller (Chicago: University of Chicago Press), p. 167.

Johnson, M. A., Betz, A. L., and Townes, C. H. 1974, Phys. Rev. Let., **33**, 1617.

Jones, T. W. and Merrill, K. M. 1976, Ap. J., **209**, 509.

Kwan, J. and Linke, R. A. 1982, Ap. J., **254**, 587.

Labeyrie, A., Kowchlin, L., Bonneau, D., Blazit, A., and Foy, R. 1977, Ap. J. (Letters), **218**, L75.

Mariotti, J. M., Chelli, A., Foy, R., Léna, P., Sibille, F., and Tchountonov, G. 1983, Astron. Ap., 120, 237.
McCarthy, D. W. 1982, Ap. J. (Letters), 257, L93.
McCarthy, D. W., Howell, R., and Low, F. J. 1978, Ap. J. (Letters), **223**, L113.
McCarthy, D. W., Howell, R., and Low, F. J. 1980, Ap. J. (Letters), **235**, L27.
McCarthy, D. W. and Low, F. J. 1975, Ap. J. Letters), **202**, L37.
McCarthy, D. W., Low, F. J., and Howell, R. 1977, Ap. J. (Letters), **214**, L85.
Merrill, K. M. and Ridgway, S. T. 1979, Ann. Rev. Astron. Ap., **17**, 9.
Michelson, A. A. and Pease, F. G. 1921, Ap. J., **53**, 249.
Natta, A., Beckwith, S., and Salpeter, E. E. 1985, in preparation.
Ney, E. P., Merrill, K. M., Becklin, E. E., Neugebauer, G., and Wynn-Williams, C.G. 1975, Ap. J. (Letters), **198**, L129.
Rowan-Robinson, M. and Harris, S. 1983, M.N.R.A S., **202**, 767.
Salpeter, E. E. 1974, Ap. J., **193**, 585.
Salpeter, E. E. 1977, Ann. Rev. Astron. Ap., **15**, 267.
Selby, M. J., Wade, R., and Sanchez-Magro, C. 1979, M.N.R.A.S., **187**, 553.
Sibille, F., Chelli, A., and Lena, P. 1979, Astron. Ap., **79**, 315.
Sutton, E. C. 1979, in IAU colloquium 50, High Angular Resolution Stellar Interferometry, ed. J. Davis and W. J. Tango (Sydney: Chatterton Astronomy Department, School of Physics, University of Sydney), p. 16.
Sutton, E. C., Betz, A. L., Storey, J. W. V., and Spears, D. L. 1979, Ap. J. (Letters), **230**, L105.
Sutton, E. C., Storey, J. W. V., Betz, A. L., Townes, C. H., and Spears, D. L. 1977, Ap. J. (Letters), **217**, L97.
Sutton, E. C., Storey, J. W. V., Townes, C. H., and Spears, D. L. 1978, Ap. J. (Letters), **224**, L123.
Toombs, R. I., Becklin, E. E., Frogel, J. A., Law, S. K., Porter, F. C., and Westphal, J. A. 1972, Ap. J. (Letters), **173**, L71.
Westbrook, W. E., Becklin, E. E., Neugebauer, G., and Wynn-Williams, C. G. 1975, Ap. J., **202**, 407.
Zuckerman, B. 1980, Ann. Rev. Astron. Ap., **18**, 263.

10 MICRON SPECKLE INTERFEROMETRY OF OH/IR STARS

Michael L. Cobb* and John D. Fix
The University of Iowa
Department of Physics and Astronomy
Iowa City, Iowa 52242

As part of a program to determine the circumstellar structure of OH/IR stars, we have obtained interferometric measurements of four OH/IR stars at a wavelength of 10 microns. The observations were carried out on 27-30 May 1984 with the 3 meter telescope of the Infrared Telescope Facility on Mauna Kea. The data were obtained by driving the image across a narrow (0".8) slit oriented either E-W or N-S. The bandwidth of the filter used was 5 microns.

We have estimated angular sizes for the objects by fitting Gaussians to the measured visibility amplitudes. The resulting angular sizes are given in the following table:

TABLE I

Object	N-S		E-W	
	θ_{FWHM}	σ_θ	θ_{FWHM}	σ_θ
IRC+10420	0".43	0".02	0".41	0".02
OH 26.5+0.6	0".32	0".04	0".30	0".04
OH 354.9-0.5	-	-	0".17	0".03
OH 1.1-0.8	0".26	0".04	-	-

For all four objects, the brightness temperature at 10 microns is about 250 K. The visibility amplitudes for OH 26.5+0.6 are shown in Figure 1 along with the visibility amplitudes of the fitted Gaussians.

For IRC+10420 and OH 26.5+0.6, Dyck et al. (1984) have obtained angular sizes at 4.8 microns. The sizes at 10 microns are about twice as large. Hydroxyl maser maps at 18 cm are available for IRC+10420 (Bowers 1984, Diamond et al. 1984), OH 26.5+0.6 (Baud 1981), and

*Visiting Astronomer at the Infrared Telescope Facility which is operated by the University of Hawaii under contract from the National Aeronautics and Space Administration.

M. Morris and B. Zuckerman (eds.), Mass Loss from Red Giants, 115–116.

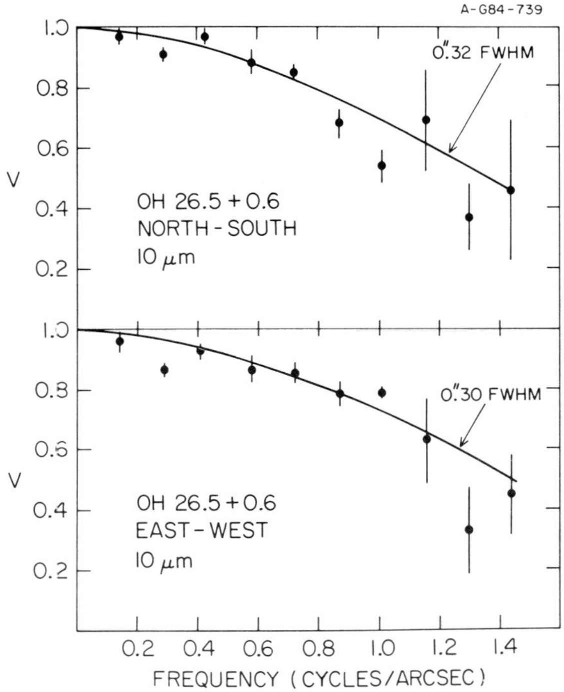

Figure 1. Visibility amplitudes for OH 26.5+0.6.

OH 354.9-0.5 (Fix and Mutel, private communication). The total extent of the OH emission is about 25 times the 10 micron angular size in each case. For IRC+10420, OH maps have led to two distinct models: a clumpy sphere model by Bowers (1984) and an E-W aligned bipolar nebular model by Diamond et al. (1984). Our observations show the distribution of warm dust, which seems to lack pronounced asymmetry, favoring Bowers' model.

REFERENCES

Baud, B. 1981, Ap.J., (Letters), 250, p. L79.

Bowers, P.F. 1984, Ap.J., 279, p.350.

Diamond, P.J., Norris, R.P., and Booth, R.S. 1983, Astr. Ap., (Letters), 124, p.L4.

Dyck, H.M., Zuckerman, B., Leinert, Ch., and Beckwith, S. 1984, Ap.J., 287, 801.

FAR-INFRARED AND SUBMILLIMETER PHOTOMETRY OF EVOLVED STARS

Michael W. Werner
Space Sciences Division
NASA-Ames Research Center
Moffett Field, CA 94035
USA

ABSTRACT. Broad-band photometric observations at far infrared and submillimeter wavelengths (30 μm < λ < 1mm) provide a unique probe of circumstellar shells in evolved objects and of the mass loss processes which produce them. The problems which these observations explore include: dust properties and dust-to-gas ratios in the outflowing material, the mass and structure of the circumstellar shell, and the pumping mechanism for maser emission. This contribution reviews the early work in this still largely unexploited field and reports results in each of the areas listed above. We also preview the dramatic advances forthcoming in this area from the IRAS satellite and the further progress which will come from the next generation of infrared and submillimeter telescopes.

1. INTRODUCTION

Compared with some of the other observational areas described in this volume, that of far infrared and submillimeter observations (30 μm <λ< 1mm) of continuum emission from mass-losing stars is very much in its infancy. The reason is that observations between 30 and 300 μm are routinely possible only from airplane altitudes and above, while the very small [compared to protostellar regions] masses of dust in the circumstellar shells make them faint objects at λ> 300 μm even for the best current instruments on the largest telescopes. The opportunities for this type of work should improve dramatically over the next few years, however; this progress began with the Infrared Astronomical Satellite (IRAS) survey and will continue with the advent of 10-m class ground-based submillimeter telescopes and the development of future space infrared and submillimeter telescopes such as SIRTF, ISO, and LDR. Therefore, this seems an appropriate juncture at which to review the progress to date in this area and to describe the future prospects.

This review emphasizes observations at > 30 μm, although reference is made as necessary to the much more numerous shorter wavelength infrared observations. The study of OH/IR stars at near infrared wavelengths has been a particularly active field in recent years, as repre-

M. Morris and B. Zuckerman (eds.), Mass Loss from Red Giants, 117–128.

sented by several contributions in this volume as well as by the work of Jones et al. (1983), Herman et al. (1984), Baud et al. (1985), and Gehrz et al. (1985), to which the reader is referred for further information and references. We also do not discuss far infrared observations of planetary and protoplanetary nebulae, which are presented by Kleinmann et al. (1978) and Moseley (1980) and will also be thoroughly covered in the IRAS data base.

2. SCIENTIFIC MOTIVATION

The continuum emission in the far infrared from an evolved star which has undergone, or is undergoing, substantial mass loss is produced by thermal emission from dust in the circumstellar shell. Even in the cases where the dust shells are quite thick, the peak luminosity of such objects appears to lie in the 5-to-30 μm region, although this may be in part a selection effect which will be overcome by the unbiased IRAS survey. Thus the far infrared observations of these objects generally result in corrections to the luminosity of less than a factor of two beyond what is measured shortward of 30 μm. The far infrared observations do bear uniquely on such important problems as the properties of the circumstellar grains and the mass of the circumstellar shell - factors of importance in assessing the contribution of a particular class of stars, or a particular phase in the stars' lifetime, to the recycling of matter to the interstellar medium. In fact, as discussed further below, the far infrared continuum measurements may be our best way of determining the total extent of a circumstellar shell and thus of estimating the duration of a rapid mass loss phase. When combined with radio observations of these systems, the far infrared observations can produce estimates of gas-to-dust ratios and tests of maser pumping schemes. Together with observations in the near infrared, the far infrared data provide a means of modeling the structure of the circumstellar shell and test important empirical relations such as whether the silicate features in these stars are of circumstellar or interstellar origin. The present paper touches briefly on each of these issues and also on the likely impact of the IRAS results on this area of research.

3. SUBMILLIMETER PHOTOMETRY OF EVOLVED STARS.

3.1 Dust Masses and Dust-to-Gas Ratios

Submillimeter ($\lambda > 300$ μm) wavelengths provide a uniquely powerful means of determining the mass of dust in a circumstellar shell because of the weak temperature dependence of the radiation at this long wavelength. Observations at this wavelength are possible from dry mountaintop sites, but with existing instrumentation and telescopes, it is possible to detect only the brightest objects. The results to date are summarized in Table 1, taken from the work of Sopka et al. (1985), which includes planetary and protoplanetary nebulae as well as both oxygen-rich and carbon-rich stars. The submillimeter emission seen from these objects is attributed to thermal emission from dust mixed with the gas in the outflowing circumstellar shell. Sopka et al. show that it is straightforward to relate the flux at 400 μm to the mass loss rate. Under the assumption that the density n(r) in the outflow scales as r^{-s} with s=2, the observed flux S can be related to the mass loss rate $\dot{M}_D$ in dust by the expression:

$$\dot{M}_D = \frac{2Sd^2v}{k\xi} \quad (1)$$

where v is the outflow velocity (known from molecular observations); k the mass opacity of the dust grains at 400 μm; d the distance to the source, and ξ is an integral over the radial temperature distribution in the source and the shape of the telescope beam. For the adopted values of v and d given in Table I, and assuming k = 20 cm^2/gm (see below), Sopka et al. find mass loss rates in dust (Table I) between $\approx 10^{-5}$ and $\approx 10^{-7}$ $M_\odot$/yr. These mass loss rates are compared in the Table with mass loss rates in gas which are derived primarily from CO observations of these objects. Given the considerable uncertainties in both estimates, it is seen that the dust-to-gas ratios in the outflowing material are generally consistent with the interstellar value, as has also been found by Werner et al. (1980) for OH/IR stars.

For some purposes, the total shell mass is of greater interest than the mass loss rate. This has been estimated and included in Table I under the assumption that all of the grains emitting at 400 μm have T=50K, which is a characteristic temperature for grains emitting at 400 μm. If, instead, T=100K were assumed, the masses would be smaller by a factor ≈ 2.5. Dust masses of order 0.01$M_\odot$ are found, which, combined with gas-to-dust ratios of order 100, lead to masses of up to a solar mass for the shells. Thus the shells observed at present could contain a substantial fraction of the mass of the progenitor star.

3.2 Grain Properties

Dust shells of this type are particularly good sites for the investigation of grain properties at far infrared wavelengths. The reason for this is that the location and character of the heating source, and the spatial distribution of the dust with respect to the heating source, are known with much less uncertainty than is usually the case for sources in the interstellar medium. The luminosity of the heating source can be determined with the help of the infrared observations and the grain properties derived from comparison with a relatively straightforward model calculation. Results of application of this technique to the submillimeter observations of Sopka et al. are shown in Figure 1 for the carbon-rich object IRC+10216 and the oxygen-rich object IRC+10011. The curves fit through the IRC+10216 data are model calculations for s=2 and for several different choices for the grain properties, as summarized in the figure caption. For this density distribution, the best fit to the long wavelength data is found for grains having long wavelength emissivity $Q(\lambda) \propto \lambda^{-1.3}$, close to the expected behavior of amorphous carbon, rather than the steeper wave-

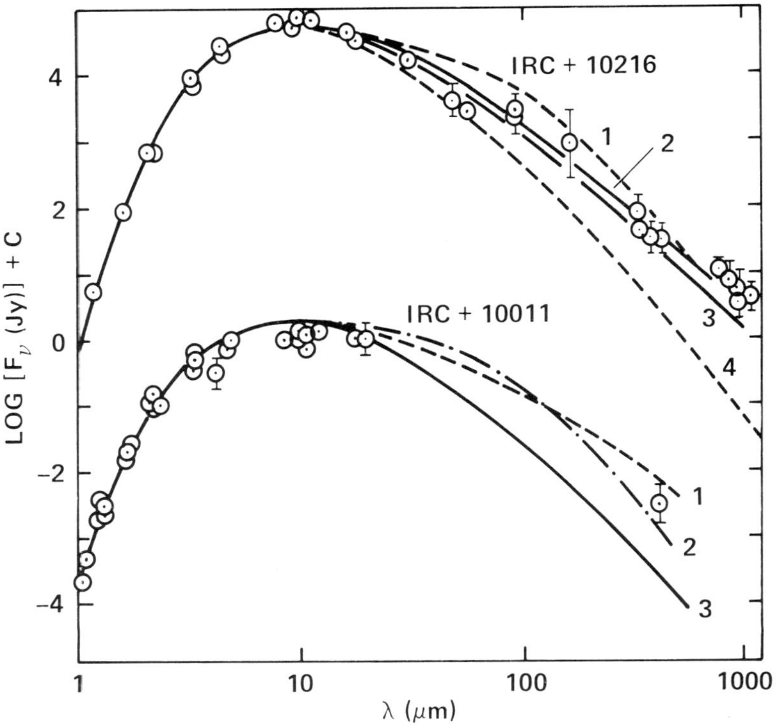

Figure 1. Model fits to the observed infrared and submillimeter spectra of IRC+10216 (top) and IRC+10011 (bottom). The value of C on the ordinate is 0 for +10216 and -3 for +10011. For IRC+10216, curves 1-4 represent, respectively, the following grain models:

1: Amorphous carbon;

2: Amorphous carbon modified so that the absorption efficiency $Q(\lambda) \propto \lambda^{-1.2}$ for $\lambda > 10$ μm;

3: The same but with $Q(\lambda) \propto \lambda^{-1.4}$ for $\lambda > 10$ μm;

4: The same, but with $Q(\lambda)$ for $\lambda > 10$ μm as expected for pure graphite grains.

An $n(r) \propto r^{-2}$ density law is assumed in each case.

For IRC+10011, curves 1-3 represent, respectively:

1: Crystalline silicate grains and an $n(r) \propto r^{-1.5}$ density distribution;

2: Crystalline silicates, modified to have $Q(\lambda) \propto \lambda^{-1}$ for $\lambda > 10$ μm, and $n(r) \propto r^{-2}$.

3: Crystalline silicates unmodified, and $n(r) \propto r^{-2}$.

Further details of the models and references to the original data can be found in Sopka et al. (1985), whence this figure is adapted.

length dependence expected for pure graphite. Similarly, the best fit to the spectrum of IRC+10011 is found for amorphous silicate grains, having $Q(\lambda) \propto \lambda^{-1}$, rather than for crystalline silicate grains which, again, have a steeper decrease in emissivity and could not account for the observed amount of submillimeter emission.

On the basis of the laboratory properties of amorphous silicate and carbon grains, Sopka et al. adopt 20 cm^2/gm for the mass opacity of the circumstellar grains at 400 μm. There is of course considerable uncertainty in this estimate, and it is somewhat higher than the value of 6 cm^2/gm suggested by Hildebrand (1983) for interstellar grains. These two results should be considered to be consistent in view of the uncertainties in both. Note, however, that a value of 6 cm^2/gm for the opacity of the circumstellar grains would lead to rather high values for the dust-to-gas ratio in the circumstellar shells (cf. Table I).

TABLE I. 400 μm OBSERVATIONS OF EVOLVED STARS

OBJECT	S(400)[a] (Jy)	d (Kpc)	$\dot{M}_D$[b] ($M_\odot$/yr)	$\dot{M}_G$[c] ($M_\odot$/yr)	$\dot{M}_G/\dot{M}_D$	M_D[d] ($M_\odot$)	L[e] ($L_\odot$)
OH231.8	22	1.3	7.6(-6)	1.0(-4)	13	0.014	4(3)
CRL2688	41	1.0	2.3(-6)	7.2(-5)	30	0.015	2.5(4)
NGC7027	15	1.1	1.3(-6)	7.8(-5)	60	0.008	6(3)
CRL618	9	1.7	8.5(-7)	1.0(-4)	120	0.01	3(4)
VYCMa	10	1.5	7.8(-7)	2.3(-4)	300	0.008	3.5(5)
+10216	32	0.3	1.7(-7)	5.7(-5)	300	0.001	5.5(4)
+10011	3	0.5	7(-8)	5.0(-5)	700	0.0003	6(3)

Notes to Table I

a. Observed 400 μm flux, corrected for free-free emission if necessary, measured with a ≈ 45" beam.
b. Mass loss rate in dust, assuming $K=20cm^2/gm$ at 400 μm. $7.6(-6) \equiv 7.6 \times 10^{-6}$, etc.
c. Mass loss rate in gas, usually from CO observations.
d. Mass of dust from 400 μm flux, assuming $K=20cm^2/gm$ and T=50K for the grains.
e. Total luminosity from broad band infrared energy distribution.

3.3 Radial Density Distribution

Figure 1 also shows that, with a suitably chosen but not unreasonable set of dust properties, an $\underline{s}=1.5$ dust density distribution fits the spectrum of IRC+10011 as well as does the s=2 distribution in combination with the amorphous silicate grains. While s=2 is perhaps most plausible, the possibility that n(r) decreases outward more slowly (corresponding to a decreasing mass loss rate or to grain growth in the outflow), or more rapidly (corresponding to an increasing mass loss rate) should be borne in mind. Werner et al. and Rowan-Robinson (1982) have found, assuming the silicate grain properties given by Jones and Merrill (1976), that energy distributions with deep silicate features such as those of the OH/IR stars shown in Figure 2 have too much far infrared emission to be well fit by an s=2 density distribution. This is illustrated in Figure 3, where it is shown that s=1.5 is required to give an adequate fit to the energy distribution of OH26.5. A similar fit can be obtained with s=2, provided that some freedom is exercised in the choice of grain properties (Schutte and Tielens 1985).

Although broadband energy distributions may not be able to distinguish between s=1.5 and s=2, the resulting shape and size of the circumstellar shell at long wavelengths would differ dramatically in the two cases. Calculations illustrating the dependence of the source size

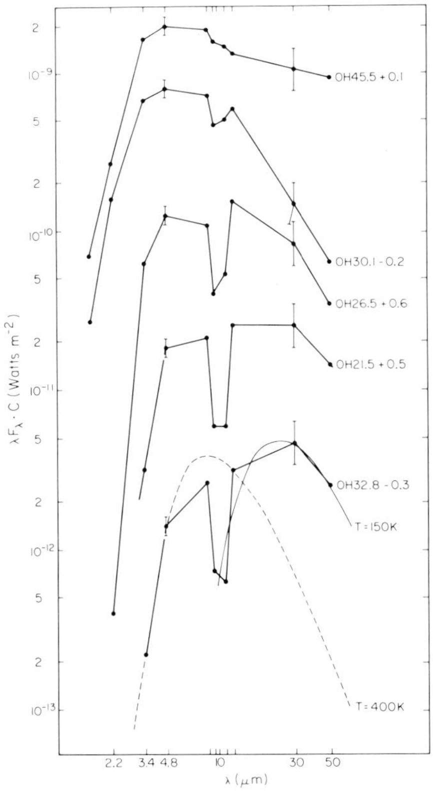

Figure 2. Infrared energy distributions for five very red OH/IR stars, 1978 July, from Werner et al. (1980). The multiplicative constant on the y-axis has the following value for the sources from top to bottom: C=265, C=42, C=1, C=2.2, and C=0.55.

on s are presented by Sopka et al. The data to investigate this question may not be present in the IRAS survey, because of its relatively large beam sizes. Measurements from the next generation of cryogenic infrared telescopes - SIRTF and ISO - which will combine 30" beams at 100 μm with very high sensitivity to low surface brightness emission, and from 10-m class submillimeter telescopes with 10" beams at 400 μm, should be able to distinguish between the two density distributions rather easily.

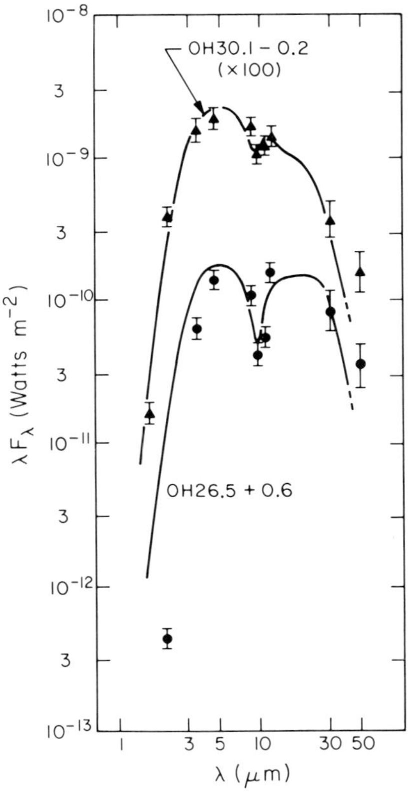

Figure 3. The observed infrared energy distributions of OH26.5 and OH30.1 are compared with two of the model calculations from Jones and Merrill (1976). The model fit to OH26.5 has density $n(r) \propto r^{-1.5}$ while that fit to OH30.1 has $n(r) \propto r^{-2}$.

4. FAR INFRARED OBSERVATIONS OF OH-IR STARS

4.1 General Results

Werner et al. (1980) carried out a set of simultaneous near infrared, far-infrared and radio observations of a group of radio-selected OH/IR stars which were originally identified by Evans and Beckwith (1977). These are very red objects without optical counterparts and appear to be stars with very thick circumstellar dust shells. As shown in Figure 2, these objects show broad energy distributions, with the radiated luminosity peaking in the 5-to-30 μm region, as well as deep silicate features in the 10 μm region. Up to 40% of the total power is radiated longward of 25 μm. The total luminosities range from a few times 10^3 to about 5×10^4 $L_\odot$.

4.2 Maser Pumping Efficiency

The results of Werner et al. were used to test the prediction of Elitzur, Goldreich, and Scoville (1976) that the 1612 MHz OH maser emission from these objects is pumped by the absorption of infrared photons in the rotational transitions at 35 μm. The pump efficiency, _e_, is defined as the ratio of the number of OH photons emitted per second, n(OH), to the number of photons available to be absorbed per second in the pumping transition, n(35 μm). A detailed discussion of the estimation of _e_ from the observations is given by Evans and Beckwith. It appears plausible in the absence of a detailed model to estimate _e_ from the observed peak OH and 35 μm flux densities by:

$$e = \frac{S_\nu(\mathrm{OH})}{S_\nu(35\ \mu\mathrm{m})} \tag{2}$$

Figure 4 compares the 35 μm and 1612 MHz fluxes for the objects observed by Werner et al., both for the simultaneous 1978 observations and for observations carried out in 1975 July by Evans and Beckwith. Observations were not carried out at 35 μm in 1975, but bounds on the 35 μm flux at that epoch were established based on the 1978 data. Note from Figure 4 that _e_ remains fairly constant as the fluxes themselves vary, and also that _e_ is close to 0.25. These results are in good agreement with the model of Eltizur et al. for the pumping of 1612 MHz maser emission. In this model, the maser is saturated, and a constant value of _e_ is consistent with this. Additionally, in the model of Elitzur et al., the pumping efficiency _e_ is predicted to be close to 0.25, as is observed for the sources shown in Figure 4. The support to this theory shown in Figure 4 has been substantiated by subsequent observations of larger numbers of OH/IR stars (Herman et al. 1984; Baud et al. 1985), in which the flux at 35 μm was estimated by extrapolation from 20 μm data. These studies show values of _e_ clustering around 0.25, although with somewhat larger scatter than shown in Figure 4. The scatter is presumably due in part to the fact that the 35 μm fluxes were not directly observed.

5. THE IMPACTS OF IRAS

5.1 The Extent of the Dust Shells

The results obtained from the IRAS all-sky infrared survey (Neugebauer et al. 1984) will have a profound influence on this field of research.

The flux levels achieved by the IRAS survey for a detection at the several sigma level in the 25 to 60 μm channels are approximately two orders of magnitude fainter than the fluxes shown in Figure 2. From this it can readily be seen that IRAS has measured most if not all of

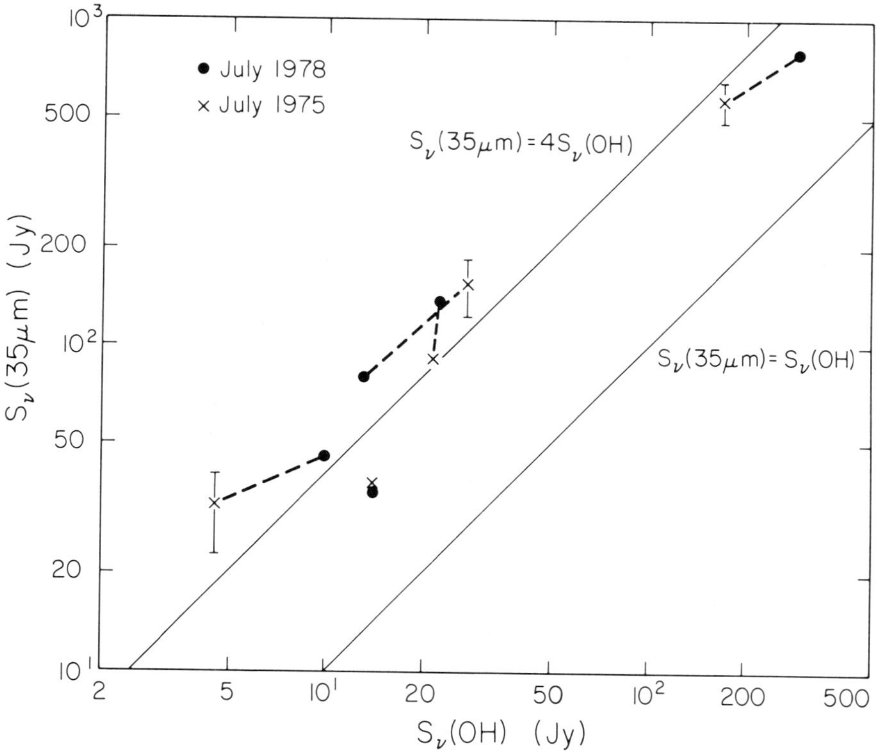

Figure 4. A comparison of the 35 μm infrared and 1612 MHz maser flux densities for the five sources studied by Werner et al. (1980) at each of two epochs. The error bar for the 1975 July data gives the uncertainties in the extrapolated 35 μm flux for that epoch. For 35 μm pumping to operate requires S(35 μm) > S(OH); S(35 μm) ≈ 4S(OH) is predicted by the detailed model of Elitzur et al. (1976).

the OH/IR stars in the Galaxy. This is particularly significant when it is borne in mind that these long-wavelength continuum observations may be the only means of determining the total mass of the circumstellar shell, which is a critical parameter in assessing the duration and significance of the high mass loss phase in which these objects are currently found.

For *s*=2, the total mass within a distance *r* of the star increases linearly with *r*; however, this distant material may be too cool to radiate appreciably shortward of 30 μm. In addition, it makes little contribution to the optical depth of the 10 μm silicate feature. Take the shell around OH26.5 as an example. Werner et al. showed for an assumed distance of 1.8 kpc that 4×10^{-4} $M_\odot$ of dust are required within a radius of 3×10^{16} cm to explain the radiation observed from this object shortward of 30 μm. The time scale (r/v) associated with this material is only 700 years. In the case of OH26.5, if it is assumed instead that the shell is as large as the projected radius of the measuring aperture, 4×10^{17} cm, and that the mass loss has been constant over the past 10^4 years at its presently inferred value of 7×10^{-5} $M_\odot$/yr, the total (dust+gas) mass of the shell could be as large as 1$M_\odot$. This would be a substantial fraction of the likely mass of the progenitor star. This material would dominate the integrated long wavelength emission from the shell but remain unseen shortward of $\approx$ 30 μm because the dust temperature at 3×10^{16} cm is only $\approx$ 100K.

The excess long wavelength radiation shown in Figure 3 suggests the presence of such an extended shell, but the limited pre-IRAS observations are not sufficiently precise to provide a definitive answer to this question. It is interesting to note that the energy distribution for OH26.5 obtained from the IRAS results by Olnon et al. (1984) shows a flux density (in Jy) at 100 μm which is only a factor of 2 less than that at 60 μm. By contrast, a Rayleigh-Jeans spectrum modified by a λ^{-1} emissivity would give a flux density at 100 μm less by a factor of 4.5 than that at 60 μm. This argument and its potential implications should be re-examined in view of the much more precise parameters given for OH26.5 in this volume by Herman (1985), but the IRAS data apparently indicate the presence of considerably more material around OH26.5 than would be inferred from the ground-based observations alone. To account for the observed 60 μm/100 μm flux ratio, much of the 100 μm emission has to come from material with $T < 75$K.

5.2 Origin of the 10 μm Silicate Feature

Another important contribution from IRAS is related to the origin of the silicate feature seen at 10 μm in the spectra of the OH/IR stars. It has been argued on various grounds both that the feature is produced by radiative transfer effects due to the temperature gradient in the circumstellar shell (Werner et al. 1980) and, alternatively, that it is principally due to interstellar reddening (Gehrz et al. 1985). The deep silicate feature seen in the nearby object OH26.5, and other arguments reviewed by Werner et al., show that the silicate feature cannot be produced entirely in the interstellar medium for all sources. However, if there is a significant interstellar contribution in some cases, a significant reddening correction must be applied to the observed energy distributions of the OH/IR stars, raising their inferred luminosities but probably reducing the inferred mass loss rates. If the 10 μm silicate feature is primarily interstellar, then the companion feature at 20 μm should be present in a fairly constant ratio to that seen at 10 μm, so this ratio can be used as a diagnostic of the interstellar contribution. On the other hand, if the origin of the 10 μm feature is circumstellar, then the 20 μm feature can vary in strength relative to that at 10 μm and perhaps even appear in emission. In this case, the relative appearance of the features is in fact an important diagnostic of the characteristics of the shell. The IRAS spectrometer, which will provide spectra between 7.5 and 23 μm with 5% resolution of sources brighter than about 5Jy, will provide spectra of numerous OH/IR stars and resolve this question directly. If the silicate features are shown to be intrinsic, then monitoring the shape of the absorption profile as the infrared brightness of the source varies is another important probe of the structure of the circumstellar shell (cf. Forrest et al. 1978).

5.3 Comparison with Radio Surveys: Galactic Structure

The far infrared observations are also much more capable than those with presently existing radio astronomical techniques of detecting these extended shells (excepting, of course, those with strong maser emission). This can be understood by a comparison of the circumstellar dust shell with the molecular cloud sources studied by both infrared and radio astronomers. The total luminosity of an OH/IR star, $\approx 10^4$ $L_\odot$, is comparable with that of a B-type star, and even a fraction of this luminosity can be detected at infrared wavelengths across the Galaxy. For example, IRAS achieves signal-to-noise ratio > 10 in a single survey scan at 60 μm for a $100L_\odot$ source at the galactic center. On the other hand, the shell mass, to which the readily available mm-wavelength radio lines are most sensitive, is no more than $1M_\odot$, which is much less than the molecular cloud masses typically associated with embedded protostellar infrared sources. High-sensitivity targeted searches with the most sensitive radio telescopes have detected circumstellar molecular clouds around only a few of the nearest mass-losing stars (Knapp et al. 1982), and the weak lines seen plus the effects of beam dilution show that it would be very difficult to detect these objects at distances $\approx$ 10 kpc.

The above discussion suggests that radio investigations of IRAS sources with appropriate energy distributions should lead to the discovery of new maser emission sources, and this has in fact been demonstrated by Engels et al. (1985). It has been recognized by many authors (e.g., Baud et al. 1981) that the radial velocity measurements made possible by the OH maser emission, together with inferences about the properties of the central star which can be drawn from the infrared and radio measurements, make OH/IR stars a powerful probe of the structure of the Galaxy and of stellar evolution throughout the Galaxy. This type of research should receive added impetus from the fact that the distribution on the sky of stars found by the IRAS with 7.5–23 μm color temperatures between 200 and 400K shows very clearly the central bulge of our Galaxy.

6. ACKNOWLEDGEMENTS

This work was supported by the Astrophysics Division of NASA. I thank Xander Tielens for comments on the manuscript, and Dora Willoughby for editorial assistance.

REFERENCES

Baud, B., Habing, H.J., Matthews, H.M., and Winnberg, A. 1981, Astr. Ap., 127, 73.
Baud, B., Sargent, A.I., Werner, M.W., and Bentley, A.F. 1985, Ap. J., in press.
Elitzur, M., Goldreich, P., and Scoville, N.Z. 1976, Ap. J., 205, 384.
Engels, D., Habing, H.J., Olnon, F.M., Schmid-Burgk, J., and Walmsley, C.M. 1985, this volume, 211, and Astr. Ap. in press.
Evans, N.J., and Beckwith, S. 1977, Ap. J., 217, 729.
Forrest, W.J., et al., 1978, Ap. J., 219, 114.
Gehrz, R.D., Kleinmann, S.G., Mason, S., Hackwell, J.A., and Grasdalen, G. 1985, Ap. J., in press (1 March 1985).
Herman, J. 1985, this volume, 215.
Herman, J., Isaacman, R., Sargent, A., and Habing, H.J. 1984, Astr. Ap., 139, 171.
Hildebrand, R.H. 1983, Q.J.R.A.S., 24, 267.
Jones, T.J., Hyland, A.R., Wood, P.R., and Gatley, I. 1983, Ap. J., 273, 669.
Jones, T.W., and Merrill, K.M. 1976, Ap. J., 209, 509.
Kleinmann, S.G., Sargent, D.B., Moseley, H., Harper, D.A., Lowenstein, R.F., Telesco, C.M., and Thronson, H.A. 1978, Astr. Ap., 65, 139.
Knapp, G.R., Phillips, T.B., Leighton, R.B., Lo, K.Y., Wannier, P.G., and Wootten, H.A. 1982, Ap. J., 252, 616.
Moseley, H. 1980, Ap. J., 238, 992.
Neugebauer, G., et al. 1984, Ap. J., (Letters), 278, L1.
Olnon, F.M., Baud, B., Habing, H.J., deJong, T., Harris, S., and Pottasch, S.R. 1984, Ap. J. (Letters), 278, L41.
Rowan-Robinson, M. 1982, M.N.R.A.S. 201, 281.
Schutte, W., and Tielens, A.G.G.M. 1985, in preparation.
Sopka, R.J., Hildebrand, R., Jaffe, D.T., Gatley, I., Roellig, T., Werner, M.W., Jura, M., and Zuckerman, B. 1985, Ap. J., in press.
Werner, M.W., Beckwith, S., Gatley, I., Sellgren, K., Berriman, G., and Whiting, D.L., 1980, Ap. J., 239, 540.

THERMAL RADIO EMISSION FROM MOLECULES IN CIRCUMSTELLAR OUTFLOWS

Mark Morris
Department of Astronomy, UCLA
Los Angeles, CA 90024

ABSTRACT. Most molecular species carried along in the outflow from a mass-losing red giant emit quasi-thermal radio radiation in their rotational lines. The shapes and intensities of the profiles of these lines permit one to infer a great deal about the physical and chemical conditions in the outflow, and about the elemental composition of the underlying star. Relative molecular abundances, for example, can be quite well determined with straightforward models describing the excitation. Since circumstellar outflows are large (up to ~ 1 pc), they can often be spatially resolved with radio spectral line observations, so one can study the chemical evolution of the outflowing gas, as well as the short-term evolution of the mass ejection process. Resolved emission regions also provide a means of determining stellar distances if the radio emission varies, as expected, with the luminosity cycle of the star. Carbon monoxide emission, now seen in > 100 stars, is a useful probe of the mass loss rate, and the surveys that have been made are beginning to provide an idea of the total rate of return of enriched stellar material to the interstellar medium and give an indication of which objects contribute significantly to the mass return rate. Not only is mass loss important for galactic evolution, but studies of individual stars indicate that their evolution is strongly affected by the loss of a major portion of their mass.

1. INTRODUCTION

For the purposes of this review, "thermal" emission refers broadly to non-maser emission; molecules in cool circumstellar outflows from mass-losing red giants are coupled to several different thermal reservoirs, so the rotational level populations, which determine the radio spectrum, are rarely Boltzmann-distributed. Nonetheless, in the absence of maser amplification, the intensities and profile shapes of the emergent "thermal" radio lines can provide a great deal of information about the state of the outflowing gas, as well as about the global properties of the outflow.

Circumstellar envelopes, in contrast to molecular clouds, are straightforward to model, at least in the outer regions where the gas

M. Morris and B. Zuckerman (eds.), Mass Loss from Red Giants, 129–148.

and dust are approaching their asymptotic state. The kinematics, the geometry, the density distribution, and the molecular excitation processes are all accessible, and can be expressed in terms of a relatively small number of determinable parameters. For those familiar with the vagaries of interstellar molecular clouds, circumstellar outflows offer a satisfying situation for modelling.

The region which dominates the thermal radio emission from molecules lies relatively far out in the flow, at radii, r, exceeding $\sim 5\times 10^{15}$ cm. Emission from smaller radii is strongly beam-diluted, and therefore is usually negligible for single-dish measurements, although it can be probed by interferometers. At such large distances from the star, known acceleration mechanisms are ineffective and the gas can be assumed to have a constant outflow velocity (inasmuch as the mass loss process has itself remained invariant).

A constant outflow velocity, coupled with a constant mass loss rate, leads to a $1/r^2$ density distribution. At radii of interest for molecular line emission, however, several processes which modify a $1/r^2$ density law for a given molecule must be considered. The most important of these are photodestruction and photoproduction by ambient galactic UV, which, in the outer envelope where dust no longer provides an effective shield, lead to a relative diminution of, among other things, H_2O, HCN, C_2H_2, and a temporary enhancement of OH, CN, and C_2H (see review by Omont in these proceedings, and references therein). Another process which affects molecular abundances is condensation onto grains (Jura and Morris 1985). SiO, for example, is probably very abundant in the innermost parts of circumstellar envelopes (indeed, SiO masers seem to require that a major fraction of the available Si be in the form of gaseous SiO [e.g., Langer and Watson 1984]), whereas the intensity of thermal radio emission from SiO in the outer envelope implies that only a small fraction remains, so SiO has presumably condensed onto grains (Morris and Alcock 1977). In the coolest and thickest envelopes, H_2O may also be depleted onto grains (Jura and Morris 1985).

Table I summarizes the molecules to which one has access at radio wavelengths. In both carbon and oxygen-rich outflows, CO is by far the most abundant (H_2 being undetectable), and therefore the most universally detected molecule. The list for carbon-rich outflows applies uniquely to IRC+10216, for which an enormous number of observations have been reported. Rather than compiling a list of such references, I refer the reader to Johansson et al. (1984) and references therein for observations carried out at millimeter wavelengths. In addition to the work discussed there, NH_3 emission at 1.3 cm was reported by Kwok, Bell and Feldman (1981) and Bell et al. (1982). The recent SiC_2 identification is due to Thaddeus, Cummins, and Linke (1984; see also Snyder et al. 1985). The popularity of this object is owed to its being nearby, its relatively large mass loss rate, and to the rich chemistry provided by such a carbon-rich environment. These characteristics account for the observation that the intensity of most molecular lines in IRC+10216 exceeds that seen in other such objects by an order of magnitude. Nonetheless, several molecules have been observed in other carbon-rich envelopes, notably

CRL2688 and CIT-6 (Zuckerman *et al.* 1976; Jewell and Snyder 1984; Sahai *et al.* 1984; Henkel *et al.* 1985; Rieu *et al.*, in these proceedings).

Table I: Molecules detected to date at radio frequencies in circumstellar envelopes

Carbon-rich (IRC+10216):

CO, CN, CS, SiS, SiO, SiC_2,
HCN, CH_3CN, HNC, NH_3
C_2H, C_3H, C_3H_2, C_4H, C_3N,
HC_3N, HC_5N, HC_7N, $HC_{11}N$,

Oxygen-rich:

CO, SiO, H_2S, (+ H_2O, OH, and SiO masers)

Oxygen-rich outflows, on the other hand, have a substantially leaner radio spectrum. Maser emission from OH, H_2O, and SiO is reviewed by Bowers in these proceedings. In terms of thermal emission, only two molecules other than CO have been detected in oxygen-rich envelopes: SiO (Buhl *et al.* 1975; Morris *et al.* 1979; Wolff and Carlson 1982; Olofsson *et al.* 1982; Nyman and Olofsson, in these proceedings; and Jewell *et al.*, in these proceedings) and H_2S, which has so far been detected in only one star, OH231.8+4.2 (Ukita and Morris 1983).

The interpretation of thermal emission measurements can provide a wealth of information about the underlying star and about the physics of the mass outflow, including the mass loss rate. Furthermore, the derived molecular and isotopic abundances can be used to draw inferences about stellar evolution. In the sections to follow, we discuss what can be learned from the shapes of the line profiles (§2), how various excitation processes affect the line intensities (§3), and how CO surveys have been used to derive mass loss rate and to infer galactic-scale consequences of the mass loss process (§4).

2. LINE PROFILES

A great deal can be learned from a quick examination of the profile of a molecular line arising in a circumstellar outflow. First, as the size of the star is negligible compared to the size of the emitting region, the velocity profile of optically thin line emission is symmetrical about the stellar radial velocity. Figure 1 illustrates that this expectation is borne out for the CO lines in IRC+10216. Thus, for the first time, one has an unambiguous means of accurately determining the radial velocities of cool giants (Reid and Dickinson 1976), a method which does not require the assumption of spherical symmetry. The velocity centroid of the line profile yields the stellar velocity as long as the distribution of molecules in the outflow is at least axially symmetric. A modest degree of caution is advised for those employing optically thick profiles to determine stellar velocities, since in that case, self-absorption can depress one side of the profile relative to the other, giving, in principle, an apparent overall velocity shift comparable to the magnitude of the local velocity dispersion in the outer envelope (Morris, Lucas, and Omont 1985). This effect has yet to be demonstrated in practice, however.

The expansion velocity, V, is also immediately evident in a typical line profile. As the edges of the profile are quite steep, especially for optically thin lines (see below), the determination of the points at which the profile edges meet the baseline is straightforward. The interval between these points, or the full width at zero intensity (FWZI), is twice the envelope expansion velocity. The accuracy of this determination is limited only by the velocity resolution and the signal-to-noise ratio and can be better than a few tenths of a kilometer per second.

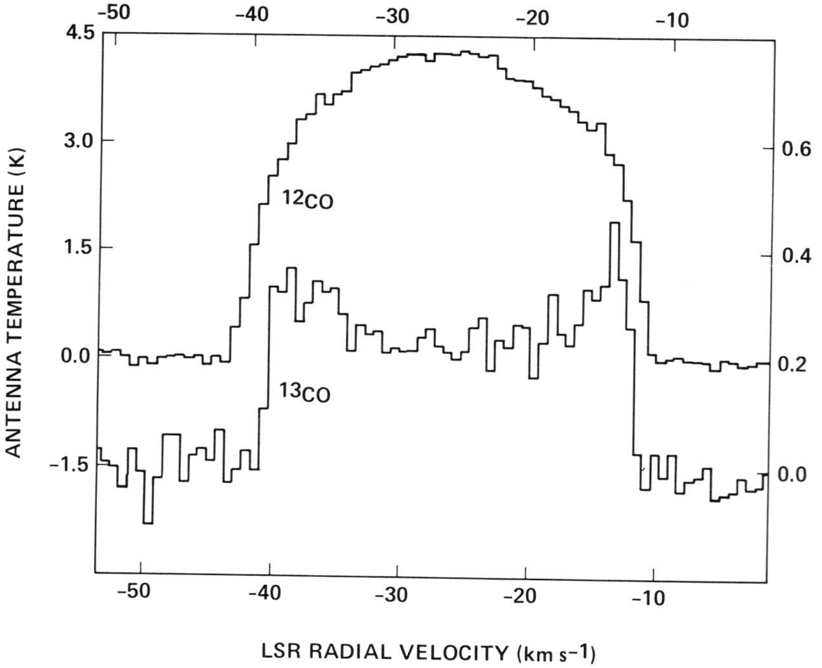

Figure 1: Profiles of J = 1 → 0 CO emission from IRC+10216, taken with the Bell Labs 7-m telescope (Morris, Jura, and Stark 1985). The horizontal axis is radial velocity with respect to the local standard of rest, the bottom one referring to the ^{13}CO spectrum and the top one to the ^{12}CO spectrum (the two spectra have the same frequency scale). The left-hand vertical scale -- antenna temperature -- refers to the ^{12}CO spectrum, while the right-hand one refers to the ^{13}CO profile.

By fitting a four-parameter profile function [equation (7) below] to high-quality data, the accuracy in the derived value of V can, in principle, be high enough to permit a study of the radial dependence (and thus the time-dependence) of V by comparing profiles of rotational lines arising at different characteristic radii (for example, the lines of HC_3N, whose characteristic emission radii decrease systematically with increasing quantum number, J, combined with lines of CO, which can arise quite far out, since CO appears to be self-shielding in circumstellar envelopes [Morris and Jura 1983]). Current evidence suggests that V is quite constant in the IRC+10216 outflow, except possibly for the largest radii, $\sim 10^{18}$ cm, where only CO emission is observed (Wannier *et al.* 1980; Olofsson *et al.* 1982; Morris, Jura, and Stark 1985), and where V may be slightly larger than it is further in (Kwan and Linke 1983). If temporal changes in V can be verified and measured accurately, then a new and important approach to the study of the effects of mass loss on the evolution of the central star will be available.

The shape of the line profile yields further information about the optical depth of the line and the angular diameter of the emitting region relative to that of the beam of the telescope used for the observation. The following derivation of the profile function (due to Morris 1975) is a generalization of arguments presented earlier by Sobolev (1960): in a spherically symmetric envelope of outer radius R, expanding at constant velocity V, the source function S (or, in temperature units, the excitation temperature) is a function of radius r from the central star. The optical depth in line $J \rightarrow J-1$ of a molecule i along a line of sight located a distance p from the central star in the plane of the sky, depends on radial velocity v_z. On the assumption that the local line width due to thermal broadening plus turbulence, ΔV, is much less than V, one can write for the optical depth:

$$\tau(p,v_z) = \tau_o(r)[1 - (v_z/V)^2]^{-1/2}/p \, , \qquad (1)$$

where

$$\tau_o(r) = \frac{2\pi^2\mu_i^2 \, f_i \dot{N} \, J \, [n^i_{J-1}(r) - n^i_J(r)]}{3hV^2} \qquad (2)$$

and

$$r = p[1 - (v_z/V)^2]^{-1/2} \qquad (3)$$

(Morris 1975). In expression (2), f_i is the fractional abundance by number of molecule i and μ_i is its permanent dipole moment. The mass loss rate is $\dot{N}$ (molecules sec^{-1}), h is Planck's constant, and $n^i_J(r)$ is the fractional population in each sublevel of rotational state J of molecule i. The assumption that $\Delta V << V$ is equivalent to the Sobolev (or local) approximation wherein the extent of the emitting region along the line of sight at a particular velocity is taken to be negligibly small.

If the observations are performed with a telescope having a normalized beam response $G(x,R_\nu)$, where x is the displacement from the beam center and R_ν is the full width at half maximum at frequency ν, then the intensity (or the brightness temperature) at velocity v_z in the line when the beam is centered on the star is:

$$T_B(v_z) = \int_0^{p_m(v_z)} G(p,R_\nu)S(r)\{1 - \exp[-\tau(p,v_z)]\}2\pi p dp , \qquad (4)$$

where $p_m(v_z) = R[1 - (v_z/V)^2]^{1/2}$ is the maximum displacement from the star at which gas is present at velocity v_z.

Consider first the case in which the envelope is unresolved. Then, G can be assumed constant over the envelope. In the optically thin limit, $1 - e^{-\tau} \rightarrow \tau$, and, when the integration variable is changed to r, equation (4) becomes:

$$T_B(v_z) = 2\pi G(0,R_\nu)\int_0^R S(r)\tau_o(r)dr . \qquad (5)$$

Because the integral is independent of v_z, T_B is constant between $\pm V$. That is, a flat-topped profile results, independent of the form of the functions $S(r)$ and $\tau_o(r)$. Note that the opacity, and thus T_B, fall off as V^{-2}, so that for a given mass loss rate, envelopes with the largest expansion velocities would be the most difficult to detect.

In the optically thick limit, $1 - e^{-\tau} \rightarrow 1$, and equation (4) reduces to:

$$T_B(v_z) = 2\pi G(0,R_\nu)[1 - (v_z/V)^2] \int_0^R S(r)rdr , \qquad (6)$$

where, again, the integration variable has been changed to r and the integral is independent of v_z. The line profile is thus parabolic in this limit.

Now, it is an oversimplification to claim that profiles are either optically thin or thick since the opacity depends on both v_z and p; an optically thick line will become optically thin at large enough p, and, as equation (1) suggests, an optically thin line will become optically thick as v_z approaches V. So the profile of thermal line emission from a realistic unresolved envelope should be intermediate between rectangular and parabolic. The following four-parameter formula, which preserves the appropriate functional dependence on v_z, can be used to describe such line profiles:

$$T_B(v_z) = T_o \{1 - [(v_z - V_c)/V]^2\} \times$$
$$[1 - \exp(-\alpha\{1 - [(v_z - V_c)/V]^2\}^{-1})]/[1 - \exp(-\alpha)]. \qquad (7)$$

Here, the reference to the stellar velocity, V_c, is explicitly included, as it is one of the parameters to be determined in fitting this function to an observed profile. The parameter T_o is the intensity at the center of the profile, $v_z = V_c$, while α is an "opacity parameter" describing the line shape. For unresolved outflows, the optical depth of the line, and thus the abundance of the emitting molecule, can be determined from α using appropriate models.

In some cases, notably IRC+10216, the envelope is clearly resolved; that is, $R_\nu < R$. When the beam is centered on the star, the effect of this is to emphasize emission near the profile edges. Emission near the center of the profile is relatively diminished because the gas emitting at these velocities in the outflow is moving in a direction perpendicular to the line of sight and has consequently moved the farthest from the center of the beam. Resolving an outflow can not only flatten a line profile, relative to what it would be in the optically thin case, but it can also cause it to be doubly peaked, as is the case with the ^{13}CO profile shown in figure 1. A number of molecules in circumstellar outflows observed with large instruments show such a profile (see for example, Kuiper et al. 1976, and, more recently, Johansson et al. 1985). The most extreme cases of doubly peaked profiles should occur for the optically thin lines of molecules created by photodissociation, such as C_2H (Olofsson et al. 1982; Huggins, Glassgold and Morris, 1984). The emitting region is then a thick spherical shell which, when adequately resolved, may show relatively little emission at the center of the profile.

A line which is optically thick at all velocities will never be doubly peaked in spherically symmetric outflows, even if the outflows are highly resolved. This may help account for the fact that, among the >100 stars so far observed to show ^{12}CO radio emission (see § 4), none of the ^{12}CO line profiles is clearly doubly peaked, with the possible exception of the $J = 2 \rightarrow 1$ line in α Ori (Knapp, Phillips, and Huggins 1980; Jura and Morris 1981; Huggins, in these proceedings). Many of the observed ^{12}CO lines must be optically thick, and most of the optically thin lines (Knapp and Morris 1985) must arise in unresolved envelopes.

When the emitting region is resolved, equation (7) cannot be applied directly to a profile, since another parameter comes into play -- the ratio of the angular size of the emitting region to the telescope beam diameter, β. A five-parameter fit to the line profile is then called for, and, for a Gaussian beam response, the right-hand side of equation (7) should therefore be multiplied by:

$$\exp\left[-\beta^2\left\{1 - [(v_z - V_c)/V]^2\right\}\right]/\exp(-\beta^2), \tag{7a}$$

where β is now the fifth parameter to be fit. The specified function then provides a good fit to flattened or doubly-peaked profiles, and the derived values of α and β can provide meaningful measures of the optical depth and the size, respectively, of the emitting region.

A more accurate description of line profiles is possible if one abandons the Sobolev approximation. Asymmetries can then occur in the profiles, although they are usually small perturbations upon the symmetric case (Rieu et al. 1984; Morris, Lucas, and Omont 1985: an expression for a line profile, derived without the local approximation, is given by both of these groups.) The predicted asymmetries account for the observation by Olofsson et al. (1982) that emission near the red edge of doubly-peaked or rectangular profiles in IRC+10216 is consistently enhanced relative to emission at the symmetrical velocity on the blue side of the profile (see the ^{13}CO profile in figure 1). Spherical symmetry need not be abandoned; the explanation of the asymmetry appears to be that the blue side is slightly self-absorbed. That is, molecules emitting at a given velocity on the red side of the profile are, on the average, closer to the central star, and thus hotter, than molecules determining the emergent intensity at corresponding velocities on the blue side. This effect is most pronounced at the extreme line edges. The magnitude of the asymmetry between the peaks of a doubly peaked profile is sensitive to ΔV, so this quantity can be determined with the intervention of models. Morris, Lucas, and Omont (1985) suggest that the value of ΔV averaged over radii between about 10^{16} and 10^{17} cm is about 0.7 km s^{-1}.

If a circumstellar outflow is observed at a position offset from the central star, then a detailed convolution of the distribution of emergent emission with the telescope beam response is required to derive the profile shapes. In general, the profile becomes narrower with increasing offset from the star (Knapp, personal communication, and Morris, Jura, and Stark, in preparation), since the emission region is increasingly extended as the radial velocity approaches that of the line center. If $R_\nu << R$, then, for beam offset p, the extreme velocities between which substantial emission can occur are $V_c \pm V_{z,max}$, where, from equation (3), we find

$$V_{z,max} = V[1 - (p/R)^2]^{1/2} . \quad (8)$$

However, for a Gaussian beam response function of finite width, some residual emission will always be present throughout the velocity interval $V_c \pm V$.

Line profiles from axisymmetric systems might be expected to differ in some characteristic way from those of spherically symmetric outflows In this case, it is difficult to make generalizations since two arbitrary new functions play a role in determining the line shape: 1) the dependence of the mass loss rate per unit solid angle on polar angle, ϕ, and 2) the dependence of the outflow velocity on ϕ. However, it is remarkable that the observed line profiles from markedly non-spherical systems -- apparently axisymmetric objects such as CRL2688, CRL618, and NGC7027 (Zuckerman et al. 1976; Lo and Bechis 1976; Thronson and Mozurkewich 1983; Beichman et al. 1983; Knapp and Morris 1985; Masson et al.,in these proceedings; Rieu, Graham and Bujarrabal, in these proceedings) -- are quite similar to profiles from spherical outflows, i.e., those that can be described by equation (7). For systems in which the mass outflow is strongly concentrated toward a plane, the line profiles should be expected to reflect this geometry, and be a strong function of the tilt of the plane with respect to the line of sight. Such a case was investigated by Morris and Bowers (1980).

Finally, observations of CO lines in a few of the most evolved objects, notably NGC7027 and CRL2688 (Masson et al., in these proceedings, and Masson, personal communication; Thronson and

Mozurkewich 1983), reveal the presence of weak, high velocity wings that extend 5 or 6 km s^{-1} beyond the nominal velocities $V_c \pm V$ that would be deduced by applying equation (7) to the profile. The J = 11 → 10 HC_3N line in CRL2688 also shows this effect (Zuckerman et al. 1985). One must conclude that, for such objects, the assumption of a constant outflow velocity breaks down. In the rapidly evolving objects where such wings are seen, a relatively rapid wind or an expanding HII region may be overtaking and shocking a slower wind (Kwok 1982). Further work remains to be done, however, before such models can be established convincingly.

3. EXCITATION OF MOLECULES IN CIRCUMSTELLAR OUTFLOWS

A number of processes must be considered in order that a thorough account of excitation in circumstellar outflows be given:

3.1. Infrared Excitation

The only molecular energy levels which are significantly populated in the extended portions of circumstellar envelopes are the rotational levels of the ground vibrational state. However, infrared radiation arising from warm dust near the central star, as well as from the star itself, can be absorbed into an excited vibrational state, and then be quickly re-emitted, leaving the molecule in an excited rotational level of the ground vibrational state (Morris 1975). The excitation of most molecules appears to be dominated by this mechanism. The v = 0 → 1 transitions, occuring predominantly between 4 and 20 μm, are by far the most important, although the v = 0 → 2 transitions, typically weaker by $\sim 10^3$, could play a role in situations where the fundamental is saturated. An illustrative and straightforward case study of an IR-excited molecule is that of HC_5N, made by Deguchi and Uyemura (1984).

For diatomic or linear molecules, the selection rule governing any vibration-rotation transition is $\Delta J = \pm 1$. Therefore, the two-step process of absorption and reemission of an IR photon connects rotational level J only to levels $J \pm 2$. This has interesting consequences: first, even J and odd J ladders are almost independent, being coupled only by the relatively slow rotational relaxation rates, or by collisions (c.f., Morris and Alcock 1977). Second, the relatively slow downward rotational cascade following the net $\Delta J = +2$ excitation of most molecules near the star by IR photons leads to inversions across the lowest rotational transitions where the cascade slows down and the population thereby stacks up. These inversions disappear for a given transition at about the radius where the excitation rate equals the spontaneous rotational decay rate. The J = 1 → 0 line of SiS, for example, is observed in IRC+10216 to be inverted and to give rise to an amplified, radially-moving beam of radiation that appears in line profiles as a spike at the edge of the profile (see paper by Claussen and Sahai in these proceedings). The SiS maser has inspired a fair amount of recent work (Grasshoff, Tiemann, and Henkel 1981; Henkel,

Matthews, and Morris 1983; Sahai, Wootten, and Clegg 1984; Rieu et al. 1984; Henkel et al. 1985).

The lowest rotational line of CO is also predicted to be inverted in a region whose size increases with decreasing mass loss rate (Morris 1980). However, for the low-J lines of CO, the magnitude of the negative optical depth is predicted to be too small, in most situations, to permit significant amplification, and the intensity is therefore characteristic of optically thin emission. Nonetheless, in a narrow range of mass loss rates around 10^{-6} $M_{\odot}/yr(8\times10^{-4}/f)$, a weak CO maser is predicted to occur if $f=[CO]/[H_2] \gtrsim 8\times10^{-4}$. No molecule other than SiS is yet known to exhibit maser action as a result of this mechanism, some (CN, C_2H) because they are produced too far from the star, some because their abundance, and thus the negative optical depth in their low-J lines, are too small, and some because their abundance is so large that maser emission gets reabsorbed before emerging from the envelope. In some oxygen-rich stars, emission from SiO in the v=0 state may have some superimposed maser components (see Jewell et al., in these proceedings), but the pumping mechanism there has not been elucidated.

Calculation of the outward transfer of infrared radiation is a difficult problem because the Sobolev approximation cannot be applied. That is, because the outflow velocity is constant, absorption at a given frequency in an infrared line can occur at all radii. The IR intensity at any point is obtained by integrating the absorption along all lines of sight to the emitting dust. For high enough abundance of the absorbing species, there occurs a radius r_s in the circumstellar outflow at which the IR lines connecting to a given rotational level are effectively squelched. Interior to this radius, the excitation of that level is dominated by the infrared process, while exterior to r_s, collisional excitation dominates. The Sobolev approximation cannot reproduce this state of affairs, so some technique must be developed to compute the transfer of infrared radiation correctly (see, for example, Sahai et al. 1984).

Of course, if the expansion velocity changes at some radius, either because of a reacceleration of the outflow, or because a change in the mass loss process at some previous time has led to a change in the momentum transfer to the envelope, then the molecules beyond that point in the outflow are subjected to a fresh segment of the infrared continuum, and infrared excitation can reestablish itself. This was offered by Kwan and Linke (1982) as one explanation for the persistence of substantial CO excitation at large radii in the outflow from IRC+10216.

Time variations: One very interesting consequence of the infrared excitation mechanism for molecules in circumstellar envelopes is that the excitation is linked to the substantial luminosity variations that accompany the pulsations of most stars having such envelopes. The intensities and the shapes of the line profiles should therefore be expected to vary in response to the variations of intensities in the infrared "pump" lines. An investigator who monitors a molecular emission line is therefore in the astronomically unusual position of having an input parameter varied in an experimental system where all other parameters are presumably held constant. Also, the potential exists for performing phase-lag measurements in a manner analogous to that employed in the study of OH masers (see papers by Bowers and by Herman in these proceedings): emission from opposite sides of a line profile arises on opposite sides of the envelope along the line of sight, and arrives with a relative time delay equal to the light travel time across the emitting region, which is a direct measure of its diameter. Coupled with high-resolution observations of the distribution of emission in the plane of the sky, such phase-lag measurements yield a measure of the distance to the star. Unlike the case of OH masers, there is no ambiguity associated with the assumption that microwave radiation emitted in a direction parallel to the radius vector from the central star has the same radial dependence as radiation emitted in a direction perpendicular to this, or tangentially. These potentialities notwithstanding, the study of time variations of thermal molecular lines in circumstellar envelopes has only just begun with the initial studies of Rieu et al. (1984), Henkel et al. (1985), Nyman and Olofsson (their v = 0 SiO paper in these proceedings), and Claussen and Sahai (these proceedings). Variations in the SiS J = 1 → 0 maser seem to be established, and there are indications that the profile shapes of higher-lying SiS lines also vary with stellar phase.

Polarization of rotational lines: The infrared photons capable of exciting a molecule at any point in the outflow beyond a few hundred a.u. almost always propagate radially outward from the vicinity of the star. The angular momentum retained by a molecule in the absorption and reemission process is therefore anisotropically distributed and, in this sense, the molecules become partially aligned. Subsequent rotational line emission is therefore partially polarized. The dependence of the degree of polarization upon the angle of emission with respect to the radial direction translates observationally into a dependence of the percentage polarization upon frequency within a line profile. This effect, which has not yet been observed, was suggested by Morris (1975), and was examined in quantitative detail by Morris, Lucas, and Omont (1985), Deguchi (these proceedings), and Deguchi and Watson (1984). These analyses suggest that the polarization predicted under the best possible conditions for some molecular lines in IRC+10216 should be readily detectable with the new generation of large millimeter-wave telescopes now being introduced. The importance of making such measurements lies in the information potentially available about depolarization mechanisms. In particular, since the total angular momentum vector of a molecule will precess about the local magnetic field direction, any non-radial magnetic field component will lead to a partial depolarization, which provides a probe of the magnetic field in circumstellar outflows.

Stellar distances: In addition to phase-lag measurements, there is another potential means of determining stellar distances. The intensity of an emission line from an infrared-excited molecule is primarily a function of three quantities: 1) the molecular abundance, which can be determined from the line profile via the parameter α (see § 2), 2) the flux in the infrared continuum at the frequencies of the vibration-rotation transitions, which can be determined by direct measurement, and 3) the distance to the mass-losing star. Since the emergent rotational line intensity does not depend linearly on the infrared flux in the envelope (Morris and Alcock 1977), there is, in principle, enough information available in the measurement of a line profile to determine the star's distance. This technique has not yet been employed, but it is well worth applying to high-quality data in view of the paucity of information on stellar distances.

3.2. Collisional Excitation

Beyond the relatively small region of a circumstellar outflow in which dust emission is substantial at the frequencies of the vibrational fundamentals, the rate of collisionally induced transitions falls off with radius in approximately the same way as the intensity of infrared continuum radiation: r^{-2}. However, whereas collisional excitation maintains this dependence at all r, the rate of infrared excitation drops abruptly at r_s, the radius at which the radial opacity in the infrared lines connecting to low-J levels, measured radially outward from the stellar surface, exceeds unity. Therefore, when infrared excitation drops out of the picture, excitation by collisions with H_2 becomes dominant. For most molecules, this never occurs because their abundance is too low to generate substantial opacity in the vibration-rotation lines, but for CO (Kwan and Hill 1977; Kwan and Linke 1982), and quite possibly for HCN (see below), r_s is quite small, so that these molecules are collisionally excited throughout the portions of the envelope accessible to radio observations. [H_2O is also subject to collisional excitation in the outer envelopes of oxygen-rich stars -- well beyond the H_2O maser region -- and although its far-IR rotational lines have not yet been observed, they are important for cooling the gas in oxygen-rich outflows (Goldreich and Scoville 1976; Tielens 1983).

An important consequence of these considerations is that, for these molecules, r_s is different for different isotopic variants of the molecule. Different variants are subject to different excitation mechanisms throughout much, if not most, of the region of interest. Isotope abundance ratios determined with molecular lines (for example, Wannier and Linke [1978] and Knapp, Chang, and Morris, in these proceedings) should therefore be regarded with caution if the most abundant isotopic variant is collisionally excited.

<u>CO</u>: Carbon monoxide is unique amongst abundant molecules in having a relatively small permanent dipole moment. A consequence of this is that the lowest rotational lines become optically thick in the tangential direction at about the same value of the CO loss rate as is required for the $v = 0 \rightarrow 1$ vibration-rotation transitions connecting to the same levels to become optically thick in the radial direction (Morris 1980). This implies two well-defined regimes for CO: 1) when $f\dot{M}(M_\odot\ yr^{-1})/V(km\ s^{-1})\Delta V(km\ s^{-1}) \lesssim 0.7 \times 10^{-10}$, where $f = [CO]/[H_2]$, CO

is IR excited and its rotational lines are optically thin at line center, and 2) when $f\dot{M}/V\Delta V \gtrsim 1.3 \times 10^{-10}$, the $V = 0 \rightarrow 1$ lines are optically thick and collisional excitation is relatively more important if not predominant (depending on $\dot{M}$ and r). At the same time, rotational lines are optically thick at all velocities. One therefore has the coincidental and utilitarian situation that the shape of a CO rotational line profile is a quick (albeit imperfect) indicator of the CO excitation mechanism.

HCN: The question of whether HCN is collisionally excited in the envelope of IRC+10216 is an important one, but one which has not yet been firmly settled. Kwan and Hill (1977) proposed a model for the $H^{12}CN$ $J = 1 \rightarrow 0$ emission line based on fluorescent IR excitation, but were unable to produce a model for the $H^{13}CN$ $J = 1 \rightarrow 0$ line that is consistent with the $[^{12}C]/[^{13}C]$ isotope ratio of ~35 in this object (Wannier and Linke 1978; Johansson *et al.* 1985, using their HC_3N data; Keady, Hall and Ridgway 1985). Also, Bieging, Chapman, and Welch (1984, see their summary in these proceedings) interpret their interferometer observations of $H^{12}CN$ in IRC+10216 in terms of IR excitation with $f = [HCN]/[H_2] \simeq 10^{-6}$, but find, with some uncertainty, a value for $[H^{12}CN]/[H^{13}CN]$ of only about 20, which is uncomfortably low.

On the other hand, Olofsson *et al.* (1982) suggest a substantially larger HCN abundance ($f \simeq 10^{-5}$), using their measured isotope intensity ratio. Such a large abundance is supported by Jura (1983a), who argues that the similar intensities of the $J = 1 \rightarrow 0$ lines of CO and HCN can be readily understood if both species are collisionally excited and both lines have a large optical depth. Jura also notes that, with such a high abundance, the opacity in the IR lines would be large enough to squelch the IR excitation mechanism so that collisional excitation would predominate. Another line of evidence is that the larger value of f is adopted in the reasonably successful models for the photoproduction of CN from HCN (Wootten *et al.* 1982; Huggins, Glassgold, and Morris 1984) and it would be difficult (though perhaps not impossible) to readjust parameters to produce enough CN if f were an order of magnitude smaller. Finally, infrared absorption line studies (Hall and Ridgway 1978) support the larger value of f.

One cannot argue that the intensity of the HCN line should differ substantially in the two cases and thus provide a means of deciding between them; the rotational line intensities can actually *decrease* as the HCN abundance increases past some point between $f=10^{-6}$ and 10^{-5} because the rotational levels uncouple from the relatively hot thermal bath represented by the IR radiation field and become coupled instead to the low kinetic temperature of the ambient gas. In other words, a given HCN line intensity can correspond to two widely differing values of f. Measurements of isotopic variants and a variety of J levels of HCN are needed to sort them out, but the problem might be complicated if the different variants are subject to different excitation mechanisms.

Kinetic temperature: Calculation of the thermal balance in circumstellar outflows is necessary for the interpretation of the intensities of lines from collisionally excited molecules. The temperature at radius r is given by:

$$\frac{dT}{dr} = -\frac{4}{3}\frac{T}{r} + \frac{8\pi r^2}{3k\dot{N}}\left[\Gamma(r) - \Lambda(r)\right] , \qquad (9)$$

in which the first term is due to cooling by adiabatic expansion, and Γ and Λ are the volume heating and cooling rates, respectively.

The dominant heating mechanism in the portions of circumstellar outflows where most molecular radio emission arises is believed to be viscous heating (Goldreich and Scoville 1976; Kwan and Linke 1982; Tielens 1983), which occurs as dust grains stream supersonically through the gas with a drift velocity, v_{dr}, found by balancing the radiation force on a grain against the drag force of the gas:

$$v_d = [QLV/(\dot{M}c)]^{1/2}. \qquad (10)$$

In equation (10), Q is the intensity-weighted average ratio of optical to geometric cross-section of the grains (or, alternatively, the momentum transfer efficiency factor), L is the luminosity of the central star, and $\dot{M}$ is the mass loss rate. For IRC+10216, Kwan and Linke suggest that Q=0.024 provides the best fit to the CO data.

If ρ is the gas mass density at radius r, then we can write, by assuming that each molecule acquires an excess velocity v_d in a collision with a dust grain,

$$\Gamma(r) = [n_g(r)\sigma v_d]^{1/2} \rho(r) v_d^2, \qquad (11)$$

where n_g is the number density of dust grains and σ is their geometrical cross section. Combining (10) and (11), we find, for the heating term in equation (9):

$$\frac{8\pi r^2}{3k\dot{N}}\Gamma = (QL/c)^{3/2} \frac{\xi}{12\pi k(\dot{M}V)^{1/2} r^2} , \qquad (12)$$

where ξ is the dust cross section per gas particle.

In the outermost regions of circumstellar envelopes (that is, beyond a few times 10^{17} cm in IRC+10216), the dust becomes optically thin in the UV. There, many of the heating mechanisms considered for diffuse interstellar clouds (photoelectric effect on grains, molecular photodissociation, CI photodissociation) can play a role in reheating the gas (Kwan and Linke 1982; Morris, Jura and Stark 1985), with the consequence that the gas temperature should go through a minimum somewhere in the envelope. The only known probe of this outer region is the self-shielded CO.

Molecular cooling is effective only for those molecules that are collisionally excited. (IR-excited molecules are, in fact, a minor heat source for the gas because the collisions they undergo tend to convert the IR-produced rotational energy into particle kinetic energy.) The emission lines of collisionally excited molecules are optically thick, so a detailed treatment of the radiative transfer in the rotational transitions is necessary for a proper calculation of the cooling rate.

The Sobolev approximation is appropriate for incorporating the effects of radiative transfer in the rotational transitions, if maser amplification does not play an important role. The formalism of Castor (1970), which is based on this approximation, has been used with profit for carrying out calculations with the coupled equations of radiative transfer and statistical equilibrium (for example, Goldreich and Scoville 1976; Morris, Lucas, and Omont 1985). In order to calculate the total cooling rate by a given molecule, a fairly major extrapolation must be made from the small number of observed transitions, because most of the energy comes out in a large number of currently inaccessible lines.

Which molecules act as coolants depends on the envelope chemistry. In oxygen-rich stars, rotational line emission from H_2O apparently controls the cooling within 2-10 $\times$ 10^{15} cm (Goldreich and Scoville 1976; Tielens 1983), but beyond that, adiabatic cooling is said to predominate. In view of the observations of CO emission in a number of oxygen-rich outflows, future studies of the thermal balance at larger radii in these objects should take account of CO cooling, as it may be comparable to the cooling rate from adiabatic expansion. Carbon monoxide is clearly important for cooling in carbon-rich outflows (Kwan and Linke 1982), but cooling by HCN, which has heretofore not been included in thermal calculations, should not be ignored if that molecule is collisionally excited, as the arguments presented above tend to indicate. Relative to CO as a coolant, the more compressed spectrum of HCN and the wider line profiles resulting from hyperfine structure should compensate at least partially for the larger dipole moment of HCN, which makes it more difficult to populate its high J states. Also, in IRC+10216, HCN is photodissociated at about 10^{17} cm (Bieging, Chapman, and Welch 1984; see summary in these proceedings), so CO, which can persist to a much greater radius because of self-shielding (Morris and Jura 1983) is, in any case, the dominant molecular coolant beyond the HCN photodissociation radius.

For the results of temperature calculations, the reader is referred to the references mentioned above. We note here that the temperature curves do not differ greatly between calculations, even when different chemistries are assumed. For example, a temperature of 100 K was found to occur in a model oxygen-rich outflow at 10^{16} cm by Goldreich and Scoville (1976) and at 3×10^{16} cm by Tielens (1983), while for the carbon-rich object IRC+10216, the same temperature occurs at 6×10^{15} cm in the model of Kwan and Linke (1982).

4. CO SURVEYS AND MASS LOSS RATES

Since the dominant constituent of circumstellar outflows, H_2, is not directly observable except under special circumstances (shocks, UV fluorescence, etc.), a readily observable probe of the matter distribution is called for. As is the case with interstellar molecular clouds, CO fulfills this role. Being the second most abundant molecule in circumstellar outflows (unless [C]/[O] >> 1), carbon monoxide is, by far, the most widely detected. It provides a reasonable estimate of the mass loss rates in two ways. First, CO is tied to H_2 by collisions (§ 3), so the intensities of its emission lines at large enough radii from a mass-losing star are linked to the density there (c.f., Knapp *et al*. 1982; Jura 1983b), and fall off as r^{-3}. The mass loss rate can be inferred by locating the radius at which the r^{-3} falloff begins, or by applying knowledge of the relative CO abundance.

Second, the relative abundance of CO is believed to be relatively well understood, so if observations can be interpreted in terms of a CO loss rate, then the total mass loss rate can be inferred. In the cool atmospheres of mass-losing red giants, where thermochemical equilibrium can be assumed, CO is calculated to incorporate almost all of the carbon or oxygen, whichever is least abundant (Tsuji 1964; Salpeter 1977). There is no evidence that its relative abundance is substantially changed in the ejection process; indeed, CO is a very stable molecule. The greatest uncertainty about its abundance lies with the difficulty of determining the stellar [C]/[H] and [O]/[H] ratios, which have, in general, been modified by CNO processing and subsequent mixing within the star.

Once values are adopted for the distance and the relative CO abundance for a particular circumstellar outflow, models can be constructed which reproduce the observed characteristics of the CO line profiles in terms of a single parameter, $\dot{M}$ (Morris 1980; Kwan and Linke 1982; Knapp and Morris 1985). The radial temperature profile of the outflowing gas is important in this calculation, and presents the largest source of uncertainty. Kwan and Hill (1977) and Kwan and Linke (1982) calculate the temperature of IRC+10216 self-consistently by including CO cooling in their model fits to CO lines, which give $\dot{M} \simeq 4 \times 10^{-5}$ (d/200 pc)^2 $M_\odot$ yr^{-1}, but with the current uncertainties in Q, in the observations, and in the importance of HCN for cooling, a future modification of this result may be warranted. In view of such uncertainties, Knapp and Morris (1985) adopted a single temperature profile for all stars, arguing that it depends only weakly on $\dot{M}$ and V. They derive approximate expressions for the CO line intensity in both the optically thin and thick regimes (§ 3) in terms of $\dot{M}$, V, f, distance and other relevant parameters. These expressions are then inverted to give $\dot{M}$ in terms of measurable and assumable parameters. The total uncertainty in the mass loss rates obtained from this procedure is a factor of 3 to 10, depending on the distance uncertainty and the quality of the CO data.

Until very recently, the Knapp and Morris survey (hereafter referred to as the KM survey) was the most extensive yet carried out.

It includes observations of the J = 1 → 0 CO line toward about 50 stars, including almost all stars detected in prior CO surveys. As had been argued previously on the basis of a smaller sample (Zuckerman et al. 1977), the carbon-rich stars (comprising almost half the sample) typically display brighter CO emission for a given bolometric magnitude than do the M-type stars. It appears that this can be accounted for in large part if the value of f in carbon stars is about three times larger, on the average, than it is in oxygen-rich stars. This is about what would be expected if the carbon abundance is roughly equal to the solar value in oxygen-rich stars, but is enhanced by the triple-α reaction to values exceeding the oxygen abundance in carbon-rich stars, even if the oxygen abundance does not change very much. The mass loss rates found for oxygen and carbon-rich stars cover the same total interval: $\sim 10^{-7}$ to $\sim 10^{-4}$ $M_\odot$ yr^{-1}. The lower end of this range is probably a result of observational selection, but the upper end of the range may correspond to an upper limit to the luminosity of AGB stars coupled with the constraint that, if radiation pressure on dust grains drives the outflow, then $\dot{M} \lesssim L/cV$. Also, there is an indication in the data that the mass loss rates for the carbon-rich stars may be concentrated toward the upper end of this range relative to oxygen-rich stars. However, the total mass loss rate by oxygen-rich stars is about the same as that of carbon-rich stars, within the accuracy of the estimates of $\dot{M}$. It is interesting that the outflow velocities found in the 50 stars of the KM survey are confined to a rather narrow range, in spite of the variation of $\dot{M}$ over three orders of magnitude. Three-fourths of the C and S stars and half of the M stars satisfy 10 km s^{-1} < V $\lesssim$ 20 km s^{-1}. Only a few exceptional M stars have V exceeding 25 km s^{-1}. This is consistent with the suggestion that, if radiation pressure on dust grains drives the outflowing matter to its terminal velocity, then V scales only as $L^{1/4}$ (Gehrz and Woolf 1971; Jura 1984a).

In the KM sample, it appears that the total rate of mass return to the interstellar medium is not dominated by objects whose mass loss rates are concentrated in any particular interval; rather, there are approximately comparable mass return rates per cubic parsec in each of the equal logarithmic intervals of $\dot{M}$ between 10^{-4} and 10^{-6} $M_\odot$ yr^{-1}, extending possibly to even smaller values of $\dot{M}$. Integrated over the Galaxy, the total mass return is estimated at $\gtrsim 0.3$ $M_\odot$ yr^{-1}, the inequality being due to the possible insensivity of the survey to stars having $\dot{M} \lesssim 10^{-7}$ $M_\odot$ yr^{-1}.

An important conclusion of the KM survey is that there appear to be very few circumstellar outflows in which the outward momentum transferred per second to the ejected matter, $\dot{M}V$, exceeds the momentum carried away per second in the stellar radiation field, L/c. Knapp et al. (1982) had argued that $\dot{M}V \gg L/c$ in a number of stars, and therefore that radiation pressure acting on dust grains is not important for driving the outflow. However, Jura (1983b) reexamined the data and argued that they are consistent in almost all cases with the hypothesis that radiation pressure on grains is important in the dynamics of the outflow. The mass loss rates derived by KM support Jura's conclusion, a few of the apparent exceptions being attributable to a recent luminosity decrease of the central star (NGC7027: Jura 1984b, or CRL618: Spergel, Giuliani, and Knapp 1983), or to the importance of mass ejection by a binary companion (OH231.8+4.2: Morris 1981 and Cohen, in these proceedings).

The stars in the KM sample were chosen on the basis of a variety of considerations, and they therefore do not constitute a well-defined sample. In order to refine the conclusions mentioned above, observations of a larger, complete sample will be necessary. Such observations are being carried out with some success by Zuckerman and Dyck (1985), who have so far detected the $J = 2 \rightarrow 1$ line of CO towards ~40 new stars, out of ~50 stars attempted, and by Zuckerman, Dyck and Claussen (1985a), who have seen $J = 1 \rightarrow 0$ emission towards ~25 additional new stars out of ~50 stars attempted. These surveys draw upon the IRAS and revised AFGL catalogs, targeting stars that are either bright at 12 and 25 μm, or very red, or both. The preliminary results of these surveys indicate that, for a given brightness and redness, carbon-rich AGB stars have the strongest CO lines, followed by oxygen-rich AGB stars, followed by M supergiants and symbiotic stars. Among the detections of Zuckerman, Dyck, and Claussen (1985), two objects -- AFGL 2233 and AFGL 2901 -- have the highest outflow velocities (~33 km s^{-1}) yet seen in carbon stars.

Finally, recent observations have led to detections of CO in well-known, mass-losing objects that previously resisted attempts at detection. These include NML Cygnus (Zuckerman *et al.* 1985b) and M1-92 (Zuckerman *et al.* 1985b; Knapp, paper on Planetary Nebulae in these proceedings), OH231.8+4.2 and IRC+10420 (KM; Zuckerman *et al.* 1985b), and VY CMa (Zuckerman and Dyck 1985). It appears that essentially all cool stars in the solar neighborhood which are losing mass at a high enough rate are accessible to CO studies. Therefore, we can soon expect to obtain a much clearer picture than now exists of the importance of mass loss for stellar and galactic evolution.

I am indebted to my colleagues Mike Jura and Ben Zuckerman for innumerable stimulating and informative discussions, and for providing helpful comments on the manuscript. This work is supported by NSF grant AST 83-18342 to UCLA.

REFERENCES

Beichman, C.A., Keene, J., Phillips, T.G., Huggins, P.J., Wootten, H.A., Masson, C., and Frerking, M.A. 1983, Ap. J., 273, 633.
Bell, M.B., Kwok, S., Matthews, H.E., and Feldman, P.A. 1982, Astro. J., 87, 404.
Bieging, J.H., Chapman, B., and Welch, W.J. 1984, Ap. J., 285, 656.
Buhl, D., Snyder, L.E., Lovas, F.J., and Johnson, D.R. 1975, Ap. J., (Letters), 201, L29.
Castor, J.I. 1970, M.N.R.A.S., 149, 111.
Deguchi, S., and Uyemura, M. 1984, Ap. J., 285, 153.
Deguchi, S., and Watson, W.D. 1984, Ap. J., 285, 126.
Goldreich, P., and Scoville, N. 1976, Ap. J., 205, 144.
Grasshoff, M., Tiemann, E., and Henkel, C. 1981, Astr. Ap., 101, 238.
Hall, D.N.B., and Ridgway, S.T. 1978, Nature, 273, 281.
Henkel, C., Matthews, H.E., and Morris, M. 1983, Ap. J., 267, 184.
Henkel, C., Matthews, H.E., Morris, M., Terebey, S., and Fich, M. 1985, Astr. Ap., in press.
Huggins, P.J., Glassgold, A.E., and Morris, M. 1984, Ap. J., 279, 284.
Jewell, P.R., and Snyder, L.E. 1984, Ap. J., 278, 176.
Johansson, L.E.B., Andersson, C., Ellder, J., Friberg, P., Hjalmarson, A., Hogland, B., Irvine, W.M., Olofsson, H., and Rydbeck, G. 1984, Astr. Ap., 130, 227.
_____ 1985, Astr. Ap. (Suppl.), in press.
Jura, M. 1983a, Ap. J., 267, 647.
____ 1983b, Ap. J., 275, 683.
____ 1984a, Ap. J., 282, 200.
____ 1984b, Ap. J., 286, 630.
Jura, M., and Morris, M. 1981, Ap. J., 251, 181.
_____ 1985, Ap. J., in press.
Keady, J.J., Hall, D.N.B., and Ridgway, S.T. 1985, preprint submitted Ap. J.
Knapp, G.R., and Morris, M. 1985, Ap. J., in press (KM).
Knapp, G.R., Phillips, T.G., and Huggins, P.J. 1980, Ap. J., (Letters), 242, L25.
Knapp, G.R., Phillips, T.G., Leighton, R.B., Lo, K.Y., Wannier, P.G., Wootten, H.A., and Huggins, P.J. 1982, Ap. J., 252, 616.
Kuiper, T.B.H., Knapp, G.R., Knapp, S.L., and Brown, R.L. 1976, Ap. J., 204, 408.
Kwan, J., and Hill, F. 1977, Ap. J., 215, 781.
Kwan, J., and Linke, R.A. 1982, Ap. J., 254, 587.
Kwok, S., Bell, M.B. and Feldman, P.A. 1981, Ap. J., 247, 125.
Kwok, S. 1982, Ap.J., 258, 280.
Langer, S.H., and Watson, W.D. 1984, Ap. J., 284, 751.
Lo, K.Y., and Bechis, K.P. 1976, Ap. J. (Letters), 205, L21.
Morris, M. 1975, Ap. J., 197, 603.
_____ 1980, Ap. J., 236, 823.
_____ 1981, Ap. J., 249, 572.
Morris, M., and Alcock, C. 1977, Ap. J., 218, 687.
Morris, M., and Bowers, P.F. 1980, Astr. J., 85, 724.

Morris, M., and Jura, M. 1983, Ap. J., 264, 546.
Morris, M., Jura, M., and Stark, A.A. 1985, in preparation.
Morris, M., Lucas, R., and Omont, A. 1985, Astr. Ap., 142, 107.
Morris, M., Redman, R., Reid, M.J., and Dickinson, D.F. 1979, Ap. J., 229, 257.
Olofsson, H., Johansson, L.E.B., Hjalmarson, A., and Rieu, N.-Q. 1982, Astr. Ap., 107, 128.
Reid, M.J., and Dickinson, D.F. 1976, Ap. J., 209, 505.
Rieu, N.-Q., Bujarrabal, V., Olofsson, H., Johansson, L.E.B., and Turner, B. E. 1984, Ap. J., 286, 276.
Sahai, R., Wootten, A., and Clegg, R.E.S. 1984, Ap. J., 284, 144.
Salpeter, E.E. 1977, Ann. Rev. Astr. Ap., 15, 267.
Snyder, L.E., Henkel, C., Hollis, J.M., and Lovas, F.J. 1985, Ap.J., (Letters), 290, L29.
Sobolev, V.V. 1960, Moving Envelopes of Stars (Cambridge; Harvard University Press).
Spergel, D.N., Giuliani, J.L., and Knapp, G.R. 1983, Ap.J., 275, 330.
Thaddeus, P., Cummins, S.E., and Linke, R.A. 1984, Ap.J. (Letters), 283, L45.
Thronson, H.A., Jr., and Mozurkewich, D. 1983, Ap. J., 271, 611.
Tielens, A.G.G.M. 1983, Ap. J., 271, 702.
Tsuji, T. 1964, Ann. Tokyo Astr. Obs., 2nd ser. 9, 1.
Ukita, N., and Morris, M. 1983, Astr. Ap., 121, 15.
Wannier, P.G., and Linke, R.A. 1978, Ap. J., 225, 130.
Wannier, P.G., Redman, R.O., Phillips, T.G., Leighton, R.B., Knapp, G.R., and Huggins, P.J. 1980, IAU Symposium 87: Interstellar Molecules, ed: B.H. Andrew (Reidel:Dordrecht), p. 487.
Wolff, R.S., and Carlson, E.R. 1982, Ap. J., 257, 161.
Wootten, A., Lichten, S.M., Sahai, R., and Wannier, P.G. 1982, Ap. J., 257, 151.
Zuckerman, B., and Dyck, H.M. 1985, in preparation for Ap. J.
Zuckerman, B., Dyck, H.M., and Claussen, M.J. 1985a, in preparation for Ap. J.
Zuckerman, B., Gilra, D.P., Turner, B.E., Morris, M., and Palmer, P. 1976, Ap. J. (Letters), 205, L15.
Zuckerman, B., Morris, M., Kaifu, N., Suzuki, H., Ukita, N. and Oishi, M. 1985b, in preparation.
Zuckerman, B., Palmer, P., Morris, M., Turner, B.E., Gilra, D.P., Bowers, P.F. and Gilmore, W. 1977, Ap. J. (Letters), 211, L97.

TIME VARIATIONS OF SiO (v=0, J=2-1) EMISSION FROM CIRCUMSTELLAR SHELLS

L-Å. Nyman and H. Olofsson
Onsala Space Observatory
S-43900 Onsala
Sweden

The SiO (v=0, J=2-1) line has been observed towards several late type stars with the Onsala 20 m telescope over a period of four years. No significant variations in the integrated intensities were detected. However, two of the sources, R Leo and R Cas, show clear changes in their line profiles during this time. Figure 1 presents two of the spectra obtained towards R Leo. The January 1984 spectrum shows a red-shifted feature at ~ 8 kms^{-1} which has completely disappeared by April 1984. The time variations and the highly redshifted velocity suggest that this feature was of maser origin. The line profile towards R Cas changed from a roughly parabolic line in 1979, to a flat topped profile with weak wings extending out to ~15 km s^{-1} from the stellar velocity in 1984. From a CO (J=1-0) spectrum we estimate the envelope expansion velocity to be ~ 11 kms^{-1}. χ Cyg and NML Tau were also observed several

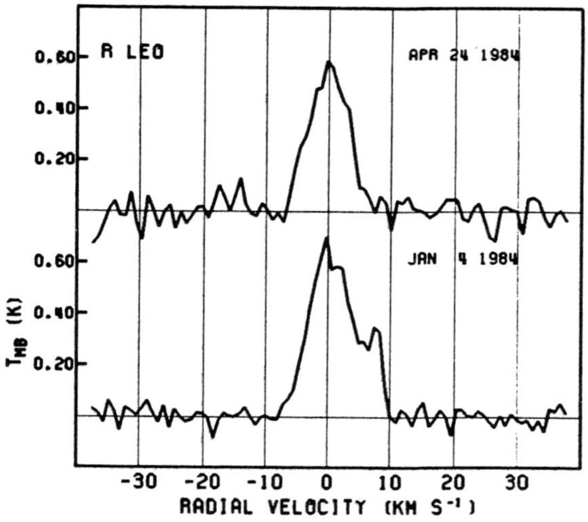

Figure 1. Time variations of the SiO (v=0, J=2-1) line profile towards R Leo.

M. Morris and B. Zuckerman (eds.), Mass Loss from Red Giants, 149–150.

times, but in these sources no time variations were detected. Their spectra are parabolic, which is the expected line profile for optically thick emission from "thermally" excited molecules in an unresolved circumstellar envelope. R Aql, o Ceti, T Cep, and U Ori were observed on one occasion, in January 1984, and SiO ground state emission was detected in the latter three sources for the first time. All four sources show evidence for anomalous excitation. The R Aql spectrum consists of a "normal" parabolic line profile with a superposed narrow feature, while the line profiles towards o Ceti, T Cep, and U Ori are narrow (FWZP $\sim$ 5 kms^{-1}) and not obviously parabolic. These observations imply that stellar radial and envelope expansion velocities may not always be correctly estimated from ground state SiO emission.

A full report of this work will be presented in Astronomy and Astrophysics.

SiS IN CIRCUMSTELLAR SHELLS

R. Sahai
Owen's Valley Radio Observatory, Caltech
Alwyn Wootten
National Radio Astronomy Observatory
R.E.S. Clegg
University College London

We have observed the spectrum of SiS toward the Mira variable IRC+10216, including the J=16-15, J=15-14, J=14-13, J=13-12, J=12-11, J=7-6, J=6-5, J=5-4 and J=1-0 lines (Fig. 1), and made a detailed model to fit these lines, incorporating a radial SiS abundance gradient due to photodissociation by interstellar UV (Sahai, Wootten, and Clegg 1984). The strength of the high J lines requires a substantial excitation of the molecule by infrared radiation near the central star; the infrared also inverts the populations in the lower levels. Variations in the 13 μm infrared flux over the 644 day period of IRC+10216 produce observable changes in some of the SiS lines. Sharp cusps in the J=5-4 line wings, observed by Olofsson et al. (1982), are explained in terms of weak maser emission at maximum light. The model predicts a different shape for this line at minimum light, in excellent agreement with Olofsson's (1985) recent observation. Line coincidences, involving 13 μm transitions of HCN and C_2H_2, which may affect the SiS vibration-rotation excitation, are identified.

We find an undissociated abundance $[SiS]/[H_2]$ in the shielded inner IRC+10216 shell of $\sim$ 2.4 10^{-7}, roughly 100 times smaller than predictions of chemical models. The SiS density falls steeply with radius, reaching one-half the undissociated value at r = 9 10^{16} cm. We show that a) Si, not S depletion is responsible for the low SiS abundance, and b) most of the available Si is in SiC grains.

Our sensitive search for SiS J=7-6 and J=6-5 lines in other carbon-rich, oxygen rich, and S-type envelopes has revealed three new sources, CIT 6, CRL 2688 and IRC+20370, all of which are carbon-rich, in agreement with the chemical equilibrium models predicting SiS to be significantly more abundant in a carbon-rich environment than in an oxygen-rich environment.

REFERENCES

Olofsson, H. 1985, this volume.
Olofsson, H., Johansson, L.E.B., Hjalmarson, A., and Nguyen-Quang-Rieu 1982, Astr. Ap., 107, 128.
Sahai, R., Wootten, A., Clegg, R.E.S. 1984, Ap. J., 284, Sep. 1 issue.

M. Morris and B. Zuckerman (eds.), Mass Loss from Red Giants, 151–152.

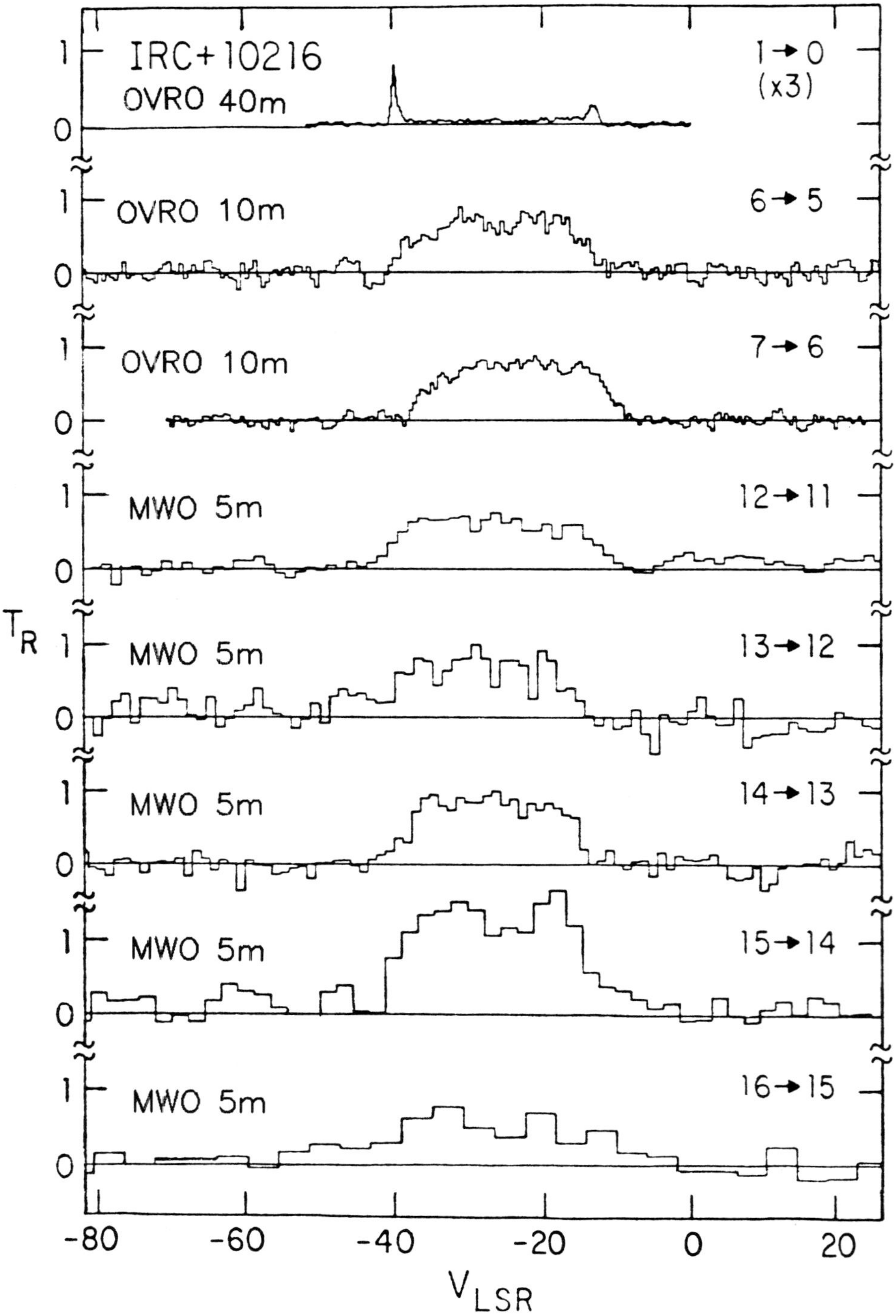

Figure 1. The rotational spectrum of SiS from IRC+10216. (For the J=5-4 line, see Olofsson et al. 1982; Sahai, Wootten and Clegg 1984.)

INFRARED PUMPING AND POLARIZATION OF MOLECULAR LINES IN IRC+10216

Shuji Deguchi
Department of Physics
University of Illinois at Urbana-Champaign
1110 W. Green Street
Urbana, Illinois 61801

In a circumstellar environment, the central star provides a source of radiation that is anisotropic at any point in the envelope. Observationally the vibrationally excited, circumstellar SiO masers are known to be quite linearly polarized and close to the star - a phenomenon that can be understood as due to the stellar radiation (Western and Watson 1983). Here, we consider polarization that is produced in the ground vibrational state emission as a result of infrared excitations in the circumstellar gas due to the star light. As a specific (and probably optimal) example, we treat the SiO molecule.

For the numerical calculation, we adopt the parameters of the envelope to fit IRC+10216. They are: a mass loss rate $\dot{M}=5\times10^{-5}$ $M_\odot$ /y, a luminosity $L=10^4$ $L_\odot$, a kinetic temperature as a function of radius R given by $T_k=75$ $(3\times10^{16}$ cm/R$)^{0.6}$ K, and a constant expansion velocity of 16 km s^{-1} for the circumstellar envelope (Kwan and Linke 1982). The abundance of SiO is treated as a free parameter and is determined by the requirement that the calculated beam averaged emission for the J= 2-1 transition should agree with the observed values (Olofsson et al. 1981). An abundance ratio $[SiO]/[H_2]=2.5\times10^{-7}$ is found by assuming that all the emission is produced within a diameter 5×10^{16} cm of the star, or within an angular diameter of 36" if IRC+10216 is at a distance of 200 pc. The escape probability method for the polarized light of the Zeeman triplet (J=1-0; Goldreich and Kylafis 1982) is readily extended for the multilevel calculation. A basic premise is that the Zeeman splitting is larger than the natural line width, but much smaller than the Doppler width. In this calculation, all rotational states J less than 9 of the v=0 state are included explicitly.

To compare the results of the calculation with observation, an average over the size of the telescope beam must be performed. This average is given in table 1 for the Onsala telescope. Because the circumstellar envelope is assumed to be spherically symmetric, the beam must be off-center to detect linear polarization. The predicted polarization is clearly undetectable by present, single dish radio telescopes for which the noise level is approximately 0.1-0.01 K. Interferometric observations with much higher resolving power might be available in the near future.

M. Morris and B. Zuckerman (eds.), Mass Loss from Red Giants, 153–154.

Table I. Calculated Linear Polarization for SiO in IRC+10216.

Transition	On-source beam		22"-off-centered beam	
	I (K)	Q (K)	I (K)	Q (K)
J= 1-0	0.88	0.0	0.53	0.009
J= 2-1	1.46	0.0	0.84	0.002

It is noteworthy that the polarization of the radiation emitted in the J=3-2 and J=2-1 transitions goes through zero at certain radii - 1.5×10^{16}cm and 3×10^{16}cm, respectively. This occurs because the direction of the net polarization changes by 90 degrees. Near the star, it is parallel to the projected radius. Two anisotropies influence the infrared radiation field at a point within the envelope - direct radiation from the star is radial whereas directions perpendicular to radii are favorable for the escape of radiation. Separately, these anisotropies tends to cause net linear polarizations in the ground vibrational state that are orthogonal. As suggested by the fact that intensity of star light varies as R^{-2}, the relative importance of the two anisotropies depends upon radius and causes the changes. When the emitted radiation is averaged over the size of the beam of the radio telescope, the polarizations tend to cancel in the higher transitions.

Conclusion

The foregoing discussion is based on a spherically symmetric matter distribution in the circumstellar envelope. Deviations from spherical symmetry would, of course, tend to increase the likelihood for detecting polarized radiation.

The extensive search performed by Wannier et al. (1983) for linear polarization in thermal emission lines utilized the J=1-0 transitions of CO and HCN and the J=2-1 transition of CS. Though noise levels of about 0.1-0.2 K (one standard deviation) were achieved in a few cases, no definite polarization was detected. We have not consider HCN explicitly, but would expect that its hyperfine splitting could lead to line overlap that would reduce the net polarization. Even though the star light causes a strong anisotropy in the radiation field at points within a (spherically symmetric) circumstellar envelope, beam averaging for single dish, millimeter radio telescopes makes it unlikely that linear polarization can be detected in the emitted radiation.

The detail is described in Astrophys. J. (Oct. 1, 1984).

REFERENCES

Goldreich, P., and Kylafis, N. D., 1982, Ap. J. 253, 606
Olofsson, H., et al., 1981, A. Ap. 107, 128
Kwan, J., and Linke, R. A., 1982, Ap. J. 254, 587
Wannier, P. G., Scoville, N. Z., and Barvainis, R., 1983, Ap. J.,267,126
Western, L. R., and Watson, W. D., 1983, Ap. J. 275, 195

THE DISTRIBUTION OF HCN IN THE CIRCUMSTELLAR ENVELOPE OF IRC+10216

J.H. Bieging, B. Chapman, and W.J. Welch
Radio Astronomy Laboratory
University of California, Berkeley

We have made aperture synthesis maps of the 88.6 GHz J=1-0 HCN emission line from IRC+10216, using the Hat Creek interferometer. The emission is well-resolved by our 8" x 10" beam and is centrally peaked on the position of the IR source. The half-peak intensity diameter is about 20".

The observations are compared with results of statistical equilibrium calculations for the HCN rotational and vibrational level populations for a uniform, constant velocity expansion model for the circumstellar envelope. The results of our model calculations lead us to these conclusions:

(1) The HCN excitation is dominated by radiative pumping through the 14 μm vibrational transitions.

(2) The best overall fit for the $1/r^2$ density model is for n(HCN) = 2 cm^{-3} $\pm$ 30% at radius 5×10^{15} cm. The HCN spectrum predicted by the models has deep absorption features at the blue-shifted edges of the hyperfine components unless the HCN density falls off sharply at radii greater than 10^{17} cm. Since these absorption features are not observed, we conclude that the HCN is cut off by photodissociation at about this radius.

(3) The observed brightness distribution does not agree very well in detail with the best $1/r^2$ model prediction, in the sense that we observe an excess of emission at about 10"-25" from the star over what the model gives. This discrepancy may be reduced by postulating that the HCN density deviates from a $1/r^2$ law. A density increase by a factor of 3 over the simple $1/r^2$ law for radii greater than 4×10^{16} cm provides a good fit to the observed model brightness distribution. This result suggests that the mass loss rate for IRC+10216 has declined by about a factor of 3 over the past 1000 years.

(4) Comparison of the model HCN spectra with observations shows good agreement over most of the velocity range of emission, but poor

M. Morris and B. Zuckerman (eds.), Mass Loss from Red Giants, 155–156.

agreement for the blue-shifted wing. This discrepancy might be explained by a non-LTE population of the J=1 hyperfine structure levels as a result of an asymmetry in the radiative excitation of the J=1-0 transition.

This work has been published in the Astrophysical Journal, Part 1, Volume 285, issue for 1984 October 15.

UPPER LIMIT FOR THE PRODUCTION OF ^{14}C IN THE CARBON STAR IRC+10216 FROM OBSERVATIONS OF THE ^{14}CO (J = 1-0) LINE

G. R. Knapp[1,2], W. D. Langer[1,3], and R. W. Wilson[1]

[1] AT&T Bell Laboratories, Holmdel, NJ 07733, USA
[2] Princeton University Observatory, Princeton, NJ 08544, USA
[3] Princeton P.P.L., Princeton, NJ 08544, USA

ABSTRACT. We report the results of an unsuccessful search for the J = 1-0 line of ^{14}CO in the envelope of the evolved carbon star IRC+10216. Comparison with models of the envelope emission shows that in the material leaving the star $^{14}C/^{13}C < 0.1$.

The ^{14}C isotope has a short half-life (~5600 years) and has never been detected outside the Earth. On Earth, its abundance ($\sim 10^{-2}$ relative to ^{13}C) is maintained via the reaction $^{14}N + n \rightarrow ^{14}C + p$ between cosmic-ray produced neutrons and ^{14}N. A recent search by Phillips and White (1983) of large molecular clouds has set 2σ limits of a few $\times 10^{-5}$ on the $^{14}C/^{12}C$ ratio and suggested that this reaction does not take place in the molecular cloud environment.

A second possible site for the production of ^{14}C has been suggested to be the interiors of evolved asymptotic giant branch stars, where s-process elements are formed [e.g., Cowan and Rose (1977), Despain (1977)]. If ^{14}C is mixed to the surface, ^{14}C molecules may be detected in the outflowing wind, since such stars are known to be undergoing copious mass loss. For IRC+10216, Barnes et al. (1977) find $^{12}CO/^{14}CO > 10^4$. However, these infrared observations measure gas close to the star, and the ^{14}C abundance may be time-variable.

The molecular envelope around IRC+10216 is known to have a radius of $\sim 10^{18}$ cm and an age of ~ 20,000 years (Morris, Jura and Stark 1984). Rodriguez-Kuiper et al. (1977) carried out an unsuccessful search for the J = 1-0 rotational lines of $H^{14}CN$ and ^{14}CO from this envelope.

We made a more sensitive, but still negative, search for the ^{14}CO J = 1-0 line at 105.87 GHz in March 1984 using the Bell Laboratories 7-m antenna. The receiver temperature was ~ 120K and the telescope beamwidth 109". The resulting spectrum measured with 1 MHz filters is shown in Figure 1. The antenna temperature scale is corrected for atmospheric attenuation and for beam efficiency, and is in units of T_A^*, the Rayleigh-Jeans equivalent brightness temperature measured by a perfect antenna above the atmosphere. The weak line near "160 km/sec"

M. Morris and B. Zuckerman (eds.), Mass Loss from Red Giants, 157–158.

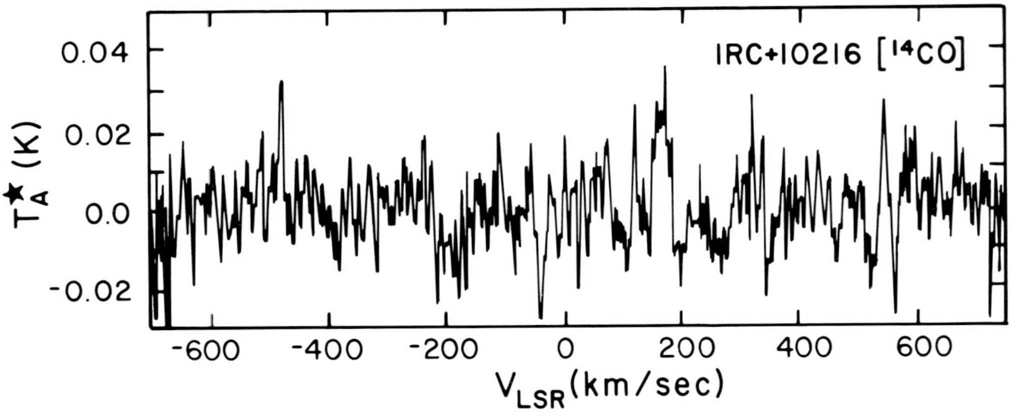

Figure 1. Spectrum of the ^{14}CO line observed towards IRC+10216

is the J = 12–11 transition of $H^{13}CCCN$ at 105.799 GHz. The 5σ upper limit for the ^{14}CO line is 0.35 K × km/sec, about ten times as sensitive as that of Rodriguez-Kuiper et al. (1977).

The CO emission from the envelope of IRC+10216 has recently been modelled (Knapp and Chang 1984) to give $\dot{M} = 5.5\times10^{-5}$ $M_\odot$ yr^{-1}, [CO]/[H_2] = 8×10^{-4}, and [^{13}CO]/[H_2] = 1.7×10^{-5}. Calculation of the ^{14}CO upper limit requires taking into account excitation by infrared radiation from the central star and the truncation of the effective radius of the ^{14}CO cloud by dissociation by interstellar ultraviolet radiation and by decay of ^{14}C. We then find a 5σ upper limit of [^{14}CO]/[H_2] < 1.7×10^{-6}, or ^{14}C/^{13}C < 0.1. This is in disagreement with the most optimistic calculations (Cowan and Rose 1977) but not with those of Despain (1977).

The receiver was built by A.A. Stark and the SIS junctions by R.E. Miller. This work was partly supported by grant AST82-13293 to Princeton University.

REFERENCES

Barnes, T.G., Beer, R., Hinkle, K.H., and Lambert, D.L. 1977, Ap.J. 213, 71.
Cowan, J.J., and Rose, W.K. 1977, Ap.J. 212, 149.
Despain, K.H. 1977, Ap.J. 212, 774.
Knapp, G.R., and Chang, K.M. 1984, in preparation.
Morris, M., Jura, M., and Stark, A.A. 1984, in preparation.
Phillips, J.P., and White, G.J. 1983, M.N.R.A.S. 202, 265.
Rodriguez-Kuiper, E.N., Kuiper, T.B.H., Zuckerman, B., and Kakar, R.K. 1977, Ap.J. 214, 394.

THE $^{13}C/^{12}C$ ISOTOPE RATIO IN CIRCUMSTELLAR ENVELOPES

G. R. Knapp and Kar Man Chang
Princeton University Observatory
Mark Morris
Astronomy Department, UCLA

A recent survey of the $^{12}CO(J = 1-0)$ emission from circumstellar envelopes produced by mass loss from evolved giants has been made using the 7 m antenna at Bell Laboratories (Knapp and Morris, 1984, submitted to Astrophysical Journal. A total of 50 stars was detected. Emission in the ^{13}CO J = 1-0 line was searched for in nineteen of the stars with the strongest ^{12}CO emission. The sample contains nine carbon stars, nine stars with [O] > [C] ("oxygen" stars) and one S star, χ Cyg. ^{13}CO emission was detected from twelve stars at a level > 3σ, and from three more stars (IRC+50096, CRL865 and χ Cyg) at the 3σ level.

The ratio of the ^{13}CO and ^{12}CO line fluxes is several times greater in oxygen stars than in carbon stars, suggesting that the $^{13}CO/^{12}CO$ ratio is greater in the oxygen stars. Derivation of the ratio of $^{12}CO/^{13}CO$ in the envelopes requires the construction of models of the line emission, taking into account collisional, radiative and infrared excitation and dissociation of CO by the interstellar ultraviolet radiation field. Infrared excitation by the central star is more important for the ^{13}CO molecule than for the ^{12}CO molecule, since the latter is usually optically thick in the circumstellar envelopes in which ^{13}CO is detectable. The outer radius of the envelope is determined by molecular self-shielding, and hence the ^{12}CO extent may be greater than the ^{13}CO extent because of the greater abundance of ^{12}CO.

The resulting values of $^{12}CO/^{13}CO$ show a marked difference between oxygen and carbon stars. For the oxygen stars, the five envelopes with detected ^{13}CO emission give values of 5-18, with a mean of 9. The non-detections give only $^{12}CO/^{13}CO \gtrsim 4$ for the other envelopes. This result does not seem to depend on mass-loss rate.

For the carbon stars, the values of $^{12}CO/^{13}CO$ range from 20 to 100, with a mean value of 50. The oxygen stars thus show the end-product abundances of the CNO cycle, while the carbon stars may show in addition the production of ^{12}C via the triple-α reaction.

The full text of this paper is in preparation for Astrophysical Journal. This work is supported by the National Science Foundation via grant AST82-13293 to Princeton University.

M. Morris and B. Zuckerman (eds.), Mass Loss from Red Giants, 159.

CO(2-1) EMISSION FROM THE CIRCUMSTELLAR ENVELOPE OF ALPHA ORI

P. J. Huggins
Physics Department
New York University
4 Washington Place
New York, NY 10003

ABSTRACT. New measurements with the NRAO 12m telescope are reported of CO 2-1 emission from Alpha Ori. The results are discussed in the context of the kinematics and mass loss rate of the circumstellar envelope. A first, tentative detection of ^{13}CO emission is also reported.

1. INTRODUCTION

Alpha Ori is the most extensively observed M supergiant undergoing mass loss. A variety of observational techniques have revealed both a chromosphere and a cooler, more extended envelope around the star, and further study offers the potential for developing a detailed picture of the circumstellar material. The presence of CO in the envelope is of particular interest as a model diagnostic of physical conditions and for understanding the association of molecules in the wind flow. CO has been detected in the envelope in absorption in the IR by Bernat et al. (1979), and in emission in the 2-1 line by Knapp et al. (1980; hereafter KPH). New observations of the CO 2-1 line are reported here.

2. OBSERVATIONS

The observations were made with the NRAO 12m telescope at Kitt Peak. They consist of position switched on-source spectra of the 2-1 lines of CO and ^{13}CO at 230 and 220 GHz, respectively. At these frequencies the beamsize (FWHM) was 36" and the pointing was estimated to be better than 8". Calibration was performed with hot and cold loads, sky tips, and observations of Mars, Jupiter, and Saturn. The temperatures reported here are main beam brightness temperatures.

There is a problem with the absolute calibration since it became apparent from other data taken during the run, and from subsequent investigations (Payne et al. 1984), that the parameters of the telescope were affected by elevation dependent effects: the beam distorts and the gain changes significantly above about 40 degrees. This problem should not seriously affect the planet calibrations which were made at lower elevations, but will affect the spectral line data. They have been

M. Morris and B. Zuckerman (eds.), Mass Loss from Red Giants, 161–164.

corrected here empirically, using the change with elevation of the peak temperature of IRC+10216 which was observed throughout the run. This is not an exact correction since the size of the two sources are different. However, an alternative procedure using a correction based on observations of point sources (Payne et al. 1984) gives results within 20% of those reported here. Internal evidence suggests the overall calibration is good to about 45%.

3. RESULTS AND DISCUSSION

The CO spectrum is shown in Fig. 1. The line shape is similar to that reported earlier by KPH. The data are consistent with a flat-topped profile characteristic of an unresolved, optically thin shell, but there is marginal evidence for double peaking, possibly suggesting partial resolution. The implications of the profile shape for the size of the emitting region have been discussed by KPH and Jura and Morris (1981). The somewhat improved S/N and velocity resolution of the present data allows some refinement of the kinematic parameters of the shell: the expansion velocity, the systemic (heliocentric) radial velocity, and the (heliocentric) velocity of the blue wing are 14.1, 19.1, and 5.0 km/s, respectively. Formal errors in these are about 1.0 km/s, but the irregular baseline makes them rather more uncertain. The radial velocity of the star itself is not accurately known due to systematic motions of its atmosphere; the mean of all photospheric values discussed recently by Goldberg (1984) is 20.6 km/s, which is slightly larger than the systemic velocity reported here, but probably not significantly so.

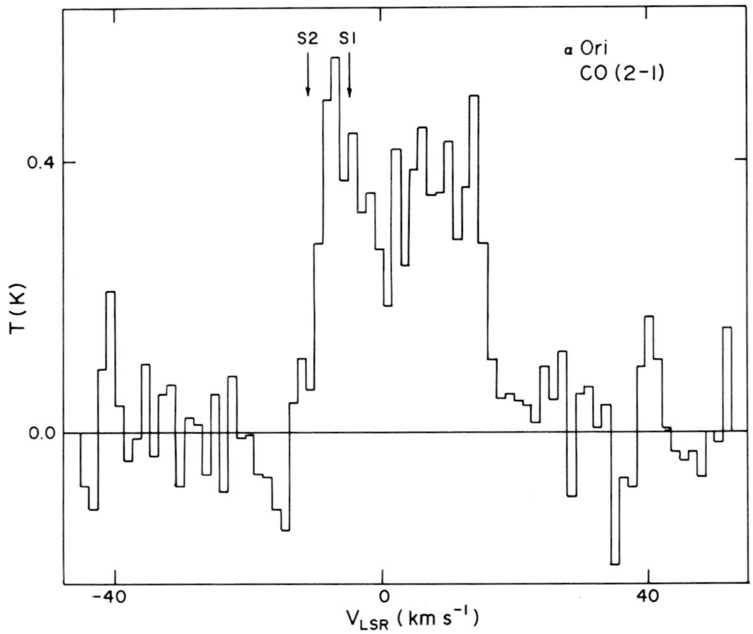

Figure 1. The CO 2-1 spectrum of Alpha Ori at a resolution of 1 MHz.

The velocity of the blue wing corresponds closely to the absorption minima of the S2 components seen in KI (5.0 km/s, Goldberg 1979) and CO (4.8 km/s, Bernat et al. 1979), and unambiguously associates the CO emission with the S2 component.

The intensity of the CO emission can be used to estimate the mass loss rate of the envelope as was done earlier by KPH and Jura and Morris (1981). The present data suggest that the 2-1 emission is weaker than that reported earlier by a factor of about two, and observations of the 1-0 line support this. Based on the earlier data and the assumptions of optical thinness, moderate excitation, and a small source, estimates for the intensity of the 1-0 line observed with the NRAO 11m, the OSO 20m, and the BL 7m telescopes are approximately 0.06, 0.2, and 0.02 K, respectively. (Relaxing the assumptions tends to increase the estimates given here.) I have searched with the first two telescopes and Knapp and Morris (1984) with the last, and we find no 1-0 emission at the levels indicated. Knapp and Morris, however, find a marginal detection of the 1-0 line at the 0.01 K level with the BL 7m telescope. The present 2-1 data, under the assumptions given above, imply an intensity of 0.013 K for this line, which is consistent with the observations. So although the calibration of the present data is not very good, as indicated in Section 2, the 2-1 and 1-0 data sets are now at least compatible. From the 2-1 data a CO loss rate of 3(39) mols/s is obtained by scaling the earlier analysis of Jura and Morris (1981). A more complete analysis, including an investigation of the effects of CO photodestruction, does not seem warranted until better data on the 1-0 and 2-1 lines have been secured.

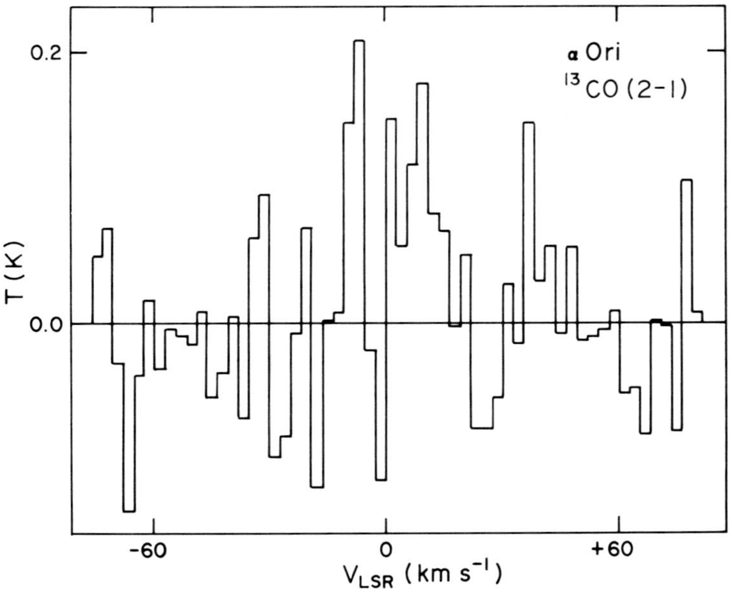

Figure 2. The ^{13}CO spectrum of Alpha Ori at a resolution of 2 MHz.

If the abundance of CO is known, then the mass loss rate of the shell can be estimated. Lambert et al. (1984) have recently determined the photospheric C/H ratio to be 2.5(-4), which probably also applies to the expanding envelope. If the CO is fully associated (i.e. all the C is in CO), this implies a mass loss rate of 4(-7) M_{o}/yr. This result is significantly less than estimates based on the KI scattering data (e.g. Mauron et al. 1984) as well as a number of other estimates. If these results are correct, then CO must be underassociated by a factor of about 10. Some support for this comes from the CO absorption measurements of Bernat et al. (1979), who find from an analysis of column densities that CO/H = 3(-5). In combination with the photospheric abundance noted above, this also implies an underassociation of CO by a considerable factor. One important consequence of this result is that free carbon may be the primary contributor of electrons in the envelope, and this possibility needs to be considered in detailed models of its ionization structure.

The ^{13}CO spectrum is shown in Fig. 2. The line is marginally detected in integrated intensity at the 3.8 sigma level. Taken at face value, it is the first detection of this species in emission in the Alpha Ori shell. The integrated intensity, divided into that of the main line, is 4.2 ± 1.1 (one sigma), and to first order is an estimate of the ^{12}CO/^{13}CO ratio in the shell under optically thin conditions. Within the large errors, this result is consistent with the value of 6, found from absorption measurements of both the shell (Bernat et al. 1979) and the photosphere (e.g. Harris and Lambert 1984), and lends some credence to the detection. Much higher S/N spectra of both lines reported here would be of considerable value for refining detailed models of the CO emitting region.

ACKNOWLEDGEMENTS. This work was partially supported by NSF (AST-8216484) and by an Alfred P. Sloan Foundation Fellowship. NRAO is operated by Associated Universities Inc., under contract with NSF.

REFERENCES

Bernat, A.P., Hall, D.N.B., Hinkle, K.H., Ridgway, S.T.: 1979, *Astrophys. J. (Letters)*, 233, L135.

Goldberg, L.: 1979, *Q.J.R.A.S.*, 20, 361.

Goldberg, L.: 1984, *P.A.S.P.*, 96, 366.

Harris, M.J., Lambert, D.L.: 1984, *Astrophys. J.*, 281, 739.

Jura, M., Morris, M.: 1981, *Astrophys. J.*, 251, 181.

Knapp, G.R., Phillips, T.G., Huggins, P.J.: 1980, *Astrophys. J. (Letters)*, 242, L25.

Knapp, G.R., Morris, M.: 1984, private communication.

Lambert, D.L., Brown, J.A., Hinkle, K.H., Johnson, H.R.: 1984, *Astrophys. J.*, 284, 223.

Mauron, N., Fort, B., Querci, F., Dreux, M., Fauconnier, T., Lamy, P.: 1984, *Astron. Astrophys.*, 130, 341.

Payne, J.M., Rhodes, P.J., Salter, C.J., Stobie, E.B.: 1984, *NRAO Mem.* 03/26/84.

CO (1-0) MAPS OF NGC7027

C. R. Masson, K. W. Cheung, G. L. Berge, M. J. Claussen,
G. M. Heiligman, R. B. Leighton, K. Y. Lo, A. T. Moffet,
T. G. Phillips, A. I. Sargent, S. L. Scott, D. P. Woody
Owens Valley Radio Observatory
California Institute of Technology

NGC7027 is a very young planetary nebula in which the ionization of the red giant wind is only partially complete. Previous measurements of the CO (1-0) emission at 115 GHz from NGC7027 (Mufson, Lyon, Marioni, 1975) did not have enough angular resolution to reveal the structure of the neutral cloud. We have recently mapped the CO (1-0) emission from NGC7027 (Masson et al., Ap. J., in preparation), using the Owens Valley Millimeter-wave Interferometer with a resolution of 7" x 11".

The main part of the CO line has a width of 33 km/s, corresponding to an outflow velocity for the gas of 16.5 km/s. The interferometer maps near the central LSR velocity of 25.5 km/s are heavily resolved. The emitting region decreases in size away from the line center, as expected from projection effects. At velocities near the edge of the line (Fig. 1a), the red-shifted and blue-shifted emitting regions are compact and coincident. At intermediate velocities (Fig. 1b), there is an asymmetry, with red-shifted emission being displaced to the NW and blue-shifted emission to the SE. This is in the opposite sense to the velocity gradient found in the HII region by Atherton et al. (1979). From the maps we deduce that the neutral cloud has an anisotropic density distribution, with a flattened shape. The higher density in the plane of the neutral cloud inhibits the spread of the HII region, accounting for the prolate shape and velocity gradient found by Atherton et al.

We have also detected weak wings of CO (1-0) and CO (2-1) emission extending to a velocity of 22.5 km/s from the line center. The similarity to the velocity structure of the HII region suggests that the CO emission arises in a shocked, neutral zone immediately outside the expanding HII region. Since the density and temperature in the HII region are known (Atherton, et al., 1979; Harris and Scott, 1976), we can use the shock velocity of 6 km/s to estimate a pre-shock density of 2.5×10^5 $(d/1\mathrm{kpc})^{-1/2}$ cm^{-3}. The radiation pressure is significantly less than the pressure within the HII region and is therefore unable to drive the shock. Therefore, some other mechanism, such as a fast wind from the central star, must be invoked.

From our estimates of the pre-shock density, we compute a mass

M. Morris and B. Zuckerman (eds.), Mass Loss from Red Giants, 165–167.

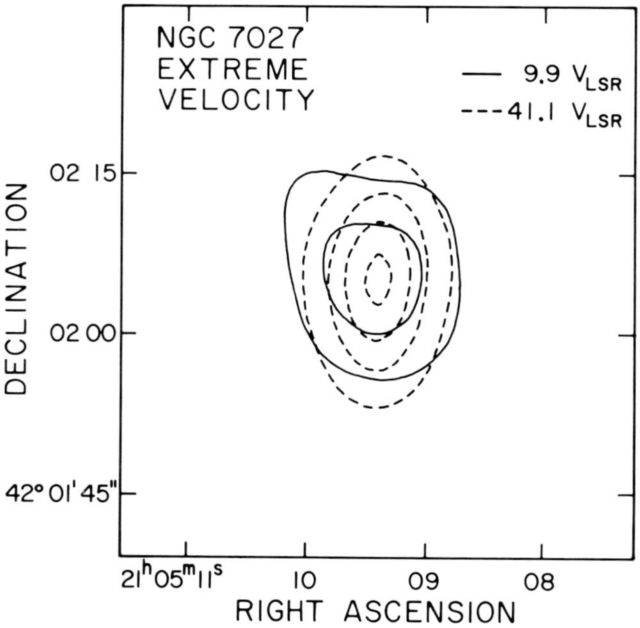

1a

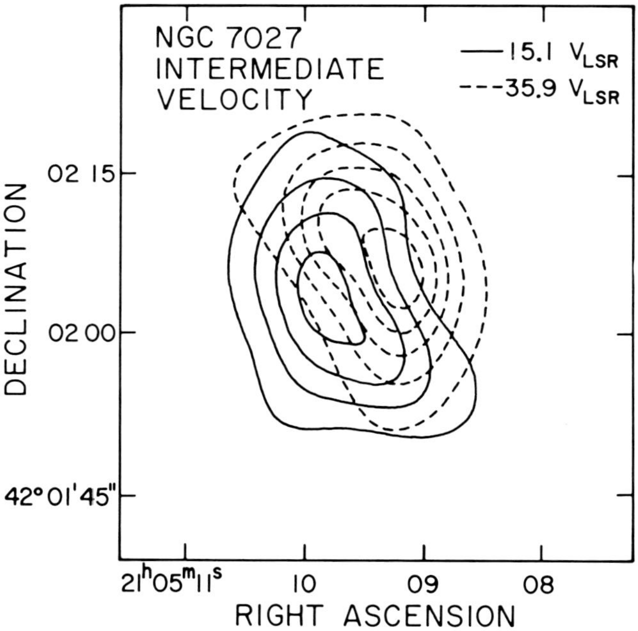

1b

Figure 1. Maps of CO emission from NGC7027: a) Extreme velocity $|\delta V|$ = 15.6 km/s. Contour interval 3Jy/beam. b) Intermediate velocity $|\delta V|$ = 10.4 km/s. Contour interval 3Jy/beam.

loss rate of 1.1 x 10^{-3} $(d/1kpc)^{3/2}$ $M_\odot$/yr. This is higher than the values estimated from consideration of molecular excitation (Knapp, et al., 1980; Jura, 1983) in the outer part of the envelope. There is some evidence, however, from measurements of scattered Hα, that the mass loss rate has increased with time (Atherton, et al., 1979), which would decrease the apparent discrepancy between the results.

REFERENCES

Atherton, P. D., Hicks, T. R., Remy, N. K., Robinson, C. J., and Warnicke, S. P.: 1979, Ap. J., 232, p. 786.

Harris, S., and Scott, P. F.: 1976, M.N.R.A.S., 175, p. 371.

Jura, M.: 1983, Ap. J., 275, p. 683.

Knapp, G. R., Phillips, T. G., Leighton, R. B., Lo, K. Y., Wannier, P. G., Wootten, H. A., and Huggins, P. J.: 1982, Ap. J., 252, p. 616.

Mufson, S. L., Lyon, J., and Marioni, P. A.: 1975, Ap. J. (Letters), 201, L85.

THE NGC 7027 MOLECULAR CLOUD

Harley A. Thronson, Jr.
Wyoming Infrared Observatory

John Bally
A.T.&T. Bell Laboratories

ABSTRACT. Emission from eight different molecules has been searched for in the planetary nebula NGC 7027. One new molecule has been detected: the N = 1 → 0 transition of CN was observed, bringing to four the number of molecules known to exist in the object. For the most part the abundances calculated for NGC 7027 agree with those found for the prototypical evolved carbon star IRC +10°216, suggesting that the two objects share somewhat similar chemical histories. One exception may be the low abundance of CS in NGC 7027.

Increasingly sensitive telescope/receiver combinations have lead to deeper searches for molecular emission from planetary nebulae and their suspected precursors (Mufson, Lyon, and Marionni 1975; Zuckerman *et al.* 1977; Davis, Seaquist, and Purton 1979; Thronson and Mozurkewich 1983; Knapp 1985). Although a few dozen highly evolved stars have been carefully searched, NGC 7027 remains the only bonafide planetary nebula with strong thermal molecular emission which can, therefore, be easily studied. As such, observations of this object are particularly important in understanding stars that have disgorged their interiors as part of the final stages of nuclear-burning evolution.

Two millimeter-wave telescopes were used to search for new molecular emission from NGC 7027: the NRAO 12 m and Bell Labs 7 m dishes. Standard observational and reduction techniques were employed and the results are presented in Tables 1 and 2. Only one new molecule was discovered in the object -- the CN radical -- bringing to four the total number known to exist in the object: H_2, CO, HCN, and CN. This pathetic collection stands in sharp contrast with the rich circumstellar cloud around the carbon star IRC +10°216 (e.g., Wannier and Linke 1978).

The emission from CN was easily detected and the abundance calculated assumed thermodynamic equilibrium, with an excitation temperature between 10 and 16 K. Huggins and Glassgold (1982) suggested that some circumstellar molecules may be photoreduced by the general interstellar radiation field. One such reaction is HCN + hν → CN + H. This mechanism appears to explain the abundance of both HCN and CN in IRC +10°216 and, by analogy, may be the likely source of CN in NGC 7027.

We were surprised by the lack of J = 2 → 1 emission from CS, since

M. Morris and B. Zuckerman (eds.), Mass Loss from Red Giants, 169–170.

TABLE 1

Observed Transitions and Line Strengths

Molecule	Transition	$\int T_R^* dv$ (K-km s^{-1})	Telescope
CO	$J = 2 \rightarrow 1$	57.0 ± 1	NRAO 12 m
CN	$N = 1 \rightarrow 0$	1.23 ± 0.12[1]	NRAO 12 m
CS	$J = 2 \rightarrow 1$	0.04 ± 0.03	BTL 7 m
$C^{18}O$	$J = 1 \rightarrow 0$	0.03 ± 0.05	BTL 7 m
HC_3N	$J = 11 \rightarrow 10$	0.15 ± 0.08	BTL 7 m
C_2H	$N = 1 \rightarrow 0,\ J_F = \frac{3}{2}_1 \rightarrow \frac{1}{2}_1$	0.07 ± 0.1	BTL 7 m
SiCC	$J_{K_-,K_+} = 6_{0,6} \rightarrow 5_{0,5}$	0.07 ± 0.06	BTL 7 m
CO^+	$N = 2 \rightarrow 1,\ J = 5/2 \rightarrow 3/2$	-0.21 ± 0.22	NRAO 12 m
H_2CO	$J = 3 \rightarrow 2,\ K_p = 1 \rightarrow 1,\ K_o = 2 \rightarrow 1$	-0.21 ± 0.35	NRAO 12 m

[1]Strongest hyperfine component

TABLE 2

Comparative Abundances

	NGC 7027[1]	IRC +10°216[2]
N(CO)	$\geq 2 \times 10^{17}$ cm^{-2}	5×10^{17} cm^{-2}
$[CO]/[H_2]$	$\sim 10^{-3}$	8×10^{-4}
[CN]/[CO]	$\leq 5 \times 10^{-4}$	1×10^{-3}
[CS]/[CO]	$\leq 5 \times 10^{-5}$	3×10^{-4}
$[^{12}C]/[^{13}C]$	≥ 43 ± 6	40 ± 14
$[^{16}O]/[^{18}O]$	≥ 180	≥ 2700

[1]Values calculated from this work and Thronson (1983).
[2]Values from Lafont, Lucas, and Omont (1982) and references therein.

this transition has been detected in highly evolved objects (e.g., Wannier and Linke). If this molecule in NGC 7027 is collisionally excited, we hypothesize that the densities in the object are not sufficient to excite the J = 2 level. A simple two-level model molecule gives $n_{H_2} < 4 \times 10^5$ cm^{-3} from our observational data. It may be that the intense radiation field from the exciting star has destroyed the molecules in the inner part of the cloud, where the densities are expected to be the highest.

Advanced nucleosynthesis is expected to alter the observed isotopic ratios of oxygen. In particular, explosive H- and/or He-burning has been predicted to increase the abundance of ^{18}O relative to the more abundant isotope. Our observational results, $[^{16}O]/[^{18}O] \geq 180$ (2σ), shows no evidence for strong enhancement above the terrestrial ratio, 490.

This work was supported in part by the University of Wyoming Division of Basic Research.

REFERENCES

Davis, L. E., Seaquist, E. R., and Purton, C. R. 1979, *Ap. J.*, 230,434.
Huggins, P. J., and Glassgold, A. E. 1982, *Ap. J.*, 252, 201.
Knapp, G. R. 1985, these proceedings, 159, 171, 175, 177.
Lafont, S., Lucas, R., and Omont, A. 1982, *Astr. Ap.*, 106, 201.
Mufson, S., Lyon, J., and Marionni, P. 1975, *Ap. J. (Letters)*, 201, L85.
Thronson, H. A. 1983, *Ap. J.*, 264, 599.
Thronson, H. A., and Mozurkewich, D. 1983, *Ap. J.*, 271, 611.
Wannier, P. G., and Linke, R. A. 1978, *Ap. J.*, 225, 130.
Zuckerman, B. *et al.* 1977, *Ap. J. (Letters)*, 211, L97.

CO EMISSION FROM PLANETARY NEBULAE

G. R. Knapp
Department of Astrophysical Sciences
Princeton University
Princeton, NJ 08544

ABSTRACT. Searches for molecular line emission from planetary nebulae generally prove negative. Observations of 10 planetaries have resulted in the detection of one of them, NGC2346. In addition, weak CO emission has been detected in M1-92, which may be a transition object between a red giant and a planetary nebula.

The detection of extensive circumstellar envelopes produced by mass loss from red giants led to the suggestion (see e.g. the review by Zuckerman 1978) that these objects are the immediate precursors of planetary nebulae. This suggestion received strong confirmation when Mufson, Lyon and Marionni (1975) detected a molecular cloud, with similar properties to a circumstellar envelope, associated with NGC7027. The second such object identified is CRL618, in which a compact HII region is embedded in a circumstellar molecular cloud (Kwok and Feldman 1981). Both of these objects appear to be carbon-rich. Vy 2-2 has recently been identified as a molecular cloud surrounding a young planetary nebula; the circumstellar cloud is an OH maser source (Seaquist and Davis 1983). This object has also been tentatively detected in the CO line (Knapp and Morris 1984).

Observations of ten planetary nebulae were made in the Spring of 1984 in the CO (J = 1-0) line using the 7 m antenna of Bell Laboratories. Only one of these objects, the bipolar planetary nebula NGC2346, was detected. Other searches for molecular emission from planetary nebulae have in general been negative (Zuckerman et al. 1978; Davis et al. 1979). In addition, the probable planetary nebula precursor, M1-92, which is a strong 1667 MHz OH maser emitter, was tentatively detected. The velocity extent of the CO emission is similar to that of the OH emission (Davis et al. 1979). The CO observations are listed in Table I. The upper limits on T_A^* are 5σ limits. The values of V_c (the central velocity of the CO profile) and V_o (the terminal wind outflow velocity) are found by fitting a parabola to the observed CO profile.

M. Morris and B. Zuckerman (eds.), Mass Loss from Red Giants, 171–173.

Table I. CO observations of planetary nebulae

PN	T_A^*(K)	
N650	<0.10	
IC418	<0.10	
N2346	0.08	V_c = 14.2 km/s, V_o = 19.7 km/s
N2371-2	<0.15	
N2440	<0.27	
N2374-5	<0.06	
IC4593	<0.07	
M2-9	<0.12	
M1-65	<0.15	
M1-92	0.02	V_c = -5 km/s, V_o = 25 km/s
Hu1-2	<0.12	

The values for $\dot{M}$ for the four detected planetary nebulae were found using models of the circumstellar envelope emission (c.f. Knapp and Morris 1984). These are listed in Table II, as is the assumed distance and the envelope chemistry for each object. The values of $\dot{M}$ for NGC7027, CRL618 and Vy 2-2 are taken from Knapp and Morris (1984).

Table II. Planetary nebulae detected in the CO line

PN	$\dot{M}$	Chemistry?	D(pc)
NGC7027	1.1×10^{-4}	C	1000
CRL618	8.0×10^{-5}	C	1300
NGC2346	1.2×10^{-5}	C	460
Vy2-2	(2.4×10^{-6})	0	1000

NGC2346 is known to be carbon-rich (Wood 1984) and its distance is estimated to be 460 pc (Walsh 1983). The model for the CO emission from NGC2346 took into account the fact that the center of the cloud is ionized - the radius of the nebula is ~ 1' (Walsh 1983). The value of $\dot{M}$ derived here is an order of magnitude larger than that found by Walsh (1983) from observations of the ionized gas. However, both values are highly uncertain - spherical symmetry is assumed in the models, but the nebula is of much more complex structure.

In addition to M1-92, two other objects are known in which circumstellar shells surround stars of type earlier than M - these are IRC+10420 and CRL2688. Mass loss rates for these three objects, which presumably represent stars in transition between the red giant and planetary nebula phases, are listed in Table III.

Table III. Protoplanetary nebulae detected in the CO line

	$\dot{M}$	Chemistry	Star	D(pc)
M1-92	$1.8x10^{-5}$	0	B0	2300
IRC+10420	$3.0x10^{-4}$	0	F8	3000
CRL2688	$1.6x10^{-4}$	C	F5	1000

From these observations, we can conclude:

(1). Most planetary nebulae are completely ionized.
(2). Only molecular clouds produced by stars with high mass-loss rates can survive after the onset of planetary nebula formation long enough to be detectable. This agrees with model predictions by Spergel et al. (1983).
(3). Both stars with [C] > [O] and with [C] < [O] are planetary nebula precursors.

I thank the staff at Bell Laboratories, A.A. Stark, R.W. Wilson, J. Bally and R.E. Miller for their support of these observations. This work was supported by the National Science Foundation via grant AST83-13293 to Princeton University.

REFERENCES

Davis, L.E., Seaquist, E.R., and Purton, C.R. 1979, Ap.J. 230, 434.
Knapp, G.R., and Morris, M. 1984, submitted to Ap.J.
Kwok, S., and Feldman, P.A. 1981, Ap.J. (Letters) 247, L67.
Mufson, S.L., Lyon, J., and Marionni, P.A. 1975, Ap.J. (Letters) 201, L85.
Seaquist, E.R., and Davis, L.E. 1983, Ap.J. 274, 659.
Walsh, J.R. 1983, M.N.R.A.S. 202, 303.
Wood, B. 1984, private communication.
Zuckerman, B. 1978, in I.A.U. Symposium 76, ed. Y. Terzian (D. Reidel).
Zuckerman, B., Palmer, P., Gilra, D.P., Turner, B.E., and Morris, M. 1978, Ap.J. (Letters) 220, L53.

THE GAS TO DUST RATIO IN CIRCUMSTELLAR ENVELOPES

G. R. Knapp
Princeton University Observatory

Mass loss rates for evolved stars, with both [O] > [C] and [O] < [C] can be found using observations of thermal emission in the CO rotational lines. Observations of the CO(1-0) line have been made of fifty stars using the Bell Laboratories 7-m antenna, and models of the circumstellar shell used to derive mass loss rates from these stars (Knapp and Morris 1984, submitted to Astrophysical Journal). Using these results and values of $\dot{M}$ derived from OH maser line observations, the gas-to-dust ratio in circumstellar envelopes has been examined. The grain loss rates are found using IR photometry between about two and twenty microns by M. Rowan-Robinson and S. Harris (M.N.R.A.S. 1982, **200**, 197; 1983, **202**, 767; **202**, 797). The results show that over 4.5 decades in mass-loss rate the gas-to-dust ratio by number in circumstellar envelopes is constant and that there is no significant difference between the carbon and oxygen stars. Using simple grain models and assuming that the grains are silicates, the gas-to-dust ratio in oxygen stars is 1:158±13 by mass. This is very close to that found in the local interstellar medium, 1:147, using the same grain model.

The full text of this paper has been submitted to Astrophysical Journal. The work is supported by the National Science Foundation via grant AST82-13293 to Princeton University.

M. Morris and B. Zuckerman (eds.), Mass Loss from Red Giants, 175.

21-CM LINE AND RADIO CONTINUUM EMISSION FROM CIRCUMSTELLAR ENVELOPES AROUND LATE-TYPE GIANTS

G. R. Knapp
Department of Astrophysical Sciences
Princeton University
Princeton, NJ 08544

ABSTRACT. Searches for atomic hydrogen in the circumstellar envelopes produced by mass loss from late-type giants and from planetary nebulae using the 21-cm line have so far proven negative, with the possible exception of the planetary nebula NGC6302. These observations show that the mass shed by red giants is overwhelmingly in molecular form, even for the hotter stars. Comparison between the upper limits and the amount of HI expected to be produced by photodissociation by interstellar UV sets an upper limit of ~ 20,000 years on the age of the envelope around IRC+10216.

Continuum observations of red giants provide a means of investigating mass-loss mechanisms, the presence of companion stars and the formation of planetary nebulae. The available observations suggest that the mass-loss mechanism may be different for cool stars (e.g. IRC+10216) than for hotter stars (e.g. α Ori). They also allow a preliminary estimate of the lifetime of the copious mass loss phase of extreme AGB stars of 20,000 -30,000 years.

1. INTRODUCTION

The large circumstellar envelopes produced by copious mass loss from red giants are observed via their infrared and radiofrequency spectral line emission. As discussed at this meeting by Werner, Morris and Bowers, these observations give values of $\dot{M}$, the mass loss rate, V_c, the stellar velocity, and V_o, the terminal outflow velocity of the wind. With the advent of the Very Large Array, observations in the 21-cm HI line and in the radio continuum are possible with greatly increased sensitivity. Even so, circumstellar envelopes are not yet detected in the HI line, and rarely in the radio continuum. These observations will be reviewed here, and their implications for mass loss mechanisms and planetary nebula formation discussed.

Throughout this contribution, it is assumed that copious mass loss immediately precedes the formation of a planetary nebula, and that the nebula is, in fact, the stellar wind ionized by the planetary nebula nucleus. In the next section, the HI observations are discussed,

M. Morris and B. Zuckerman (eds.), Mass Loss from Red Giants, 177–183.

particularly from the point of view of the limits set on the duration of the mass loss phase. In §III the continuum observations are described and used to estimate a time scale for mass loss and planetary nebula formation.

2. HI IN CIRCUMSTELLAR ENVELOPES AND PLANETARY NEBULAE

HI is very difficult to observe in circumstellar envelopes and planetary nebulae, which are generally low-mass (< 1 $M_\odot$) objects, because of (1) the intrinsic extreme weakness of the line (2) the small size (a few arc minutes or less) of the objects and (3) the ubiquity of strong emission from galactic HI (which can also absorb any emission from distant objects). Aperture synthesis instruments, in particular the VLA, help considerably with the last two problems for nearby objects.

Circumstellar envelopes are studied using their molecular line emission and planetary nebulae their ionized continuum. The transition time between the two stages is short, ~ 10^3 years (see later in this paper) so that we do not expect a long period during PN formation when the envelope is atomic. Nevertheless, some CS envelopes may be expected to be partly, or completely, atomic, either because of equilibrium abundances in the stellar photosphere or because of dissociation by interstellar UV as the envelope expands into the ISM. The expected abundances of HI and H_2 in CS envelopes have recently been calculated in detail by Glassgold and Huggins (1983).

The only such object in which HI may have been detected is the bipolar planetary nebula NGC6302 (Rodríguez and Moran 1983). These authors see an HI component in absorption which may be due to blueshifted gas in the surrounding envelope. This component is not seen in other nearby continuum sources. In NGC6302, Rodríguez and Moran find M(HI) ~ 0.06 $M_\odot$, which probably represents a few percent of the total envelope mass.

HI line absorption observations have also been made by Pottasch et al. (1982) for NGC7027. HI absorption is seen at the correct velocity for the surrounding neutral shell, but this absorption component is also seen in other nearby continuum sources, and may readily be ascribed to intervening Galactic HI. From this observation, Pottasch et al. (1982) find M(HI) < 0.04 $M_\odot$, or M(HI)/M(total) < 0.02. Limits of ~ 0.01 $M_\odot$ are found for several other planetaries (Rodríguez and García-Barretto 1984).

Searches for HI in several other PN, both with the VLA and with Arecibo, have proven negative (Zuckerman, Terzian and Silverglate 1980; Knapp and Bowers 1983). These authors have also searched mass-losing stars, again without success. For α Ori, Knapp and Bowers (1983) find $\dot{M}$(HI) < 5 x 10^{-6} $M_\odot$ yr^{-1}. Since it is expected that most of the hydrogen leaving this relatively hot star is in atomic form, this sets a limit on $\dot{M}$ in agreement with most recent values (e.g. Knapp and Morris 1984).

For IRC+10216, the observed upper limit ($\dot{M}$ (HI) < 4 x 10^{-6} $M_\odot$ yr^{-1}) shows that the mass lost by this star is > 90% molecular. HI is also expected in the envelope because of dissociation by interstellar UV, and we expect M(HI) ~ T^3, where T is the age of the envelope. Using the UV flux of Jura (1982) and the upper limit on the HI content of 0.01 $M_\odot$, the envelope age is found to be < 2.1 × 10^4 years, about the same as the value measured in the CO line of ~ 1.9 × 10^4 years (Morris, Jura and Stark 1984).

3. CONTINUUM EMISSION FROM CIRCUMSTELLAR ENVELOPES

The VLA makes possible extremely sensitive searches for continuum sources, with sources of ~ 0.5 mJy being detectable in about thirty minutes of observing time. Normally, stars are undetectable at radio wavelengths, (Johnston et al. 1983), but there are at least four ways in which continuum emission might be expected to arise from late-type giants: (a) ionization by an external source (b) an extended ionized stellar chromosphere (c) ionization produced by the stellar white-dwarf core as the object evolves to a planetary nebula. A few tens of objects have been surveyed to the above sensitivity; a sample of carbon stars and OH/IR stars by Spergel et al. (1983), and a sample of OH/IR stars by Baud and Herman (1983; described by Herman 1983). In most cases, no continuum emission has been detected. The implications of these results for mass loss mechanisms and planetary nebula formation will be discussed below.

3.1. Mass-Losing Stars with External Ionization Sources

In the general interstellar medium (HI regions) the flux of ionizing photons is very low. However, a circumstellar envelope may be ionized if the object is in an HII region or near an OB association. The only current example of this phenomenon is NML Cyg, whose continuum emission is offset by ~ 0'.5 from the star (Habing et al. 1982). Morris and Jura (1983) interpret this observation as due to ionization by stars in the Cygnus OB2 association, giving a value of 2 kpc for the distance of NML Cyg. Several other luminous mass-losing stars may be members of such associations, and the morphology of the OH maser emission suggests that the envelopes are affected by external radiation (Bowers, Johnston and Spencer 1983). For one such object, VY CMa, an upper limit of 0.3 mJy at λ 6 cm has been found for any continuum emission (Spergel et al. 1983).

3.2. Stars With Extended Chromospheres

Several late-type giants have now been detected as continuum sources at the VLA. These include α Ori (Newell and Hjellming 1982), α Sco (Hjellming and Newell 1983), o Cet (?) and IRC+10216 (?) (Spergel et al. 1983), and α^1 Her and α Boo (Drake and Linsky.1983). If the continuum fluxes are used to calculate values of M (H^+), using the formulae of Wright and Barlow (1975) and Panagia and Felli (1975), the winds of α Ori, α Sco, o Cet and α^1 Her are found to be ~ 1% ionized. These stars thus have extended (R ~ 1 -4 R_*, Newell and Hjellming 1982), warm (3-4,000K) chromospheres. These observations lend support to the models of Alfvén-wave driven mass loss (Hartmann and McGregor 1980, Hartmann and Avrett 1984). However, many other late-type giants with considerably higher values of $\dot{M}$ do not show continuum emission and therefore are likely to be losing mass by some other mechanism. In these properties, these late-type stars are similar to pre-main sequence stars embedded in molecular clouds, which are expelling large amounts of molecular material from their vicinity, yet show only very weak radio continuum emission (e.g., Moran et al. 1983).

3.3. Stars With Companions Embedded in the Circumstellar Material

Several stars undergoing copious mass loss are known to have hot companions embedded in the circumstellar shell. These include R Aqr, α Sco and o Cet. High resolution VLA observations of α Sco (Hjellming and Newell 1983) and R Aqr (Sopka et al. 1982; Spergel et al. 1983; Hollis et al. 1984) show the radio continuum emission from these stars to have multiple thermal components.

Hjellming and Newell (1983) find that the double source structure of α Sco is well fit by an extended chromosphere (n_e/n_H ~ 0.01) around the primary and an ionized cavity in the wind produced by the B2.5V secondary. These observations allow the derivation of a value for the total mass loss rate from the primary of ~ 2×10^{-6} $M_\odot$ yr^{-1}.

For o Cet, it is not known whether the weak radio continuum is associated with the star or with its companion - the separation is only 0".6. For R Aqr, three thermal radio sources are found, one coincident with the star and two with structure in the optical nebulosity surrounding the star. It is likely that the ionization of all three sources is due to the hot, unseen companion; that the wind is neutral as it leaves R Aqr is shown by the presence of an SiO maser (Zuckerman 1979). However the multiple source structure is explained, it strongly suggests that in this star, and by inference in other late-type stars, the mass loss takes place both as clumps and as a steady wind.

The peculiar properties of the companions of o Cet and R Aqr, including the extremely low X-ray fluxes from the systems, have been discussed by Jura and Helfand (1984) who suggest that the companions are low-mass main sequence stars rather than evolved compact dwarfs. The ionizing radiation is produced by accretion from the envelope of the primary. This suggestion is attractive because main-sequence companions are more likely than evolved companions.

3.4. Formation of Planetary Nebulae

After the first molecular-line observations demonstrating the existence of copious mass loss from red giants, reviewed here by Morris, it was suggested (see e.g., the review by Zuckerman 1978) that these objects are the immediate precursors of planetary nebulae. Planetary nebula formation is then imagined to occur at the end of the mass loss phase, when the central star forms a very hot white dwarf and ionizes the surrounding circumstellar material. The density distribution in the forming planetary nebula can be modified by the action of a fast wind from the central star. This picture received strong confirmation when Mufson, Lyon and Marionni (1975) detected CO emission from a molecular cloud surrounding the compact PN NGC7027.

A list of suggested protoplanetary nebulae is given by Zuckerman (1978). Since the observations by Mufson, Lyon and Marionni (1975) some tens of planetary nebulae and possible transition objects have been surveyed for CO and/or OH maser emission (Davis et al. 1979; Marionni 1977; Zuckerman et al. 1977, 1978; Knapp 1984). To date, circumstellar molecular clouds have been found associated with the following planetary nebulae and transition objects (Table I):

Table I. Planetary and proto-planetary nebulae in circumstellar molecular clouds.

Object	$\dot{M}$	CO?	OH?	Central *	S(6cm) (mJy)	
N7027	1.1×10^{-4}	✓	x	pnn	110	BPN
CRL618	8×10^{-5}	✓	x	B0	16	BPN
M1-92	1.8×10^{-5}	✓	✓	B0	-	BPN
N2346	1.2×10^{-5}	✓	-	pnn	-	BPN, binary
IRC+10420	3×10^{-4}	✓	✓	F8	-	BPN?
CRL2688	1.6×10^{-4}	✓	x	F5	<0.3	BPN
Vy 2-2	(2.4×10^{-6})	(✓)	✓	pnn	15	

NGC7027, CRL618, NGC2346 and Vy 2-2 are unmistakably compact planetary nebulae. M1-92, IRC+10420 and CRL2688 may be transition objects. Thermal radio emission is seen from NGC7027, CRL618 and Vy 2-2. All of the above objects are bipolar in appearance except for Vy 2-2, and the nucleus of NGC2346 is binary (Walsh 1983; Kohoutek 1983).

The expansion of the ionization front as the planetary nebula forms has been directly observed for CRL618 by Kwok and Feldman (1981) and by Kwok and Bignell (1984). High-resolution observations with the VLA by these authors have shown that the inner ionized structure is bipolar.

The existence of the molecular envelope around Vy 2-2 has been shown by the detection of OH maser emission (Davis et al. 1979) and, tentatively, CO emission (Knapp and Morris 1984). The CO observations give a value for $\dot{M}$ of $\sim 2 \times 10^{-6}$ $M_\odot$ yr^{-1} and an outflow velocity of 12 - 18 km/sec. Seaquist and Davis (1983) have made a beautiful set of high-resolution VLA observations which suggest that the thermal continuum source is optically thick and is surrounded by the molecular cloud.

Vy 2-2 is unusual in that it has only a single 1612 MHz maser emission spike. The shape of this line and its position with respect to the continuum source suggest that it is the blue-shifted (near-side) component. Its velocity (-62 km/sec) compared with that of the CO (-44 km/sec) supports this hypothesis. Since the nebula is optically thick at 1.6 GHz, Seaquist and Davis suggest that the redshifted component, on the far side of the nebula, is absorbed.

If this model of Vy 2-2 is more generally applicable, an estimate of the number of planetary nebulae forming in the envelopes of OH/IR stars can be made, using the unbiassed surveys of the galactic plane made by several groups in a search for OH/IR stars via their 1612 MHz maser emission. The relative numbers of double- and single- peaked sources can be compared, when the observations have been made with enough sensitivity that both components might reasonably have been expected to be seen. The available published surveys are not on the whole presented in a form to do this well (single- peaked sources having generally received short shrift from the authors), so the following estimates of the ratio of single- to double- line sources are somewhat uncertain:

Dwingeloo Survey (Baud et al. 1978):	3/55
Onsala Survey (Johansson et al. 1977):	3/57
NRAO Survey (Bowers 1978):	2/45
Parkes Survey (Caswell and Haynes 1975):	1/21
NRAO OH Survey (Turner 1979):	7/111

Interestingly, all the single -line source profiles presented with adequate sensitivity and velocity resolution resemble those of blueshifted maser spikes. The best candidates for Vy 2-2 type objects from these surveys are OH337.3-0.2 (Caswell and Haynes 1975), OH12.6+0.2 (Baud et al. 1978), OH28.3-0.0 (Johansson et al. 1977), OH31.3-0.2 (Turner 1979) and OH12.3+0.1 and OH56.1+2.1 (Bowers 1978). The hypothesis could be confirmed by a VLA search for continuum emission from these objects.

The above surveys all suggest that 3-5% of OH/IR stars are currently evolving to the planetary nebula stage. Typical values of $\dot{M}$ for these stars are 3×10^{-5} $M_\odot$ yr^{-1} (Bowers, Johnston and Spencer 1983; Knapp and Morris 1984), while the OH shell radius is typically 5×10^{16} cm (Bowers, Johnston and Spencer 1983). The models of Spergel et al. (1983) suggest that the propagation time of an ionization front to this radius is $\sim 10^3$ years. Thus, the lifetimes of the stars in the OH/IR phase are $2 - 3 \times 10^4$ years (which is similar to the upper limit found for the lifetime of IRC+10216). The total mass lost is then ~ 1 $M_\odot$, suggesting that the stars have progenitor masses of ~ 2 $M_\odot$.

This work is supported by NSF grant AST82-13293 to Princeton University.

REFERENCES

Baud, B., Habing, H.J., Matthews, H.E., and Winnberg, A. 1978, Astr. Ap. Suppl. 36, 193.
Bowers, P.F. 1978, Astr. Ap. Suppl. 31, 127.
Bowers, P.F., Johnston, K.J., and Spencer, J.H. 1983, Ap.J. 274, 733.
Caswell, J.L., and Haynes, R.F. 1975, M.N.R.A.S. 173, 649.
Davis, L.E., Seaquist, E.R., and Purton, C.R. 1979, Ap.J. 230, 434.

Drake, S.A., and Linsky, J.L. 1983, Ap.J. (Letters) 274, L77.
Glassgold, A.E., and Huggins, P.J. 1983, M.N.R.A.S. 203, 517.
Habing, H.J., Goss, W.M., and Winnberg, A. 1982, Astr. Ap. 108, 412.
Hartmann, L., and Avrett, E.H. 1984, Ap.J. (in press).
Hartmann, L., and McGregor, K.B. 1980, Ap.J. 242, 260.
Herman, J. 1983, Ph.D. Thesis, Leiden University.
Hjellming, R.M., and Newell, R.T. 1983, Ap.J. 275, 704.
Hollis, J.M., Kafatos, M., Michalitsianos, A.G., and McAlister, H.M. 1984, Ap.J. (in press).
Johnston, K.J., Wade, C.M., and Florkowski, D.R. 1983, in I.A.U. Colloquium 76, ed. A.G. Davis Philip and A.R. Upgren, p.69.
Johansson, L.E.B., Andersson, C., Goss, W.M., and Winnberg, A. 1977, Astr. Ap. Suppl. 28, 199.
Jura, M. 1982, private communication.
Jura, M., and Helfand, D.J. 1984, Ap.J. (in press).
Knapp, G.R. 1985, this meeting, 159, 171, 175, 177.
Knapp, G.R., and Bowers, P.F. 1983, Ap.J. 266, 701.
Knapp, G.R., and Morris, M. 1984, submitted to Ap.J.
Kohoutek, L. 1983, M.N.R.A.S. 204, 93p.
Kwok, S., and Feldman, P.A. 1981, Ap.J. (Letters) 247, L67.
Kwok, S., and Bignell, R.C. 1984, Ap.J. 276, 544.
Marionni, P.A. 1977, Ph.D. Thesis, University of Maryland.
Moran, J.M., Garay, G., Reid, M.J., Genzel, R., Wright, M.C.H., and Plambeck, R.L. 1983, Ap.J. (Letters) 271, L31.
Morris, M., and Jura, M. 1983, Ap.J. 267, 179.
Morris, M., Jura, M., and Stark, A.A. 1984, in preparation.
Mufson, S.L., Lyon, J., and Marionni, P.A. 1975, Ap.J. (Letters) 201, L85.
Newell, R.T., and Hjellming, R.M. 1982, Ap.J. 263, L85.
Panagia, N., and Felli, M. 1975, Astr. Ap. 39, 1.
Pottasch, S.R., Goss, W.M., Arnal, E.M., and Gathier, R. 1982, Astr. Ap. 106, 229.
Rodríguez, L.F., and García-Barreto, J.A., 1984, Ap.J. (in press).
Rodríguez, L.F., and Moran, J.M. 1983, Nature 229, 323.
Seaquist, E.R., and Davis, L.E. 1983, Ap.J. 274, 659.
Sopka, R.J., Herbig, G., Kafatos, M., and Michalitsianos, A.G. 1982, Ap.J. (Letters) 258, L35.
Spergel, D.N., Giuliani, J.L., and Knapp, G.R. 1983, Ap.J. 275, 330.
Turner, B.E. 1979, Astr. Ap. Suppl. 37, 1.
Walsh, J.R. 1983, M.N.R.A.S. 202, 303.
Wright, A.E., and Barlow, M.J. 1975, M.N.R.A.S. 170, 41.
Zuckerman, B. 1978, in I.A.U. Symposium 76, ed. Y. Terzian (D. Reidel).
Zuckerman, B. 1979, Ap.J. 230, 442.
Zuckerman, B., Terzian, Y., and Silverglate, P. 1980, Ap.J. 241, 1014.
Zuckerman, B., Palmer, P., Morris, M., Turner, B.E., Gilra, D.P., Bowers, P.F., and Gilmore, W.S. 1977, Ap.J. (Letters), 211, L97.
Zuckerman, B., Palmer, P., Gilra, D.P., Turner, B.E., and Morris, M. 1978, Ap.J. (Letters) 220, L53.

20 GHZ CONTINUUM AND SiS MASER EMISSION FROM IRC+10216

M.J. Claussen and R. Sahai
Owens Valley Radio Observatory
California Institute of Technology
Pasadena, CA USA

We are monitoring the SiS (1-0) maser line at 18.19 GHz from the late-type carbon star IRC+10216 (CW Leo) over the infrared light cycle (644 days) using the OVRO 40-m radio telescope and a supercooled K-band maser receiver. We have also searched for, discovered, and are monitoring continuum emission at 20 GHz from the star using the same telescope system. The detection of the SiS maser lines in IRC+10216 (Grashoff et al. 1981) is the first evidence of maser activity in any carbon-rich stellar envelope and the interpretation of the masers is important in understanding the physics and chemistry of the circumstellar shell.

The results of the spectral line monitoring so far are shown in Figure 1, and cover about 80% of a full light cycle. The peak intensity of the blue-shifted feature, in particular, shows evidence for time variations which correlate rather well with the infrared light variations in the sense that at maximum light the strength of the blue-shifted peak is also at a maximum. The variations of the red-shifted feature is less obvious, but the ratio of the blue-shifted peak to the red-shifted peak is constant, within the errors of measurement over the range of observations. These observational results agree with the independent observations of Matthews (1984). These results imply that the SiS masers are pumped by infrared radiation from the central star.

We have also detected 20 GHz continuum emission from IRC+10216 and have begun a monitoring program in conjunction with the SiS observations. So far, we have not detected any significant variation in the 20 GHz flux; it has remained constant at a level approximately 6 mJy. This measurement supplements observations at other frequencies in the radio continuum: 145 mJy at 90 GHz by Schwartz and Spencer (1977) and 0.42 mJy at 5 GHz by Spergel et al. (1983).

Schwartz and Spencer make the very plausible argument that the 90 GHz flux is due to dust emission. Based on this and Spergel et al.'s argument that there is no chromospheric emission from the star, we conclude that the centimeter emission is due to a very compact HII region surrounding the star. The diameter of this HII region is calculated to be about 60 milliarcseconds ($\sim 10^{14}$ cm). The discovery

M. Morris and B. Zuckerman (eds.), Mass Loss from Red Giants, 185–186.

of such an HII region has interesting implications for the evolutionary status of IRC+10216.

REFERENCES

Grashoff, M., Tiemann, E., and Henkel, C.: 1981, Astr.Ap., 101, p. 238.
Matthews, H.: 1984, this conference.
Schwartz, P.R. and Spencer, J.H.: 1977, Monthly Notices of R.A.S., 280, p. 297.
Spergel, D.N., Giulani, J.L., and Knapp, G.R.: 1983, Ap.J., 275, p. 330.

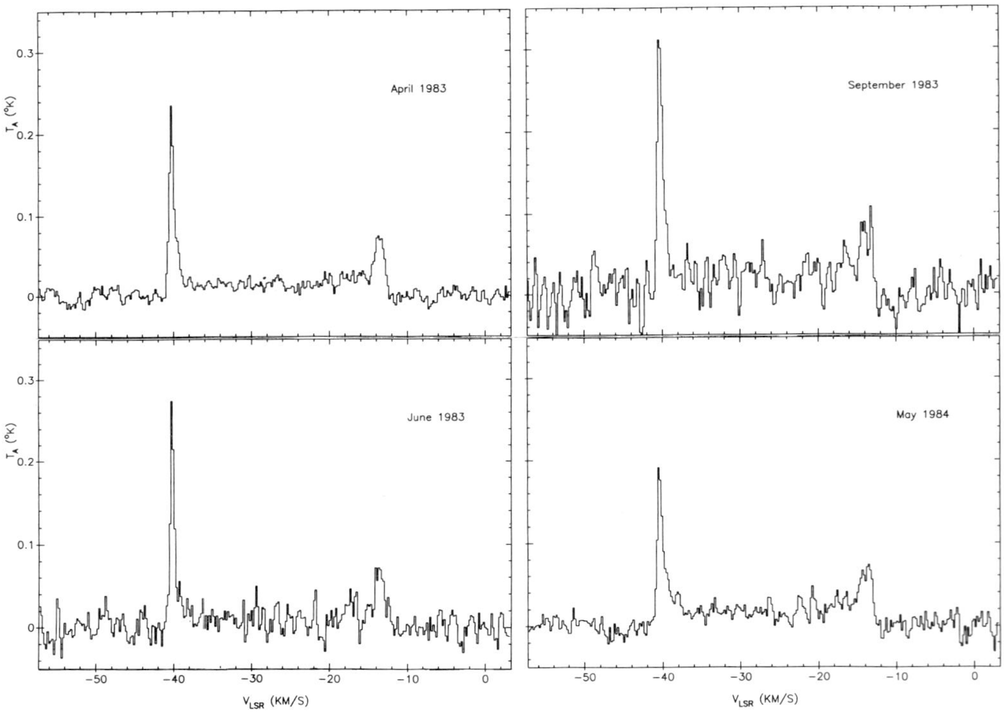

FIGURE 1. Spectra of the SiS (J=1-0) maser toward IRC+10216 at different epochs.

RADIO CONTINUUM OBSERVATIONS OF G-M GIANTS AND SUPERGIANTS AND INFERRED IONIZED MASS LOSS RATES

S. A. Drake
Joint Institute for Laboratory Astrophysics, University of Colorado and National Bureau of Standards, Boulder, CO 80309

We have conducted a 6 cm survey of 33 nearby G-M giants and supergiants with the NRAO VLA.* We have detected α^1 Her (M5 II) at 0.9 mJy and α Boo (K2 IIIp) at 0.27 mJy, and obtained upper limits for the remaining 31 stars significantly lower than most previous values. If we interpret both detections as being due to radio emission from partially ionized stellar winds, we derive ionized mass loss rates of 2×10^{-9} $M_{\odot}$ yr^{-1} (α^1 Her) and 7×10^{-11} $M_{\odot}$ yr^{-1} (α Boo). We infer an ionization fraction for the wind of α^1 Her of 1.4%, significantly higher than the upper limits of 0.045 and 0.012% for the two other M5 stars (RX Boo and W Hya) that were observed. The weak radio source identified with α Boo appears to share the high proper motion of the star (~2".5 per year), confirming its association with the K giant. This is probably another example of a stellar wind-type source, but we cannot rule out at this time the possibility that the radio emission is from a static, extended chromosphere ($T_e \sim 10^4$ K) of angular diameter about 1.7 times the photospheric angular diameter.

31 out of 33 (94%) red giants and supergiants were not detected as quiescent continuum radio sources, with typical 3σ upper limits of 0.1-0.3 mJy. Since the sample was chosen on the basis of proximity and large angular sizes, we do not expect that more than a few additional sources can be found, given the present sensitivity level of the VLA of ~0.1 mJy at 6 cm. The general trend of our results suggests that the Reimer's scaling formula for red giant mass loss, frequently used to calculate the significance of mass loss on the evolution of red giants, tends to overestimate the rate of mass loss for many of these stars by up to an order of magnitude. For coronal and hybrid giants, where the winds are presumably completely ionized,

*The National Radio Astronomy Observatory is operated by Associated Universities, Inc., under contract with the National Science Foundation.

M. Morris and B. Zuckerman (eds.), Mass Loss from Red Giants, 187–188.

our results suggest that the Reimers' (1975, 1977) mass loss rate formula overestimates the actual mass loss rates by factors of at least two to three. For the wind-type G and K giants and supergiants we derive ionized mass loss rates $\lesssim$5% of the Reimers' values. Given the strength of emission lines like Ly-α and the O I λ1304 triplet seen in many of these stars, which suggest that there exists substantial ionized material, we suspect that the total mass loss rates are again lower than the Reimers' formula values, by as much as an order of magnitude.

For M giants and supergiants, there is a sharp decrease in ionization fraction from ~1% for the earlier types to $\lesssim$0.03% for the later types. This is presumably related to the anticorrelation of chromospheric emission with a high dust-to-gas ratio reported by Hagen et al. (1983). In particular, none of the stars with SiO maser emission were detected as radio continuum sources, in agreement with the SiO maser models of Elitzur (1981).

REFERENCES

Elitzur, M. 1981, in Physical Processes in Red Giants, eds. I. Iben and A. Renzini, p. 363.
Hagen, W., Stencel, R. E., and Dickinson, D. F. 1983, Ap. J., 274, 286.
Reimers, D. 1975, Mem. Roy. Soc. Sci. Liège, 8, 369.
_____. 1977, Astr. Ap., 61, 217 [1978, Astr. Ap., 67, 161 (Erratum)].

MASER EMISSION AS A TOOL TO STUDY MASS LOSS FROM EVOLVED STARS

P. F. Bowers
E. O. Hulburt Center for Space Research
Naval Research Laboratory, Washington, D. C. 20375-5000
and
Sachs/Freeman Associates, Bowie, Maryland 20715

ABSTRACT. The properties of circumstellar OH, H_2O, and SiO masers which can be used to examine the mass loss process in red giant stars are discussed from an empiricist's point of view. Two aspects of mass loss are considered: namely, the mass loss mechanism and the mass loss rate.

Maser emission now provides good evidence that at radii $\gtrsim$ 1000 AU the circumstellar envelopes of most oxygen-rich red giants can be modelled to first order as predominantly spherical shells whose dominant kinematic mode is radial expansion at a nearly constant velocity. This result implies that the mass loss mechanism does not produce large deviations from sphericity when averaged over the expansion timescale ($\sim$ 1000 y). More specific constraints are difficult to obtain because the observed distribution of maser emission depends not only on the distribution of gas produced by the mass loss process but also on many other factors, ranging from interaction of the outflowing gas with the ambient interstellar medium or radiation field to limitations imposed by variables of the observing program. An assessment is made of the impact of these factors on the interpretation (and associated uncertainties) of the data.

The large data base provided by the many known circumstellar OH/H_2O/SiO maser stars also can be used to examine possible connections between certain properties of maser emission and the stellar mass loss rate. Evidence for two such connections is presented: a) the outer radius of the OH, H_2O, or SiO maser region appears to be related to the mass loss rate; b) the maser luminosity also can be influenced by the mass loss rate, but the dependence is different for each of these molecular species, providing a simple observational method to obtain a distance-independent mass loss rate by determining the flux ratio for the maser lines. These correlations support the scale of mass loss obtained from IR/CO data and can be used to derive an order-of-magnitude estimate of the mass loss rate for a given star. By combining these results with estimates of the number density of OH, H_2O, and SiO maser stars in the solar neighborhood, it is concluded that long-period, oxygen-rich, Mira-type variable stars contribute at least 35% of all material recycled to the interstellar medium.

M. Morris and B. Zuckerman (eds.), Mass Loss from Red Giants, 189–209.

1. INTRODUCTION

Soon after the discovery of OH and H_2O maser emission from "infrared" stars (Wilson and Barrett 1968; Knowles et al. 1969), optical, infrared, and microwave data led to a schematic model of a circumstellar maser source for which a variable, oxygen-rich M giant or supergiant star undergoes significant mass loss and consequently produces an expanding envelope of gas and dust which emits predominantly in the infrared (see reviews of Caswell 1974; Hyland 1974). Much qualitative support for this model was accumulated throughout the 1970's [including the discovery of circumstellar SiO masers (cf. Snyder and Buhl 1975)], but little progress was made in understanding the structure of the envelopes or the evolutionary nature of the stars. A wealth of infrared and microwave data obtained in the last few years has significantly enhanced our ability to examine both these aspects. In this paper emphasis is placed on the extent to which OH, H_2O, and SiO masers presently can be used to determine the nature and degree of the mass loss process. Recent reviews encompassing other aspects of maser emission are given by Snyder (1980), Goldreich (1980), Downes and Genzel (1980), Reid and Moran (1981), and Elitzur (1981, 1982). Implications for the evolutionary nature of the central star are broached, but this is an active area of research with several competing scenarios and a thorough discussion is beyond the scope of this paper. The interested reader is encouraged to see the references in the last paragraph of this section.

Much of the usefulness of circumstellar masers is based on their common association with late-type stars. Maser emission produced by rotational transitions in the ground vibrational states of OH and H_2O and the excited vibrational states of SiO is frequently detected from the circumstellar envelopes of long-period ($\gtrsim$ 250 days), oxygen-rich (O > C) Mira variables and, to a lesser extent, from supergiant stars and short-period (< 250 days) semiregular variables. If the spectral type of the star is known, it is usually later than M3 at maximum light; some notable exceptions include the well known S star χ Cyg (Wolff and Carlson 1982), the F8 supergiant IRC+10420 (Bowers 1984), three short-period ($\sim 70^d$) G variables (Fix and Claussen 1984), and several early-type emission line stars (Davis et al. 1979; Allen et al. 1980). A list of late-type circumstellar maser stars discovered through 1978 is given by Engels (1979). Additional detections which supplement his list have been made for OH and H_2O by Olnon et al. (1980), for OH and SiO by Ukita and Goldsmith (1984), for OH by Silverglate et al. (1979), Bowers (1981), Caswell et al. (1981), Habing et al. (1983), Ukita and LeSqueren (1984), and Fix and Claussen (1984), for H_2O by dos Santos et al. (1979), Dickinson and Dinger (1982), Crocker and Hagen (1983), and Bowers and Hagen (1984), and for SiO by Zuckerman (1979), Spencer et al. (1981), Wolff and Carlson (1982), and Jewell et al. (1984). OH has been detected from about 350 stars, H_2O from 150, and SiO from > 100.

Particularly valuable as a tool to study long-period variables is the ground-state rotational ($^2\pi_{3/2}$, J=3/2, F=1-2) transition of OH at

1612 MHz and much (although not all) of the discussion in this paper is biased toward this emission. Its utility can be attributed to two observational properties. First, the emission seldom is strongly polarized and the profile structure usually varies smoothly and periodically with the stellar light cycle (Harvey et al. 1974; Fillit et al. 1977; Jewell et al. 1979; Herman 1983), so the spatial and velocity distributions of the maser emission are not strongly dependent on the epoch of the observations. Two exceptions for which significant polarization and flare-like behavior have been observed are IRC+10420 and U Ori (Benson et al. 1979; Cimmerman 1979; Reid et al. 1979; Jewell et al. 1981). [See Claussen and Fix (1982) and Ukita and LeSqueren (1984) for recent discussions of polarization in the main-line, 1665/1667 MHz transitions]. For the most part, however, OH (1612) emission can provide relatively accurate estimates of various stellar and circumstellar properties, including the position, systemic radial velocity, period, and distance of the star as well as the morphology, radius, and velocity structure of the maser region. Recent results and implications for the mechanism of mass loss are discussed in Section 2.

The second, extremely useful property of circumstellar OH (1612) emission is its characteristic double-peaked velocity signature which provides almost unequivocal identification of the affiliated star as an oxygen-rich, late-type variable star. The profile structure of no other maser transition is so well defined. Indeed, most (> 200) OH maser stars have been discovered by this velocity signature in 1612 MHz surveys of the galactic plane region (Winnberg et al. 1973; Kerr and Bowers 1974; Caswell and Haynes 1975; Johansson et al. 1977a; Bowers 1978a; Baud et al. 1979a,b; Caswell et al. 1981; Habing et al. 1983). This emission also has been used to identify stars discovered by other maser transitions (e.g., Ukita and Goldsmith 1984).

The large number of "radio-luminous OH/IR stars" (those discovered in the galactic OH surveys) has played an especially important role in unveiling the range of properties attributable to long-period variable stars and in establishing the importance of mass loss in the red giant phase of evolution. Because of their high maser luminosities, these stars can be detected on a transgalactic distance scale, thus providing a valuable probe of the galactic distribution and kinematics of an evolved population, unhindered by interstellar obscuration. Their population characteristics range from Extreme Population I to Young Disk, indicating ages $\sim 10^7$ to 10^9 y (Caswell 1974; Johansson et al. 1977b; Bowers 1978b; Olnon et al. 1981; Baud et al. 1981a,b; Habing et al. 1983). Although a small percentage ($\sim$ 10%?) are probably late-type supergiants, the majority of these stars appear to be long-period, Mira-type variables based on the similarity of their bolometric luminosities ($\sim 3 \times 10^3$ to 3×10^4 $L_\odot$), their infrared and OH light curve shapes, and their OH profile structures to those of known Mira variables in the solar neighborhood (Bowers and Kerr 1978; Herman 1983; Engels et al. 1983; de Jong 1983; Jones et al. 1983a,b). However, periods of variability range up to 2000 days (Engels et al. 1983; Herman 1983), and many of these stars are undergoing extensive mass loss with derived rates $\sim 10^{-5}$ to 10^{-4} $M_\odot$ y^{-1}. In comparison, most nearby ($\lesssim$ 1 kpc), optically identifiable Mira variables have periods $\lesssim$ 600 days and mass loss rates $\lesssim 10^{-6}$ $M_\odot$ y^{-1}.

From analysis of galactic survey data, the radio-luminous OH/IR stars are spatially rare, comprising perhaps only 1 or 2 percent of the Mira population by number density (Bowers 1978b; Baud and Habing 1983; Herman 1983; de Jong 1983; Jones et al. 1983b). Thus only a small percentage of Mira variables may undergo copious mass loss, depending on the initial stellar mass, or else all Miras may briefly experience this phase (Engels et al. 1983; de Jong 1983; Baud and Habing 1983; Jones et al. 1983b). In either case it is now well recognized that mass loss by a steady wind in the asymptotic red giant phase clearly is an important parameter in models of stellar evolution (Iben and Renzini 1983) and of the formation of planetary nebulae (Kwok 1983). Known properties of OH, H_2O, or SiO masers which relate to and serve as an indicator for the mass loss rate are discussed in Section 3.

2. MASS LOSS MECHANISM

The mass loss mechanism for red giant stars is not yet known (e.g., Zuckerman 1980). Models of the outflow usually assume it to be spherically symmetric with radiation pressure on dust grains responsible for acceleration of the gas at distances $\gtrsim$ 10 AU from the star (Kwok 1975; Goldreich and Scoville 1976). Such models are consistent with radiation-driven or shock-driven mass loss. However, nonspherical circumstellar distributions of material seen for closely related evolved objects such as bipolar nebulae and some planetary nebulae are obvious exceptions and might indicate other mechanisms, including nonradial pulsations, rotationally-forced ejection, mass loss driven by Alfvén waves, or mass ejection in close binary systems.

Because maser emission simultaneously provides information about the spatial and velocity structure of the envelope, it can provide important constraints for the mass loss mechanism. Additionally, each molecular maser transition arises from a different portion of the envelope because of different excitation energies, so the radial structure at distances ranging from a few to a few thousand AU from the star can, in principle, be determined. However, the observed distribution of masers depends on many factors. The output intensity at a given velocity depends on the abundance of inverted molecules, the line-of-sight velocity coherence which determines the pathlength of amplification, the available pump energy, and the intensity of the background radiation field at the transition frequency of the maser. In addition to the distance, the apparent intensity depends on variables of the observing program, including the spatial and spectral sensitivity, the polarization characteristics, the velocity resolution, and the epoch of the observations. Interpretation of the data for individual cases therefore can be difficult. Despite these problems, recently acquired interferometric data for a large number of stars have provided valuable insight into the structure of the circumstellar envelopes and the nature of the mass loss mechanism.

2.1 Outer Envelope Structure (OH Results)

Attempts throughout the 1970's to examine the spatial distribution of OH emission with Very Long Baseline Interferometry (VLBI) have been useful to determine the relative locations and sizes of individual masers spots, but these data are sensitive only to brighter, small scale ($\lesssim$ 0.″1) features and typically over-resolve much of the emission (e.g. Wilson et al. 1970; Reid et al. 1977; Bowers et al. 1980). Lower limits to the angular size of the masing regions demonstrate indirectly that the double-peaked velocity structure must be produced by an expanding shell, but the distribution of maser features usually appears to be an almost random collection of spots at various velocities. Recently developed intermediate baseline interferometers such as the Very Large Array (VLA) and the Multi-Element Radio-Linked Interferometer (MERLIN) are sensitive to weaker, extended emission present in these sources. To date, these instruments have been used to map the OH emission at 1612, 1665, or 1667 MHz from about 30 stars (Booth et al. 1981; Baud 1981; Bowers et al. 1981, 1983 = BJS; Morris et al. 1982; Norris et al. 1982; Herman 1983; Seaquist and Davis 1983; Diamond et al. 1983; 1984a,b; Bowers 1984; Chapman et al. 1984; Chapman and Cohen 1984; Spencer et al. 1984) and to determine accurate positions for an additional 70 stars (Reid et al. 1981; Bowers and de Jong 1983; Fix and Mutel 1984). These data provide direct evidence that the outer ($\gtrsim$ 1000 AU) regions of the envelopes for many oxygen-rich, long-period variables can be modelled, to first order, as spherical shells expanding at a nearly constant velocity.

One of the better examples confirming the expanding, spherical shell (ESS) model is OH127.8-0.0. This source, coincident with AFGL 230, was discovered in an OH survey of the galactic plane (Kerr and Bowers 1974; Bowers 1978a), and it is the brightest known OH/IR star in the anticenter region of the Galaxy. The emission at 1612 MHz has been mapped with the VLA (BJS), MERLIN (Booth et al. 1981; Norris et al. 1982) and VLBI techniques (Bowers et al. 1980; Norris et al. 1984). The velocity profile and selected VLA maps are shown in Figure 1. The points of agreement with the predictions of the ESS model are as follows:

a) The unresolved (< 1") peak emission features (A, E) are positionally coincident to within 0.″1. In the ESS model these strong features are produced along the line of sight to the star and have the largest gain because the velocity coherence is largest in this direction. The absolute position of these features unambiguously defines the stellar position to within 0.″1 (BJS). VLBI data indicate compact (< 0.″1) structure in Feature A which may be amplifying the stellar thermal emission (Norris et al. 1984), so it ultimately may be possible to measure the stellar position at a level of a few milliarcsec for this star.

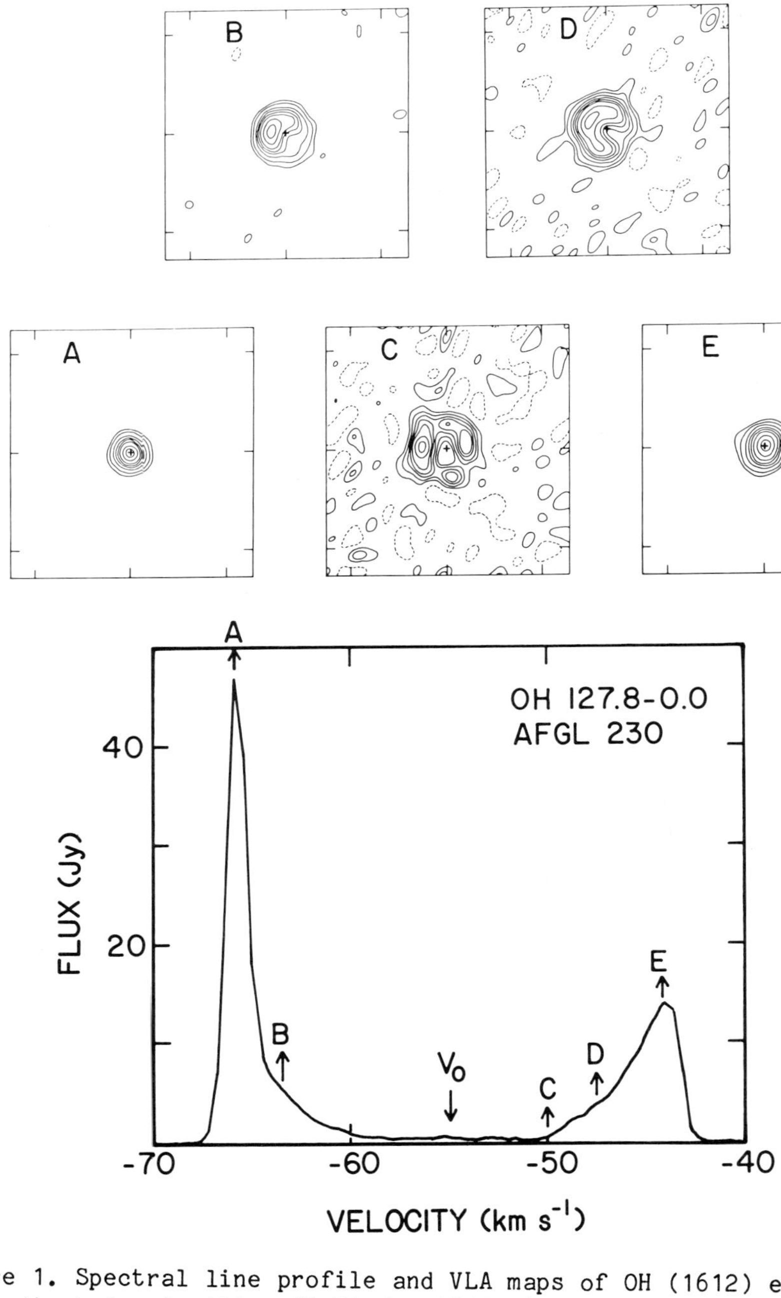

Figure 1. Spectral line profile and VLA maps of OH (1612) emission at the indicated velocities (A-E) for OH127.8-0.0. Tic marks on the maps are separated by 5" in right ascension (abscissa) and declination (ordinate) and contour levels range from 5 to 90% of the peak flux. The cross on each map indicates the inferred position of the star. The data are taken from BJS.

b) As predicted by the ESS model, the emission at intermediate velocities (B,C,D) arises from the outer portion of the maser shell and is centered at the stellar position derived from Features A and E. The radius of the emission region increases as the velocity approaches the stellar velocity V_o, and an underlying spherical geometry is indicated by the low-level circular contours in these three maps. Some additional examples for which a well-resolved, approximately circular ring structure is seen near the stellar velocity include OH26.5+0.6 (Baud 1981; BJS; Herman 1983), OH32.8-0.3 (BJS; Herman 1983), and OH30.1-0.7 (Herman 1983).

c) For these and many other OH/IR stars, a phase lag between the light curves of the low and high velocity emission features is detected (Schultz et al. 1978; Jewell et al. 1980; Morris and Bowers 1980; Herman 1983; Bowers and Morris 1984). Because the OH emission varies in response to the varying light cycle of the central star, the sense of the phase difference is further proof that the dominant kinematic mode is expansion. Additionally, the magnitude of the phase lag, which depends on the light travel time across the maser region, directly yields the linear size (for a spherical geometry) and confirms that the OH is located at radii which range from 500 to 10^4 AU.

In summary, these sources provide good evidence for a radially expanding, predominantly spherical mass distribution in the outer envelope. However, they also indicate that significant deviations between the observed and predicted maser distributions can occur. Some examples are: clumping (e.g., Figure 1), radial asymmetries in the plane of the sky or along the line of sight for virtually all OH/IR stars, significant positional separation (> 0.″1) of peak emission features (e.g., OH26.5+0.6), asymmetric, extended structure of peak features (e.g., OH32.8-0.3), and profile broadening due to emission outside the peak features, generally $\lesssim$ 2 km s^{-1} but sometimes ranging up to 7 km s^{-1} (BJS; Diamond et al. 1984b).

Such deviations may be related to the mass loss mechanism or to interaction of the circumstellar gas with the ambient interstellar medium or radiation field; it sometimes can be difficult to determine the cause of the deviations. For example, small positional offsets of the peak features and profile broadening can be produced in a tilted, expanding disk, or in an ESS model with either random velocity fluctuations or a rotational component superimposed on radial expansion. It is not possible to distinguish between these models if the maser emission is unresolved.

Features B and D in Figure 1 provide another example. These features have brightness distributions at 1612 MHz which are notably asymmetric relative to the star. Interestingly, their similar asymmetric shapes, approximately equidistant from the stellar velocity, indicate a front-back symmetry. This might reflect a) fluctuations in the mass density distribution (e.g., a ring-like segment of higher density in the eastern portion of the shell) which are caused by the mass loss process or by interaction of the circumstellar and interstellar gas, or b) fluctuations in the OH or velocity distributions caused, for example, by an asymmetry in the interstellar UV radiation field (cf. Huggins and Glassgold 1982; Deguchi 1982). Diamond et al. (1984b) indicate that the OH distribution at 1667 MHz is complementary (i.e., strongest in the western portion of the shell), further illustrating that the physical conditions responsible for the maser emission at each frequency vary throughout the envelope. More extreme cases in which the maser emission is asymmetric relative to the star at many velocities are IRC+10420 (Bowers 1984) and NML Cyg (Diamond et al. 1984a; Spencer et al. 1984).

Because the aforementioned deviations are present in the sources for which a circular shell structure is detected near the stellar velocity, BJS propose that the outer envelopes of most OH/IR stars can be described by a modified ESS (MESS?) model in which there are OH (and possibly mass) density fluctuations ($\sim 10^{15}$ to 10^{16} cm) and velocity fluctuations due to subsonic turbulence and small streaming motions (usually $\lesssim 2$ km s^{-1}). For many stars it is not possible to determine the envelope structure unambiguously because of an extremely clumpy or complex emission distribution, inadequate spatial resolution, or a poor signal-to-noise ratio. However, with the exception of VY CMa, the positions of the extreme (and often the peak) low and high velocity features of all double-peaked OH/IR stars thus far mapped are coincident to within 10% of the angular radius of the maser region and usually to within 0.″1, consistent with a modified ESS model.

The apparent validity of the ESS model at radii $\gtrsim$ 1000 AU suggests that the mass loss mechanism for most oxygen-rich red giants does not produce large deviations from spherical outflow when averaged over the expansion timescale ($\sim$ 1000 y). This conclusion supports a time-averaged isotropic mass loss mechanism. It does not eliminate the possibility of sporadic (or episodic) ejection which distributes material in different directions at different times (Bowers 1984) or of steady mass loss confined to a moderate (but not small) opening angle.

More specific constraints on the mass loss mechanism are difficult to obtain because of the inherent difficulty in relating the observed distribution of maser emission to the envelope structure. For the Mira-type variables some interesting possibilities which require further investigation include a time-varying mass loss rate for OH26.5+0.6 (Baud 1981, but see BJS and Diamond et al. 1984b), a small rotational component in the velocity field for U Ori (Chapman and Cohen 1984) and OH104.9+2.4 (Norris et al. 1982, but see Herman 1983), and significant radial velocity gradients (perhaps due to continued acceleration of the gas) for various stars (BJS; Bowers 1984; Chapman and Cohen 1984). Another interesting case is the bipolar nebula OH231.8+4.2, the central star of which is identified as an M9III Mira variable (Cohen 1981; Bowers and Morris 1984). The OH (at 1667 MHz) from this object clearly displays properties of both an expanding sphere and disk (Morris et al. 1982). There are several reasons to suspect that this is a close binary system (Morris 1981; Cohen 1984), in which case the overall geometry of mass loss may change with time (BJS).

It is also noteworthy that the structures of the OH (1612) maser regions for the three well-known supergiant stars VY CMa, NML Cyg, and IRC+10420 are very complex and apparently very different from each other. Models include: an expanding thick disk or ellipsoid geometry for VY CMa (BJS), time-varying mass loss in which the inner part of the maser region is expanding faster than the outer part for NML Cyg (Diamond *et al.* 1984a), and a bipolar nebula (Diamond *et al.* 1983) or a clumpy sphere model (Bowers 1984) for IRC+10420.

Deviations of the envelope structure from the simple ESS model introduce an intrinsic uncertainty in estimates of the various stellar and circumstellar properties listed in Table 1 which can be determined from OH (1612) profiles and maps. The best method to measure each parameter (in the opinion of this author), an estimate of the smallest uncertainty currently obtainable with this method for "proto-type" ESS sources like OH127.8-0.0, and references for further discussion are included in Table 1.

The uncertainties in Table 1 apply only to optimum cases with small deviations from the ESS model. Uncertainties may be significantly larger, depending on peculiarities associated with individual sources and on the quality of the data. For example, the line-of-sight stellar velocities and circumstellar expansion velocities derived for the complex OH sources VY CMa, NML Cyg, and IRC+10420 may have uncertainties of 2 or 3 km s^{-1} (Table 2 of BJS; Diamond *et al.* 1984a; Bowers 1984). Similarly, the position, stellar velocity, expansion velocity, and radius of the maser region derived for U Ori from the OH (1612) emission cannot be considered to be reliable because of the anomalous time-dependent behavior (Jewell *et al.* 1981; BJS; Chapman and Cohen 1984).

Another example is OH26.5+0.6, for which maser maps have been obtained by a number of authors. Like most OH/IR stars, this source exhibits evidence for both transverse and front-back radial asymmetries of the maser region. These influence the uncertainties for the linear (phase-lag) radius (since the value derived from this technique assumes a symmetric distribution of the maser emission relative to the star) and for the angular radius. In the latter case the average angular radius is estimated to be 2".2 (Herman 1983), 2".8 (Diamond *et al.* 1984b), 3".5 (BJS), and 4" (Baud 1981). Obviously, significantly different values are possible, depending on the spatial and spectral sensitivity used to obtain the data and on the approach used to measure the angular radius. Angular radii derived by measuring the positions of knots of emission (Herman 1983; Diamond *et al.* 1984b) are systematically smaller than those derived by accounting for low-level extended emission (Baud 1981; BJS). With the linear radius derived by Herman (1983), corresponding distance estimates range from 980 to 550 pc, yielding an uncertainty of about 30% relative to the mean value (Table 1). The uncertainty ultimately may be improved by a better understanding of the structure of the envelope and of how to relate the angular radius derived from emission

Table 1. Parameters Obtainable for OH/IR Stars with Double-Peaked 1612 MHz Emission

Parameter	Range of Values	Suggested Method of Measurement	Optimum Uncertainty	References
		A. Stellar		
Position	--	Mean position at extreme velocities (position of low-velocity peak in some cases)	0".1	a,b,c,d,e
Radial velocity	--	Mean velocity of emission profile or of outer peaks	0.1 km s^{-1}	a,f,g,h
Period	$250-2000^d$	Amplitude variations of peak features	1%	i,j,k
Distance	--	Combined linear and angular sizes of maser region	30%	a,b,k,l
		B. Circumstellar		
Expansion velocity of maser region	2-40 km s^{-1}	Half full-width of emission profile	0.5 km s^{-1}	a,b,g,h,k
Linear radius of maser region	$\sim 500-10^4$ AU	Phase lag between low and high velocity peak features	10%	j,k,m
Angular radius of maser region	$\lesssim$ 4"	Aperture synthesis maps as a function of velocity	30%	a,b,k,l

Reference key: a)BJS; b) Diamond et al. 1984b; c) Norris et al. 1984; d) Bowers and de Jong 1983; e) Fix and Mutel 1984; f) Diamond et al. 1984a; g) Chapman and Cohen 1984; h) Bowers 1984; i) Harvey et al. 1974; j) Jewell et al. 1979; k) Herman 1983; l) Baud 1981; m) Jewell et al. 1980.

in the plane of the sky to the linear radius derived from emission along the line of sight to the star. However, for stars where emission is detected only at velocities near the peak features, errors for the estimated angular radii (and distances) may easily approach 50 to 100% (BJS).

2.2. Inner Envelope Structure (H_2O, SiO)

H_2O and SiO masers provide potentially powerful probes of the mass loss mechanism because they are produced closer to the star than is OH, but they are as yet of limited usefulness for understanding the inner envelope structure. Unlike OH (1612), there is no characteristic H_2O or SiO velocity signature. H_2O masers may, in fact, have four classes of profile structure, depending on the expansion velocity (Olnon et al. 1980). Periodic and secular variations with time are observed for both H_2O (Schwartz et al. 1974; Cox and Parker 1978; Johnston et al. 1984) and SiO (Spencer and Schwartz 1975; Hjalmarson and Olofsson 1979; Lane 1982; Wolff and Carlson 1982; Clark et al. 1984a,b), and the emission from both species can be significantly linearly polarized (Knowles and Batchelor 1978; Troland et al. 1979; Clark et al. 1984a,b), so it can be difficult to compare data obtained with different telescopes at different times.

Despite these complications, single-dish H_2O and SiO profiles for some sources (most notably VY CMa) can exhibit striking, exceedingly complex symmetric velocity structure, as might be caused by edge-on rotating rings (Knowles and Batchelor 1976; Van Blerkom 1978; Zhou Zhen-pu and Kaifu 1984). Unfortunately, the interpretation is not unique. Interferometric data do not support this or any other simple model. These data generally are interpreted to be consistent with an expanding shell model (Rosen et al. 1978; Spencer et al. 1979; Moran et al. 1979; Lada et al. 1981; Lane 1982; Johnston et al. 1984), but the agreement is qualitative at best and most authors are quick to point out the possible influence of turbulent or radially accelerative motions which are likely to be present in the inner envelope region. Interferometric observations closely spaced in time are needed to overcome the severe problems introduced by the complex temporal behavior of these masers.

3. MASS LOSS RATE

Maser emission requires a suitable combination of various physical conditions, not all of which are related to the mass loss rate. If, however, such emission is detected, it can, in principle, be used to derive a mass loss rate. The major uncertainty in such a derivation is the estimate of the mass density in the maser region. A value for the density can be derived from interferometric measurements of the brightness temperatures, observed sizes, and angular distribution of maser spots combined with a model of radiative transfer and knowledge of the pumping requirements. Such estimates are sensitive to clumping in the maser region and to the assumed molecular abundance which can vary not only throughout the maser region but also from star to star because of differences in the ambient interstellar environment. Mass loss rates have not been commonly estimated by this method, partly because of the many uncertainties in the derivation of the density and partly because of a paucity of the required interferometric data.

An alternative (empirical) approach is to determine what, if any, connection exists between maser emission and the mass loss rate by considering the sample of maser stars whose mass loss rates are independently derived from infrared or CO data. This approach has the advantage that it is not dependent on understanding the physics of the masers, and can, in fact, be used to provide constraints on the pump mechanism. Moreover, the ubiquity of known maser stars provides that any such connection can be a powerful tool to examine many individual cases.

Several methods based on infrared photometry have been used to estimate $\dot{M}$ for a considerable number of stars; these include a) measuring the depth of the 10 micron silicate feature (Gehrz and Woolf 1971), b) model-fitting to the infrared excess (Hyland et al. 1972 and BJS), and c) estimating the bolometric luminosity and assuming the momentum in the envelope is obtained from the stellar radiation (Werner et al. 1980; Engels et al. 1983; de Jong 1983). There are many sources of error intrinsic to each method, among the more significant being the adopted distance to the star and time variability of the infrared spectrum. However, it is encouraging that the "infrared" mass loss rates from the above references (normalized for distance) agree to within about a factor of 10 and often to within a factor of 3 with values of $\dot{M}$ derived from CO data (Knapp et al. 1982; Jura 1983) for the few ($\sim$ 10) stars where both determinations are available.

3.1 Radius of the Maser Region

Correlations of certain maser properties with the IR/CO mass loss rates support the possibility that the scale of mass loss is correct, in a statistical sense, although there may be order-of-magnitude errors for individual stars. An example is shown in Figure 2, where the radius R of the maser region for the ground-state rotational transitions of OH (1612 MHz), H_2O (22 GHz) and the vibrationally excited rotational (v=1, J=1-0) transition of SiO (43 GHz) is plotted as a function of the IR or CO mass loss rate taken from the references in the preceding paragraph. For OH an attempt has been made to minimize scatter along both axes by preferentially selecting sources (filled circles) for which estimates of both phase lag radii (Herman 1983) and angular radii (Baud 1981; BJS; Herman 1983; Diamond et al. 1984b) are available, so that an average "phase-lag distance" can be obtained and the mass loss rate can be scaled accordingly. The few data points available for the H_2O and SiO masers clearly illustrate the need for further interferometric observations of these masers.

The figure shows two interesting results. First, it is immediately

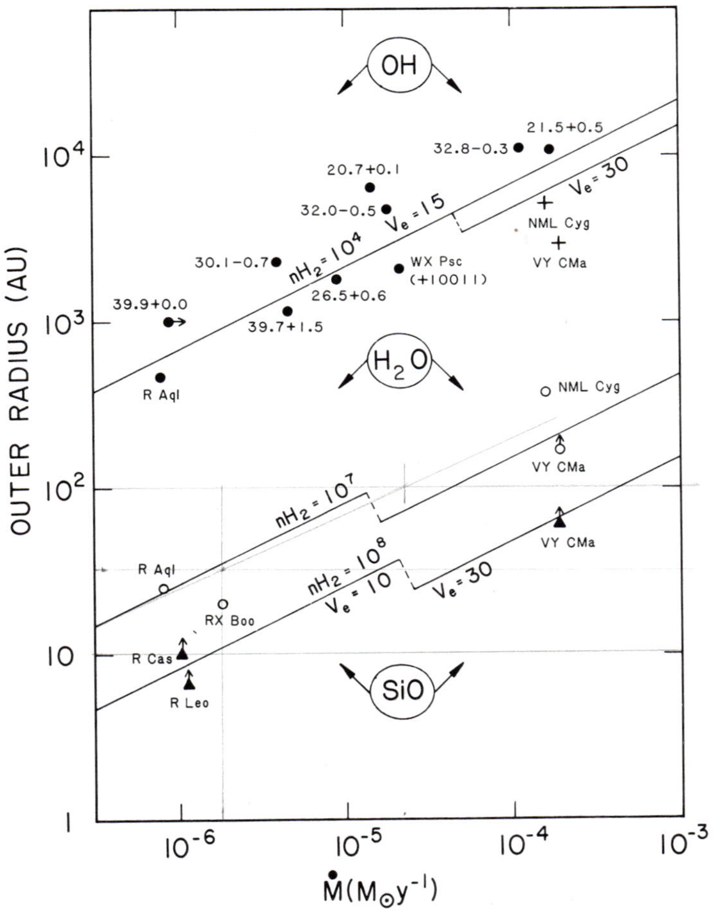

Figure 2. Radius of the maser region as a function of the IR or CO mass loss rate for OH/H_2O/SiO maser stars. The figure includes all H_2O (empty circles) and (v=1, J=1-0) SiO masers (triangles) for which there is an estimate of the angular radius of the maser region (Johnston *et al.* 1984, Rosen *et al.* 1978; Lane 1982; 1984), the distance, and the mass loss rate (cf. Bowers and Hagen 1984). Filled circles denote OH (1612) sources for which a phase-lag distance is available (see text). Linear OH radii for the supergiants NML Cyg and VY CMa (crosses) are derived using the angular sizes given by BJS and Diamond *et al.* (1984a) and the distances given by Bowers and Hagen (1984). The IR/CO mass loss rate for each star is scaled to the adopted distance with no additional correction for interstellar obscuration. For the indicated expansion velocity V_e (km s^{-1}), diagonal lines show the number density (cm^{-3}) of molecular hydrogen in the case where it is assumed to be the primary gaseous component of the envelope. Vertical arrows indicate VLBI measurements which are only lower limits if the emission is resolved. The horizontal arrow for OH39.9+0.0 indicates that the OH region for this source may be unresolved.

evident that the maser emission for each molecule arises from a different part of the envelope, as expected based on different excitation energies required for each transition. The characteristic density $nH_2 \simeq 10^4$ cm^{-3} found for the OH maser regions is comparable to that expected in the pumping model of Elitzur et al. (1976), supporting the absolute scale of mass loss. Note, however, that the SiO occurs much farther from the star than predicted by Elitzur (1981); this may indicate clumping in the maser region or a higher temperature than is commonly assumed for the inner envelope region.

Second, the radius R of the maser region tends to increase with the mass loss rate for each of these species. The R-$\dot{M}$ relation for OH masers previously has been discussed by Bowers et al. (1981; 1983), Deguchi (1982), and Huggins and Glassgold (1982). There is considerable scatter for the OH; the few points plotted for H_2O and SiO masers show a similar relation but additional data obviously are necessary. Nevertheless, Figure 2 suggests that the radius of the maser region for any of these species may provide an estimate of $\dot{M}$ to within about an order of magnitude because of the tendency for each maser to be produced in a region whose density is relatively constant from star to star, if allowance is made for the different expansion velocities of the envelopes.

Some of the scatter may be due to errors in the quantities plotted, but much of it also may be caused by intrinsic differences in the physical properties of the envelopes. For example, the OH radius for the supergiant VY CMa is smaller than might be expected from the high mass loss rate derived from infrared data (cf. BJS), but this star has a complex envelope structure and is subject to a larger than average ambient UV radiation field capable of influencing the OH radius (see Lada and Reid 1978; Huggins and Glassgold 1982; BJS). Individual, star-to-star peculiarities therefore may limit the accuracy of the R-$\dot{M}$ relation as a mass loss indicator.

3.2 Maser Luminosity

It has long been known that the OH luminosity and the infrared excess are correlated (e.g., Wilson et al. 1972; Nguyen-Q-Rieu et al. 1979; Bowers 1981; Engels et al. 1983), but it is only recently that quantitative relationships between the maser luminosity and the mass loss rate have been established for OH (Baud and Habing 1983; Bowers and Hagen 1984) and H_2O (Bowers and Hagen 1984). The dependences are shown in Figure 3. The broken line for OH indicates that strongest emission usually occurs at 1612 MHz for $\dot{M} \gtrsim 3 \times 10^{-6}$ $M_\odot$ y^{-1} and at 1665 or 1667 MHz for lower mass loss rates, although the bipolar nebula OH231.8+4.2 is a prominent exception (Bowers and Morris 1984). The filled circles represent data for the (v=1, J=1-0) SiO maser, and the dashed line indicates a median value (biased toward maximum light) of about 3.6×10^{43} s^{-1} (see figure caption). This is not particularly the strongest SiO maser (e.g. Schwartz et al. 1982), but data for it are available for a sample of stars with a reasonably wide range of mass loss rates. The

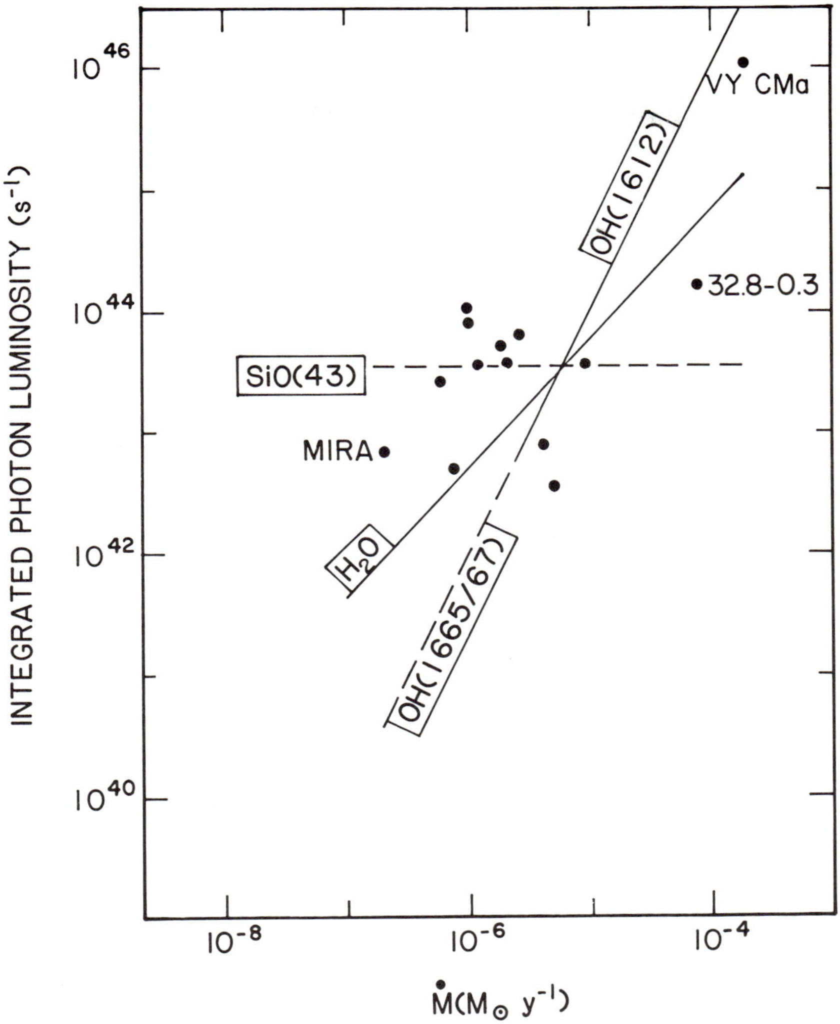

Figure 3. Integrated photon luminosity as a function of the IR/CO mass loss rate for OH, H_2O, and SiO masers. For H_2O and OH masers the indicated lines are plotted using equations 4 and 7, respectively, from Bowers and Hagen (1984); these lines represent least squares fits to the data discussed in that paper and refer to the integrated maser luminosity at maximum light. The filled circles represent SiO (v=1, J=1-0) luminosities for sources whose distances and mass loss rates are given in Table 4 of Bowers and Hagen and whose SiO flux densities are given by Cahn (1981), Lane (1982), or Jewell et al. (1984). For most cases several flux determinations are available, and the largest SiO luminosity is plotted. The dashed line represents the median of the SiO values for all the sources except the supergiant VY CMa and indicates no strong dependence of the SiO luminosity on the mass loss rate for the Mira-type stars.

luminosity for this transition appears to depend more strongly on the bolometric luminosity of the star (Cahn 1981) than on the mass loss rate (Figure 3; Jewell 1984), which may explain the position of the supergiant VY CMa in the upper right-hand portion of the diagram. Further efforts to determine how this and other SiO emission lines are related to the mass loss rate and stellar luminosity could provide valuable constraints on the pump mechanism.

The different dependences illustrated in Figure 3 suggest a method to use the flux ratio of maser lines (corrected for time variability) to obtain a distance-independent estimate of the mass loss rate. From consideration of this technique for OH and H_2O masers, Bowers and Hagen (1984) conclude that it can be used to provide order-of-magnitude estimates of the mass loss rate. Like the R-$\dot{M}$ relations in Figure 2, it is limited by a) its calibration which is subject to uncertainties of the presently available IR/CO mass loss rates and b) different star-to-star physical conditions which, for a given mass loss rate, may influence the relative line strengths.

The different slopes of the lines in Figure 3 also summarize the importance of $\dot{M}$ in determining the detectability of the various maser species. For example, nearby ($\lesssim$ 1 kpc), optically identifiable Mira variables usually have $\dot{M} \lesssim 5 \times 10^{-6}$ $M_{\odot}$ y^{-1} (see Table 4 of Bowers and Hagen 1984). Figure 3 indicates that the SiO flux for such stars should typically exceed the H_2O flux by a factor of 10 and the OH flux by a factor of 100. Indeed, SiO surveys of nearby stars yield high detection rates, and Cahn (1977) has suggested that perhaps all Miras with sufficiently late spectral types (> M5) may have SiO maser emission. H_2O emission from these stars also is commonly detected. Bowers and Hagen (1984) find that at least 60% of 30 M-type Mira variables known to be within 400 pc of the Sun have H_2O emission; the average maximum luminosity is 14.4×10^{42} s^{-1}, corresponding to an average mass loss rate of about 2.5×10^{-6} $M_{\odot}$ y^{-1}.

In comparison, the percentage of Mira variables which are OH maser stars is relatively small. Only ten Mira variables within about 400 pc of the Sun are known OH masers [compare Table 3 of Bowers and Hagen (1984) to the list of Engels (1979)], suggesting that about 30% of M-type Mira variables may have significant (> 10^{41} s^{-1}) OH emission. This result is consistent with other estimates (Bowers and Kerr 1977; Herman 1983). However, further efforts are needed to determine if this percentage is due to a minimum column density required for the OH maser or to sensitivity-limited OH surveys because of the low OH luminosities (and mass loss rates) of the nearby stars. As indicated in Section 1, the percentage of Mira-type variables which are radio-luminous OH/IR stars is very small ($\sim$ 1%). Their luminous OH emission serves as a tracer of stars with mass loss rates $\gtrsim 10^{-5}$ $M_{\odot}$ y^{-1}.

These results can be used to update previous estimates of the amount of material returned to the interstellar medium by long-period variables (e.g., Gehrz and Woolf 1971; Bowers and Kerr 1977; Cahn and

Wyatt 1978). Wyatt and Cahn (1983) estimate a local surface density of 1.6×10^{-4} pc^{-2} for Mira variables, most of which are likely to be spectral class M (cf. Blanco and McCarthy 1981). If at least 60% of these stars are undergoing moderate mass loss ($\simeq 2.5 \times 10^{-6}$ $M_\odot$ y^{-1}) and $\sim$ 1% are undergoing high mass loss ($\simeq 3 \times 10^{-5}$ $M_\odot$ y^{-1}), then the total contribution is at least 2.9×10^{-10} $M_\odot$ y^{-1} pc^{-2}, or approximately 35% of all material returned to the interstellar medium (cf. Alloin et al. 1976). Because of their small number density, the stars with the highest mass loss rates apparently make only a minimal contribution ($\lesssim$ 10%) to the recycled material.

4. CONCLUSION

Of the commonly detected circumstellar masers, the 1612-MHz transition of OH is currently the most useful. Indeed, because it is a sensitive tracer of stars with $\dot{M} \gtrsim 10^{-5}$ $M_\odot$ y^{-1}, it has been an extremely important tool for the discovery of stars with high mass loss rates and thus for the recognition of the significance of mass loss in red giant evolution. This emission can provide relatively accurate estimates of the position, systemic radial velocity, period, and distance of the star, as well as the morphology, radius, and velocity structure of the maser region at radii ranging from 500 to 10^4 AU. These data yield valuable information about the mass loss mechanism and the interaction of the circumstellar material with the ambient interstellar environment.

It is much more difficult to relate observations of H_2O and SiO masers to basic parameters of the star and envelope because of complexities introduced by time variability and polarization of the emission. It appears that these masers can serve as a tool to examine the mass loss rate, but their usefulness to study the envelope structure is presently limited. Further efforts are needed to determine the properties of these masers as a function of time and over a wide range of mass loss. Because they are produced relatively close to the star at radii between about 10 and 100 AU, H_2O and SiO ultimately may provide the most powerful maser probes of the mass loss mechanism.

REFERENCES

Allen, D. A., Hyland, A. R., and Caswell, J. L. 1980, M.N.R.A.S., 192, 505.
Alloin, D., Cruz-Gonzáles, C. and Peimbert, M. 1976, Ap. J., 205, 74.
Baud, B. 1981, Ap. J. (Letters), 250, L79.
Baud, B. and Habing, J. H. 1983, Astr. Ap., 127, 73.
Baud, B., Habing, H. J., Matthews, H. E., Winnberg, A. 1979a, Astr. Ap. Suppl., 35, 179.
___________. 1979b, Astr. Ap. Suppl., 36, 193.
___________. 1981a, Astr. Ap., 95, 156.
___________. 1981b, Astr. Ap., 95, 171.
Benson, J. M., Mutel, R. L., Fix, J. D., and Claussen, M. J. 1979, Ap. J. (Letters), 229, L87.
Blanco, V. M. and McCarthy, M. F. 1981, in Physical Processes in Red Giants, eds: I. Iben, Jr. and A. Renzini (Dordrecht:Reidel), p. 147.
Booth, R. S., Kus, A. J., Norris, R. P., and Porter, N. D. 1981, Nature, 290, 382.
Bowers, P. F. 1978a, Astr. Ap. Suppl., 31, 127.
___________. 1978b, Astr. Ap., 64, 307.
___________. 1981, A. J., 86, 1930.
___________. 1984, Ap. J., 279, 350.
Bowers, P. F. and de Jong, T. 1983, A. J., 88, 655.
Bowers, P. F. and Hagen, W. 1984, Ap. J., 285, 637.
Bowers, P. F., Johnston, K. J., and Spencer, J. H. 1981, Nature, 291, 382.
___________. 1983, Ap. J., 274, 733 (BJS).
Bowers, P. F. and Kerr, F. J. 1977, Astr. Ap., 57, 115.
___________. 1978, A. J., 83, 487.
Bowers, P. F. and Morris, M. 1984, Ap. J., 276, 646.
Bowers, P. F., Reid, M. J., Johnston, K. J., Spencer, J. H., and Moran, J. M. 1980, Ap. J., 242, 1088 and Ap. J., 248, 880 (Erratum).
Cahn, J. H. 1977, Ap. J. (Letters), 212, L135.
___________. 1981, A. J., 86, 1935.
Cahn, J. H. and Wyatt, S. P. 1978, Ap. J., 221, 163.
Caswell, J. L. 1974, in Galactic Radio Astronomy, IAU Symp. No. 60, eds: F. J. Kerr and S. C. Simonson, (Dordrecht:Reidel), p. 423.
Caswell, J. L. and Haynes, R. F. 1975, M.N.R.A.S., 173, 649.
Caswell, J. L, Haynes, R. F., Goss, W. M., and Mebold, U. 1981, Austr. J. Phys., 34, 333.
Chapman, J. M. and Cohen, R. J. 1984, M.N.R.A.S., 212, 375.
Chapman, J., Cohen, R. J., Norris, R. P., Diamond, P. J., and Booth, R. S. 1984, M.N.R.A.S., 207, 149.
Cimmerman, M. 1979, Ap. J. (Letters), 228, L79.
Clark, F. O., Troland, T. H. Pepper, G. H., and Johnson, D. R. 1984a, Ap. J., 276, 572.
___________. 1984b, Ap. J., 283, 174.
Claussen, M. J. and Fix, J. D. 1982, Ap. J., 263, 153.
Cohen, M. 1981, P.A.S.P., 93, 288.
___________. 1985, These proceedings, 291.

Cox, G. G. and Parker, E. A. 1978, M.N.R.A.S., 183, 111.
Crocker, D. A. and Hagen, W. 1983, Astr. Ap. Suppl., 54, 405.
Davis, L. E., Seaquist, E. R., and Purton, C. R. 1979, Ap. J., 230, 434.
Deguchi, S. 1982, Ap. J., 259, 634.
de Jong, T. 1983, Ap. J., 274, 252.
Diamond, P. J., Norris, R. P., and Booth, R. S. 1983, Astr. Ap. Letters, 124, L4.
__________. 1984a, M.N.R.A.S., 207, 611.
Diamond, P. J., Norris, R. P., Rowland, P. R., Booth, R. S., and Nyman L-A. 1984b, M.N.R.A.S., in press.
Dickinson, D. F. and Dinger, A. S. 1982, Ap. J., 254, 136.
dos Santos, P. Lépine, J. R. D., and Balboa, A. M. 1979, A. J., 84, 787.
Downes, D. and Genzel, R. 1980, in Interstellar Molecules, IAU Symp. No. 87, ed: B. H. Andrew (Dordrecht:Reidel), p. 565.
Elitzur, M. 1981, in Physical Processes in Red Giants, eds. I. Iben, Jr. and A. Renzini (Dordrecht:Reidel), p. 363.
__________. 1982, Rev. Mod. Phys., 54, 1225.
Elitzur, M., Goldreich, P., and Scoville, N. 1976, Ap. J., 205, 384.
Engels, D. 1979, Astr. Ap. Suppl., 36, 337.
Engels, D., Kreysa, E., Schultz, G. V., and Sherwood, W. A. 1983, Astr. Ap., 124, 123.
Fillit, R., Proust, D., and Lépine, J. R. D. 1977, Astr. Ap., 58, 281.
Fix, J. D. and Claussen, M. J. 1984, Ap. J. (Letters), 287, L35.
Fix, J. D. and Mutel, R. L. 1984, A. J., 89, 406.
Gehrz, R. D. and Woolf, N. J. 1971, Ap. J., 165, 285.
Goldreich, P. 1980, in Interstellar Molecules, IAU Symp. No. 87, ed: B. H. Andrew (Dordrecht:Reidel), p. 551.
Goldreich, P. and Scoville, N. 1976, Ap. J., 205, 144.
Habing, H. J., Olnon, F. M., Winnberg, A., Matthews, H. E., and Baud, B. 1983, Astr. Ap., 128, 230.
Harvey, P. M.,Bechis, K. P., Wilson, W.J., and Ball, J. A. 1974, Ap. J. Suppl., 27, 331.
Herman, J. 1983, Ph.D. Thesis, University of Leiden.
Hjalmarson, A. and Olofsson, H. 1979, Ap. J. (Letters), 234, L199.
Huggins, P. J. and Glassgold, A. E. 1982, A. J., 87, 1828.
Hyland, A. 1974, in Galactic Radio Astronomy, IAU Symp. No 60, eds: F. J. Kerr and S. C. Simonson, (Dordrecht:Reidel), p. 439.
Hyland, A. R., Becklin, E. E., Frogel, J. A., and Neugebauer, G. 1972, Astr. Ap., 16, 204.
Iben, I. and Renzini, A. 1983, Ann. Rev. Astr. Ap., 21, 271.
Jewell, P. R. 1985, These proceedings.
Jewell, P. R., Batria, W., Walmsley, C. M., and Wilson, T. L. 1984, Astr. Ap. Letters, 130, L1.
Jewell, P. R., Elitzur, M., Webber, J. C., and Snyder, L. E. 1979, Ap. J. Suppl., 41, 191.
Jewell, P. R., Webber, J. C., and Snyder, L. E. 1980, Ap. J. (Letters), 242, L29.
__________. 1981, Ap. J., 249, 118.
Johansson, L. E. B., Andersson, C., Goss, W. M., and Winnberg, A. 1977a, Astr. Ap. Suppl., 28, 199.
__________. 1977b, Astr. Ap., 54, 323.

Johnston, K. J., Spencer, J. H., and Bowers, P. F. 1984, Ap. J., in press.
Jones, T. J., Hyland, A. R., and Gatley, I. 1983a, Ap. J., 273, 660.
Jones, T. J., Hyland, A. R., Wood, P. R., and Gatley, I. 1983b, Ap. J., 273, 669.
Jura, M. 1983, Ap. J., 275, 683.
Kerr, F. J. and Bowers, P. F. 1974, Astr. Ap., 36, 225.
Knapp, G. R., Phillips, T. G., Leighton, R. B., Lo, K. Y., Wannier, P. G., and Wootten, H. A. 1982, Ap. J., 252, 616.
Knowles, S. H. and Batchelor, R. A. 1976, M.N.R.A.S., 174, 69P.
__________. 1978, M.N.R.A.S., 184, 107.
Knowles, S. H., Mayer, C. H., Cheung, A. C., Rank, D. M., and Townes, C. H. 1969, Science, 163, 1055.
Kwok, S. 1975, Ap. J., 198, 583.
__________. 1983, in Planetary Nebulae, IAU Symp. 103, ed. W. D. Flower (Dordrecht:Reidel), p. 293.
Lada, C. J., Blitz, L., Reid, M. J., and Moran, J. M. 1981, Ap. J., 243, 769.
Lada, C. J. and Reid, M. J. 1978, Ap. J., 219, 95.
Lane, A. P. 1982, Ph.D. Thesis, University of Massachusetts.
__________. 1984, in VLBI and Compact Radio Sources, IAU Symp. 110, eds. R. Fanti et al. (Dordrecht:Reidel), p. 329.
Moran, J. M., Ball, J. A., Predmore, C. R., Lane, A. P., Huguenin, G. R., Reid, M. J., and Hansen, S. S. 1979, Ap. J. (Letters), 231, L67.
Morris, M. 1981, Ap. J., 249, 572.
Morris, M. and Bowers, P. F. 1980, A. J., 85, 724.
Morris, M., Bowers, P. F., and Turner, B. E. 1982, Ap. J., 259, 625.
Nguyen-Q-Rieu, Laury-Micoulaut, C., Winnberg, A., and Schultz, G. V. 1979, Astr. Ap., 75, 351.
Norris, R. P., Booth, R. S., Diamond, P. J., Nyman, L-A., Graham, D. A., and Matveyenko, L. I. 1984, M.N.R.A.S., 208, 435.
Norris, R. P., Diamond, P. J., and Booth, R. S. 1982, Nature, 299, 131.
Olnon, F. M., Walterbos, R. A. M., Habing, H. J., Matthews, H. E., Winnberg, A., Brzezinska, H., and Baud, B. 1981, Ap. J. (Letters), 245, L103.
Olnon, F. M., Winnberg, A., Matthews, H. E., and Schultz, G. V. 1980, Astr. Ap. Suppl., 42, 119.
Reid, M. J. and Moran, J. M. 1981, Ann. Rev. Astr. Ap., 19, 231.
Reid, M. J., Moran, J. M., and Johnston, K. J. 1981, A. J., 86, 897.
Reid, M. J., Moran, J. M., Leach, R. W., Ball, J. A., Johnston, K. J., Spencer, J. H., and Swenson, G. W. 1979, Ap. J. (Letters), 227, L89.
Reid, M. J., Muhleman, D. O., Moran, J. M., Johnston, K. J., and Schwartz, P. R. 1977, Ap. J., 214, 60.
Rosen, B. R., Moran, J. M., Reid, M. J., Walker, R. C., Burke, B. F., Johnston, K. J., and Spencer, J. H. 1978, Ap. J., 222, 132.
Schultz, G. V., Sherwood, W. A., and Winnberg, A. 1978, Astr. Ap. Letters, 63, L5.
Schwartz, P. R. , Harvey, P. M., and Barrett, A. H., 1974, Ap. J., 187, 491.

Schwartz, P. R., Zuckerman, B., and Bologna, J. M. 1982, Ap. J. (Letters), **256**, L55.
Seaquist, E. R. and Davis, L. E. 1983, Ap. J., **274**, 659.
Silverglate, P., Zuckerman, B., Terzian, Y., and Wolff, M. 1979, A. J., **84**, 345.
Snyder, L. E. 1980, in Interstellar Molecules, IAU Symp. No. 87, ed: B. H. Andrew (Dordrecht:Reidel), p. 525.
Snyder, L. E. and Buhl, D. 1975, Ap. J., **197**, 329.
Spencer, J. H., Bowers, P. F., and Johnston, K. J. 1984, Preprint.
Spencer, J. H., Johnston, K. J., Moran, J. M., Reid, M. J., and Walker, R. C. 1979, Ap. J., **230**, 449.
Spencer, J. H. and Schwartz, P. R. 1975, Ap. J. (Letters), **199**, L111.
Spencer, J. H., Winnberg, A., Olnon, F. M., Schwartz, P. R., Matthews, H. E., and Downes, D. 1981, A. J., **86**, 392.
Troland, T. H., Heiles, C. E., Johnson, D. R., and Clark, F. O. 1979, Ap. J., **232**, 143.
Ukita, N. and Goldsmith, P. F. 1984, Astr. Ap., **138**, 194.
Ukita, N. and LeSqueren, A. M. 1984, Astr. Ap., **138**, 343.
Van Blerkom, D. 1978, Ap. J., **223**, 835.
Werner, M. W., Beckwith, S., Gatley, I., Sellgren, K., Berriman, G., and Whiting, D. L. 1980, Ap. J., **239**, 540.
Wilson, W. J. and Barrett, A. H. 1968, Science, **161**, 778.
Wilson, W. J., Barrett, A. H., and Moran, J. M. 1970, Ap. J., **160**, 545.
Wilson, W. J., Schwartz, P. R., Neugebauer, G., Harvey, P. M., and Becklin, E. E. 1972, Ap. J., **177**, 523.
Winnberg, A., Goss, W. M., Hőgland, B., and Johansson, L. E. B., 1973, Ap. Letters, **13**, 125.
Wolff, R. S. and Carlson, E. R. 1982, Ap. J., **257**, 161.
Wyatt, S. P. and Cahn, J. H. 1983, Ap. J., **275**, 225.
Zhou Zhen-pu and Kaifu, N. 1984, Astr. Ap., **138**, 359.
Zuckerman, B. 1979, Ap. J., **230**, 442.
____________. 1980, Ann. Rev. Astr. Ap., **18**, 263.

A SEARCH FOR OH AND H_2O MASER EMISSION FROM UNIDENTIFIED IRAS SOURCES

D. Engels
Astron.Institute, SFB 131, Auf dem Hügel 71, D-5300 Bonn 1
H.J. Habing, F.M. Olnon
Sterrewacht, Postbus 9513, NL-2300 RA Leiden
J. Schmid-Burgk, C.M. Walmsley
MPI f. Radioastronomie, Auf dem Hügel 69, D-5300 Bonn 1

OH and H_2O maser emission has been searched for in 33 IRAS sources (IRAS circular 9) having far infrared colors approximating those of OH/IR stars. The observations were carried out with the 100m radiotelescope in Effelsberg. Eleven Type II OH masers and six H_2O masers were found, with two sources common to both groups. Nine of the OH spectra are characteristic of OH/IR stars. These OH/IR stars are all located in the region of OH/IR variables in the IRAS color-color plot (see figure on the next page). Of the remaining detected OH masers, one is suggested to be also an OH/IR star while the second shows circularly polarized features at 1612 MHz and is probably another type of object. The water masers which are not OH masers too are probably related to star forming regions. One is peculiar in its large velocity spread of more than 110 km s^{-1}.

Thus about 50% of the unidentified IRAS sources from IRAS circular 9 could be identified with galactic objects, with planetary nebulae, HII regions and carbon stars possibly making up a sizeable fraction of the remainder (cf. Engels et al., 1984).

REFERENCES

Engels, D., Habing, H.J., Olnon, F.M., Schmid-Burgk, J., Walmsley, C.M.: 1984, Astron. Astrophys. (Letters), in press

IRAS circular 9: 1984, Astron. Astrophys. 132, C2

Olnon, F.M., Baud, B., Habing, H.J., De Jong, T., Harris, S., Pottasch, S.R.: 1984, Astrophys. J. (Letters) 278, L41

M. Morris and B. Zuckerman (eds.), Mass Loss from Red Giants, 211–212.

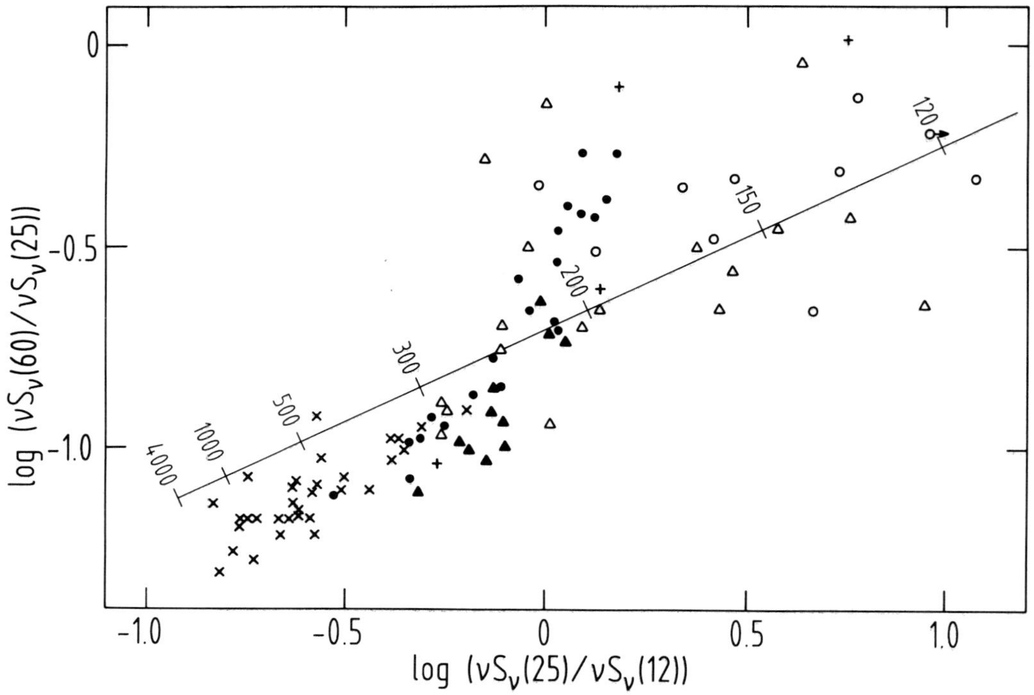

Color-color diagram of (νS_ν) IRAS flux ratios between 12, 25, and 60μm. Indicated are Mira variables, IRC and AFGL sources with OH maser emission (x), OH/IR stars (large amplitude variables (●), small amplitude or non-variables (o)) (from Olnon et al., 1984) and IRAS circular 9 objects (OH masers (▲), H_2O masers (+), without OH or H_2O maser emission (Δ)) (from Engels et al., 1984).

RADIO AND INFRARED OBSERVATIONS OF OH/IR STARS

Anneila I. Sargent
Owens Valley Radio Observatory 320-47
California Institute of Technology
Pasadena, California 91125

Boudewijn Baud
Space Research Department
University of Groningen, P. O. Box 800
9700 AV Groningen
The Netherlands

Combined infrared and OH spectral line observations are essential to understanding the nature of OH/IR stars, the properties of their circumstellar envelopes and the OH maser pump mechanism. We have used the Very Large Array of the National Radio Astronomy Observatory to determine very precise radio positions for a selection of such objects near the tangential point and at the galactic center and have subsequently made simultaneous infrared and OH observations of both samples. From the infrared measurements, luminosities of these stars and masses of the thick dust shells which surround them have been derived, while the double-peaked OH emission line profiles have yielded both the stellar radial velocities, v_R, and the expansion velocities of the envelopes, v_e, as well as the maser luminosities. The selection criteria and the observations are described in detail elsewhere (Baud et al. 1984). The results are consistent with these stars lying at the top of the Asymptotic Giant Branch (AGB) of the Hertzsprung-Russell diagram; the dense circumstellar shells suggest that these objects are evolving rapidly through a phase of high mass loss.

As a result of i)the homogeneity of the samples ii)the radio and infrared observations being obtained simultaneously and iii)the new measurements at wavelengths longward of 10μm, it has been possible to carry out a quantitative analysis of the relations between the OH maser luminosity, L_{OH}, the bolometric luminosity, L_*, and the mass loss rate, $\dot{M}$ (Baud et al. 1984). An improved understanding of the influence of $\dot{M}$ on the maser pump efficiency has also been achieved.

Briefly, the time averaged value of L_{OH} appears to be independent of L_* and of v_e, a quantity which is statistically related to the main-sequence mass (Baud et al. 1981). Thus, sources with similar stellar parameters can have very different circumstellar properties. Stars with the same L_* but higher L_{OH} have redder energy distributions and

M. Morris and B. Zuckerman (eds.), Mass Loss from Red Giants, 213–214.

deeper silicate absorption features, indicating cooler circumstellar shells and higher mass loss rates. In fact, the ratio L_{OH}/L_* is heavily dependent on the degree of reddening of the circumstellar shell and, by implication, on $\dot{M}$. We demonstrate quantitatively that the mass loss rate is the dominant parameter governing the maser flux. As expected from observations, the OH maser output follows the time variations in L_* fairly closely but, because of the steep dependence on the degree of reddening of the shell, there will be no direct correlation between L_{OH} and L_*. In addition, a quantitative relation between L_{OH} and the optical depth in the 10μm silicate feature, τ_{sil}, has been derived; the observed increase in τ_{sil} with decreasing L_* is a natural consequence of this relation.

To summarize, it appears that, in terms of mass loss and maser characteristics, OH/IR stars of different masses on the AGB follow similar evolutionary paths. Since L_{OH} is strongly dependent on $\dot{M}$, high OH luminosity OH/IR stars probably represent the end point of Mira evolution on the AGB (c.f. Jones et al. 1983; Baud and Habing 1983).

REFERENCES

Baud, B., Habing, H., Matthews, H. J., and Winnberg, A., 1981, Astron. Astrophys., 95, 156.

Baud, B., and Habing, H., 1983, Astron. Astrophys., 127, 73.

Baud, B., Sargent, A. I., Werner, M. W., and Bentley, A. F., 1984, Astrophys. J., submitted.

Jones, T. J., Hyland, A. R., Wood, P. R., and Gatley, I., 1983, Astrophys. J., 273, 669.

MASS LOSS FROM OH/IR STARS

J. Herman
Space Science Department of ESA
P.O.Box 299
2200 AG Noordwijk
The Netherlands

ABSTRACT. Mass loss rates for OH/IR stars can be estimated in several independent ways, each with its own drawbacks or rather arbitrary assumptions. For a number of OH/IR stars 'geometrical' distances are available, that have an accuracy of 10%. In that case it is possible to determine the total envelope mass, i.e. all the mass the star has lost since it left the Zero Age Main-Sequence. Consequently, these stars might be used to calibrate the various mass loss formulae, or alternatively to determine the Asymptotic Giant Branch lifetimes.

Subject headings: stars - late type, long period variables, OH/IR, masers

1. INTRODUCTION

OH/IR stars are an evolved stellar population at the tip of the Asymptotic Giant Branch (AGB). Most of these stars appear to be (very) long period variables, with pulsation periods that can be as long as 2000 days (see Herman and Habing, 1985a). Their mass loss rates are the highest known for late-type stars and consequently a substantial fraction of their Zero Age Main-Sequence (ZAMS) mass, which typically was 2-5 $M_{\odot}$, is lost in this 'superwind' phase. In this contribution some aspects concerning the superwind phase will be considered. A part of the results presented here can also be found in Herman and Habing (1985b).

2. MEASURED QUANTITIES

In 1979 a monitor program was started with the 25-m Dwingeloo Radio Telescope. The program includes 60 so-called type IIb, i.e. strongest 1612 MHz, OH masers. Among the more prominent results from this program are i) radio flux densities ii) pulsation periods and iii) radii of the OH shells (see Herman and Habing, 1985a). The radii can be determined directly from differences in light travel time towards the observer

M. Morris and B. Zuckerman (eds.), Mass Loss from Red Giants, 215–220.

between various, synchronously varying, parts of the shell. Typical values for the phase lag are 1-20 days. With a mean pulsation period for the OH/IR stars of 1000 days, it is clear that the determination of the radii is not easy, but can be achieved by frequent and sensitive observing.

Comparison of these phase lag radii with apparent angular sizes, that can be obtained from interferometric measurements, yields geometric distances with an accuracy of 10-20% (see Herman et al., 1985). The geometrical distances depend on one crucial assumption only: spherical symmetry. When one compares the observations with a model of a uniformly expanding circumstellar shell (see Reid et al., 1977), it is found that the deviations from spherical symmetry are quite small (10%).

Once such accurate distances are available, life becomes much easier. Herman et al. (1984) made infrared observations of a number of OH/IR stars, getting good absolute bolometric magnitudes. Because ~20% of the flux comes from wavelengths longward of 20 μm it is even better to use the IRAS data: all these OH/IR stars have been readily picked up (see Olnon et al., 1984).

3. MASS LOSS RATES

The Reimers formula (see Reimers, 1975), which is valid for ordinary red giants, is not applicable for the OH/IR stars; their mass loss rates appear to be a factor 100 higher. But for the 25 stars in our sample with accurate distances and IR observations available, the mass loss rate, determined as described in the following sections, is found to have the same dependence on stellar luminosity, radius and mass as in the Reimers equation. We may write a modified Reimers formula:

$$\dot{M} = -(1.00 \pm 0.12)\ 10^{-11}\ L_*R_*/M_* \qquad M_\odot\ yr^{-1} \qquad (1)$$

where L_*, R_*, and M_* are in solar units. This might be compared with $-(1.3$ to $4)\ 10^{-13}$: the constant in the 'normal' Reimers formula.

There are four different, independent, ways to estimate the mass loss rates, and so arrive at equation (1), that will be discussed briefly.

3.1 $\dot{M}$ based on infrared observations

For a mass outflow driven by radiation pressure on the dust grains one may write (see Elitzur, 1981)

$$\dot{M} = -2.0\ 10^{-8}\tau_d L_*/v_e \qquad M_\odot\ yr^{-1} \qquad (2)$$

where τ_d is the optical depth corresponding to the dust response to radiation pressure and v_e the expansion velocity of the shell in kms^{-1}. A slight complication is that the mean luminosity for these stars is rather poorly known.

The best studied ones, such as OH 26.5+0.6 have only a few measurements of their total flux. Certainly $\tau_d>1$, and following Werner et al. (1980) we choose $\tau_d=2$, probably a slight underestimate for the OH/IR stars with the thickest envelopes.

3.2. $\dot{M}$ based on CO observations

Knapp et al. (1982) used CO observations at 230 GHz to estimate the mass loss rates for a number of OH masers. For the few stars we have in common there is reasonable agreement with the results from equations (2) and (3) (see for an example Table 1), but the expansion velocity as found from the CO profiles is systematically ~25% larger than v_e based on the OH single-dish observations.

3.3. $\dot{M}$ based on radio observations

Baud and Habing (1983) used the fact that OH masers have a constant surface brightness (see Herman and Habing, 1985b) to derive

$$\dot{M} = -1.8\ 10^{-7}\ v_e L_{OH}{}^{0.5} \qquad M_\odot\ yr^{-1} \qquad (3)$$

where L_{OH} is the OH luminosity in Jy kpc^2. The constant is determined by assuming a minimum OH column density needed for a saturated maser to operate (see Elitzur et al., 1976), consequently overestimating the mass loss rate.

3.4. The mass of the circumstellar envelope

When it is possible to find out what the mass of an OH/IR star once was, and what its present mass is, per force it is known how much mass the star has lost during its evolution. The ZAMS mass can be determined in several ways: i) Paczynski's relation (see Paczynski, 1970) allows us to convert the observed luminosity into a value for the core mass, M_{ct}. The initial core mass, M_{co} which is not too different from the 'observed' M_{ct}, at the moment when hydrogen is reignited on the AGB, is solely a function of ZAMS mass (see Iben and Truran, 1978). ii) The balance at large distances from the star, e.g. at R_{OH}, between radiation pressure on the one hand and gravitational and drag forces on the other, makes it possible to determine the total mass confined within that radius (see Olnon, 1981). iii) According to Baud and Habing (1983) the number distribution of the OH/IR stars over expansion velocity reflects the initial mass function. This leads to a simple relation between initial mass and v_e. The three different methods give reasonably well the same values for the ZAMS mass (see Table 1). Statistically, the derived masses can be checked by comparing the scale heights for groups of OH/IR stars with the z-distribution of main-sequence stars (see Herman and Habing, 1985b), giving completely consistent results.

The pulsation period and radius of a star give us the present stellar mass (see Fox and Wood, 1982). This still is a rather uncertain procedure, because the pulsation mode of OH/IR stars is subject to discussion and the Fox and Wood relations have to be extrapolated to very long periods. It is reassuring, however, that $M_{ct} < M_{pul} < M_{ZAMS}$ is always found, when fundamental mode pulsation is assumed.

The percentage in the OH phase of the total population on the AGB, combined with the lifetime on the AGB, yields a canonical duration for the superwind phase of 5 10^4 yr. With envelope masses of 1-2 $M_\odot$ again mass loss rates of $> 10^{-5}$ $M_\odot yr^{-1}$ are deduced. When theoretically the lifetime on the AGB can be firmly established, the envelope masses may be used to determine the dust properties (equation 2) and the densities in the shell (equation 3). The reverse possibility, which seems more likely at present, is to assume that the calibration of the mass loss formulae is not too far off, and then to deduce the AGB lifetime.

4. RESULTS AND DISCUSSION

Two OH stars, for which much information is available, are presented in Table I. They are IRC+10011, which is identified with the Mira variable WX Psc, and OH 26.5+0.6, identified with the infrared source AFGL 2205. All quantities are given with their formal errors between parentheses. The first 6 lines give the measured, or directly inferred parameters (see section 2).

Table 1.

		IRC+10011	OH 26.5+0.6	Remarks
1. D	kpc	0.47 (0.06)	0.98 (0.09)	Geometrical distance
2. v_e	km s^{-1}	17.91 (0.003)	14.07 (0.005)	Expansion velocity of the shell
3. L_{OH}	Jy kpc^2	7.33 (1.99)	264 (51)	Mean OH luminosity
4. R_{OH}	10^{16} cm	2.97 (0.06)	3.29 (0.26)	Phase lag radius
5. P_r	days	632.0 (17.6)	1566 (30)	Pulsation period
6. $L_\star$	$10^4 L_\odot$	1.08 (0.29)	1.00 (0.26)	Bolometric luminosity
7. T_{eff}	K	2969 (180)	2745 (125)	Effective temperature
8. $R_\star$	$R_\odot$	405 (50)	456 (56)	Stellar radius
9. M_{co}	$M_\odot$	0.655 (0.047)	0.632 (0.032)	Core mass at start AGB
10. M_{ct}	$M_\odot$	0.677 (0.049)	0.664 (0.033)	Present core mass
11. M_{pul}	$M_\odot$	1.77 (0.58)	0.76 (0.34)	Present stellar mass
12. M_{dyn}	$M_\odot$	2.10	2.68	Mass within R_{OH}
13. M_{ZAMS}	$M_\odot$	3.36	3.19	Based on M_{ct} (Iben & Truran)
14. M_{v_e}	$M_\odot$	4.16	2.40	Based on v_e (Baud & Habing)
15. $M_{\star i}$	$M_\odot$	3.21 (0.60)	2.76 (0.23)	Mean of 12, 13, and 14
16. M_{env}	$M_\odot$	1.44 (0.50)	2.00 (0.45)	$M_{\star i} - M_{pul}$
17. $\dot{M}_R$	$10^{-7} M_\odot$ yr^{-1}	-3.73 (0.96)	-3.46 (0.80)	Reimers formula (start AGB)
18. $\dot{M}_{RM}$	$10^{-5} M_\odot$ yr^{-1}	-2.47 (1.07)	-6.00 (1.48)	Reimers modified, equation (1)
19. $\dot{M}_{IR}$	$10^{-5} M_\odot$ yr^{-1}	-2.42 (0.65)	-2.85 (0.56)	Equation (2)
20. $\dot{M}_{OH}$	$10^{-5} M_\odot$ yr^{-1}	-0.86 (0.22)	-4.04 (0.37)	Equation (3)
21. $\dot{M}_{CO}$	$10^{-5} M_\odot$ yr^{-1}	-2.21		Knapp et al. (1982)
22. t_{AGB}	10^5 yr	3.22 (0.23)	5.49 (0.39)	Total life time on the AGB
23. Δt_{OH}	10^5 yr	0.74 (0.27)	0.53 (0.13)	Length of the superwind phase

The effective temperatures are found by assuming that on the average these stars lie on the AGB, but they have been corrected for the large mass loss rates (see for a detailed discussion: Herman and Habing, 1985b). The pulsation mass, which is the present stellar mass, has been calculated from the Fox and Wood (1982) relations, assuming fundamental mode pulsation. The three different ways to estimate the main-sequence mass have been discussed in section 3.4., and M_{*i} is the mean of the three values. The mass loss rates $\dot{M}_{IR}$ and $\dot{M}_{OH}$ are described in the text. The Reimers wind was calculated using the luminosity, mass and radius at the start of the AGB evolution; the modified Reimers wind, $\dot{M}_{RM}$, with the slightly higher luminosity and radius and the smaller present day stellar mass, M_{pul}. The growth rate of the core mass determines the length of the thermal pulse phase, that is roughly half of the total lifetime on the AGB (t_{AGB} see Iben, 1981). For both stars t_{AGB} = 4 10^5 yr. The length of the OH phase (superwind phase) is ~15% of the thermal pulse phase (see Herman and Habing, 1985b), and thus will be ~3 10^4 yr. With the derived envelope masses of ~1.5 $M_\odot$ a mass loss rate of 5 10^{-5} $M_\odot yr^{-1}$ is deduced, if one assumes that prior to the superwind phase no significant mass loss has taken place. Because this assumption is not strictly true, the actually measured values for the mass loss rates are slightly lower.

Notice that, although for both stars the geometrical distances are very well determined, still an uncertainty of 20% in the luminosity results, leading to a 40% uncertainty in the present stellar mass. Therefore, it still seems better not to look at individual objects, but to derive mean properties for the OH/IR stars as a population (see Herman and Habing, 1984b for these group characteristics).

5. CONCLUSIONS

The possibility to determine very accurate 'geometric' distances, make the OH/IR stars outstanding probes for the theory of the last stages in stellar evolution. Probably the most promising approach is to obtain a good calibration of the mass loss rates, which means a good understanding of the dust properties and the mass loss mechanism, followed by a derivation of such fundamental parameters as the AGB lifetimes, pulsation modes, and various stellar quantities. In this way the OH/IR stars may be a valuable calibration point for all stellar evolution models.

The good agreement between the various methods of determining the mass loss rates implies i) that the AGB life time is ~5 10^5 yr ii) that $\tau_d \gtrsim 2$ and iii) that the OH column density is 2-3 times the minimum column density needed to get a saturated maser.

ACKNOWLEDGEMENTS

J. Herman acknowledges receipt of an ESA Research Fellowship.

REFERENCES

Baud, B., Habing, H.J.: 1983, Astron. Astroph. **127** ,73

Elitzur, M.: 1981, in Physical Processes in Red Giants, Eds. I. Iben and A. Renzini, Reidel, pp 363-382

Elitzur, M., Goldreich, P., Scoville, N.: 1976, Astroph.J. **205** ,384

Fox, M.W., Wood, P.R.: 1982, Astroph.J. **259** ,198

Herman, J.: 1983, The Nature of OH/IR Stars, Ph.D. thesis, RUL

Herman, J., Habing, H.J.: 1985a, Astron. Astroph. in press (see Herman, 1983: Chapter II)

Herman, J., Habing, H.J.: 1985b, in Reviews on Progress in Physics, in press (see Herman, 1983: Chapter V)

Herman, J., Baud, B., Winnberg, A., Habing, H.J.: 1985, Astron. Astroph. **143** , 122

Herman, J., Sargent, A., Isaacman, R., Habing, H.J.: 1984, Astron. Astroph. **139** , 171

Iben, I.: 1981, in Physical Processes in Red Giants, Eds. I. Iben and A. Renzini, Reidel, pp 3-24

Iben, I., Truran, J.W.: 1978, Astroph.J. **220** ,980

Knapp, G.R., Phillips, T.G., Leighton, R.B., Lo, K.Y., Wannier, P.G., Wootten, H.A., Huggins, P.J.: 1982, Astroph.J. **252** ,616

Olnon, F.M.: 1981, in Physical Processes in Red Giants, Eds. I. Iben and A. Renzini, Reidel, pp 237-240

Olnon, F.M., Baud, B., Habing, H.J., de Jong, T., Harris, S., Pottasch, S.R.: 1984, Astroph.J.Lett. **278** ,L41

Paczynski, B.: 1970, Acta Astron. **20** ,47

Reid, M.J., Muhlemann, D.O., Moran, J.M., Johnston, K.J., Schwartz, P.R.: 1977, Astroph.J. **214** ,60

Reimers, D.: 1975, Mém. Soc. Roy. Sci. Liège, 6e Ser., **8** ,369

Werner, M.W., Beckwith, S., Gatley, I., Sellgren, K., Berriman, G.: 1980, Astroph.J. **239** ,540

WHAT CIRCUMSTELLAR SiO AND OH MASERS TELL US ABOUT MASS LOSS FROM RED GIANTS

Charles Alcock
Department of Physics, M.I.T., Cambridge, MA 02139

Randy R. Ross
Department of Physics, College of the Holy Cross
Worcester, MA 01610

We advocate an alternative model of mass loss from red giant and supergiant stars. We suggest that the stars are ejecting dense knots of cool molecular gas; the typical size of a knot is <0.1 x the radius of the star and typical density $n_{H_2} < 10^{12} cm^{-3}$ (Alcock and Ross, 1985a). The filling fraction of these knots is low (~10% at the photosphere, decreasing outward) and once the knots become optically thin in the dust they move out radially at constant solid angle (subtended at the star). Some knots fall back onto the star. The mean outflow is spherically symmetric, but close to the star and density and velocity structure is chaotic and rapidly changing. This model is not well approximated by a wind.

The origin of the stochastic asymmetry will lie in the envelope of the star. Schwarzschild (1975) suggested that convection in red giant envelopes might be dominated by a few giant cells. Salpeter (1974) suggested a scenario in which grains might form in local, temporarily cool patches on the surface of the star. Polarization studies (Hayes 1984) are consistent with this picture.

The cool (~800K) material observed by Hinkle et al (1982) in χCyg probably comprises one or two knots, located a few stellar radii above the photosphere. The timescale observed (several years) corresponds to the dynamical time in this region.

This model may resolve significant problems in the interpretation of the maser emission. The SiO masers are significantly (~100x) more luminous than the best numerical pump models predict (Langer and Watson 1984). Many of these difficulties could be resolved if the densities in the maser region were ~100x greater than predicted by the wind models (Alcock 1985). Furthermore, the chaotic VLBI maps (Lane 1982) and the observation of SiO emission at many stellar radii are more consistent with emission from a collection of blobs than from a wind.

The maps of OH 1612 MHz emission (e.g. Bowers et al 1983; Diamond et al 1984) are not convincingly modelled by a wind-like outflow; the

M. Morris and B. Zuckerman (eds.), Mass Loss from Red Giants, 221–222.

reason is that the characteristic OH maps are reproduced only for extremely "finely tuned" choices in the parameter space of density and pump rate (Alcock and Ross 1985b). The wide range of luminosities spanned by this class of sources argues strongly that "fine tuning" does not occur. These problems are alleviated in a blob-like model.

REFERENCES

Alcock, C. R. 1985, in preparation.

Alcock, C. R. and Ross, R. R. 1985a, Ap.J., in press.

Alcock, C. R. and Ross, R. R. 1985b, submitted to Ap.J.

Bowers, P. F., Johnston, K. J. and Spencer, J. H. 1983, Ap.J., 274, 733.

Diamond, P. J., Norris, R. P., and Booth, R. S. 1984, M.N.R.A.S., 207, 149.

Hayes, D. P. 1984, Ap.J. Suppl., 55, 179.

Hinkle, K. H., Hall, D. N. B., and Ridgway, S. T. 1982, Ap.J., 252, 697.

Lane, A. P. 1982, Ph.D. thesis, University of Massachusetts.

Langer, S. H. and Watson, W. D. 1984, Ap.J., in press.

Salpeter, E. E. 1974, Ap.J., 193, 585.

Schwarzschild, M. 1975, Ap.J., 195, 137.

TIME MONITORING OF SiO (v=1, J=2-1) MASER EMISSION FROM LATE TYPE STARS

L-Å. Nyman and H. Olofsson
Onsala Space Observatory
S-43900 Onsala
Sweden

The time variations of the SiO (v=1, J=2-1) maser line towards eight late type stars and Orion A have been observed with the Onsala 20 m telescope. R Leo, o Ceti, and Orion A were observed at monthly intervals between December 1977 and mid 1979 (Hjalmarson and Olofsson, 1979). During 1979 and 1980 six more sources were added (χ Cyg, R Cas, NML Tau, U Ori, R Aql, and T Cep), and the observations were then performed every second to third month, occasionally with longer breaks, until January 1984. Only the late type stars will be discussed below.

All objects, except R Leo and possibly R Aql, show a correlation between the optical variations and the SiO integrated fluxes. The maximum SiO intensity can vary by a large amount from cycle to cycle and the same is true for the dynamic range, i.e. the ratio between maximum and minimum SiO flux during one period. The integrated flux variation towards o Ceti is shown in Figure 1. This is the source with the highest dynamic range, and therefore the object where the periodicity is best seen. In all sources with periodic SiO variations the maximum SiO intensity occurs at stellar optical phase 0.1 to 0.2 and minimum intensity at phase 0.5 to 0.7. This is in excellent agreement with the infrared 1.04 μm variations (Lockwood and Wing, 1971). However, there is no agreement between the average 1.04 μm amplitudes and the average SiO dynamic ranges. On the other hand the 1.04 μm region is free from molecular absorptions, but if the SiO masers are radiatively pumped this is done by photons at 8 μm (Kwan and Scoville, 1974; Bujarrabal and Rieu, 1981) where the amplitude can be affected by molecular absorptions.

The integrated flux towards R Leo did not show any correlation with the optical variations, instead the flux decreased steadily between 1978 to 1982 with superposed oscillations not varying with the stellar luminosity. The observations by Clark et al. (1984) show that at the beginning of 1979 the R Leo integrated flux was dominated by a single feature whose flux did not vary together with the visual intensity, but the SiO fluxes of the components with lower intensities were in fact correlated with the optical variations.

M. Morris and B. Zuckerman (eds.), Mass Loss from Red Giants, 223–225.

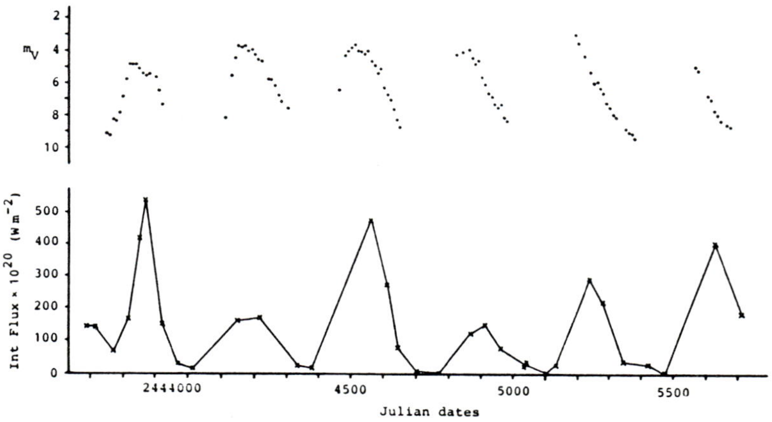

Figure 1. Visual magnitude and SiO flux variations towards o Ceti.

Simultaneous variations of the infrared and SiO fluxes do not necessarily imply that SiO masers are radiatively pumped. During the increase in stellar luminosity the temperature in the atmosphere may also increase, an effect that would enhance collisional pumping. Thus it is not clear from these observations whether the SiO masers are pumped by radiation or collisions.

The SiO grand averages, i.e. the average of all spectra obtained towards one source, show some interesting properties. The total width of the SiO maser emission varies only between 12.5 and 17 km s^{-1} among the different sources while the expansion velocities of the circumstellar shells range from 5 to 19 km s^{-1}. Thus in some cases $\Delta v_{SiO} > 2 \times v_{exp}$ suggesting that the SiO masers are located inside the circumstellar envelope, in a region having roughly the same velocity structure in all these stars. The centre velocities of the grand average profiles fall within 1 - 2 km s^{-1} of the stellar radial velocities (as opposed to the SiO v=1 and 2, J=1-0 lines, which are systematically redshifted, Lane (1982)). This implies that they can be used to estimate stellar velocities but observations over several periods are required.

We have also produced an unweighted average of the grand averages of the eight sources, which can be regarded as a global SiO (v=1, J=2-1) maser profile for Mira variables. It is apparent that the emission is not fully symmetric with respect to the stellar velocity: the blueshifted side is much steeper than the redshifted side, which results in a blueshifted peak ($\sim$ 1 km s^{-1}), but a slight redshift skew in total.

Long-term changes (i.e., longer than an optical period) of the total width and centre velocity of the SiO lines can be seen in some of the sources. The best example is o Ceti where the centre velocity has increased steadily from 45 to 50 km s^{-1} from 1978 to 1984. This may be an effect of o Ceti being a binary star. Similar variations (but not as pronounced) are also seen in some of the other stars. A full report of this work will be presented in Astronomy and Astrophysics.

REFERENCES

Bujarrabal, V., Rieu, N.-Q. 1981, Astr. Ap. 102, 65.
Clark, F.O., Troland, T.H., Pepper, G.H., Johnson, D.H. 1984, Ap.J. 276, 572.
Hjalmarson, Å., Olofsson, H. 1979, Ap. J. 234, L199.
Kwan, J., Scoville, N. 1974, Ap. J. 194, L97.
Lane, A.P. 1982, Ph.D. dissertation.
Lockwood, G.W., Wing, R.F. 1971, Ap. J. 169, 63.

NEW 43 GHz SIO OBSERVATIONS WITH THE MPIFR 100m TELESCOPE

P. R. Jewell, C. Henkel, C. M. Walmsley, and T. L. Wilson
Max-Planck-Institut fur Radioastronomie
L. E. Snyder
Max-Planck-Institut fur Radioastronomie and
the University of Illinois
D. Engels
Astronomisches Institut der Universitat Bonn

Approximately 50 new sources of SiO maser emission have been detected with the MPIfR 100m telescope at a frequency of 43 GHz. These discoveries, made during 1983 December and 1984 January, increase the number of known SiO maser sources by ~66%. We have additionally observed 42 of the 76 previously known sources. Our observations were primarily in the v=1, J=1→0 SiO line (43.122 GHz), although we were able to simultaneously observe the v=0, J=1→0 line (43.424 GHz) in most cases. Several examples of masering in the v=0 line of Mira variables were found. We also found evidence for strong linear polarization in the v=1 line of R Leo. Thermal, v=0, J=1→0 SiO emission from the carbon star IRC +10216 was studied and a comparison between the emission angular extent and profile shapes of other species was made.

The new detections were of Mira variables (27 definite, 6 possible detections), OH/IR stars (19 definite, 2 possible detections), and miscellaneous long period variables (3). A total of 185 objects were searched for SiO emission. Our search list included stars with a variety of temperatures, luminosities, and mass loss rates. In this preliminary report we have not attempted to remove selection effects, but for Mira variables, there does appear to be a strong peak in the distribution of SiO sources near spectral type ~M7. Miras with spectral types as early as M4 were detected, however. The OH/IR stars on our search list are principally those whose coordinates have been accurately determined by infrared or OH maser interferometric measurements. Including observations we have recently reported elsewhere (Jewell *et al.* 1984), the detection rate of SiO masers in OH/IR stars was ~50%. We thus conclude that SiO maser emission is quite a common phenomenon in OH/IR stars.

SiO masers have now been detected in classical Mira variables, IRC stars, OH/IR stars, late-type supergiants, and in Orion-KL -- objects encompassing a range of physical properties. Relations between such properties as SiO photon luminosity L_{SiO}, stellar bolometric luminosity L_*, and stellar mass loss rate $\dot{M}$ can

M. Morris and B. Zuckerman (eds.), Mass Loss from Red Giants, 227–228.

provide insight into the SiO maser mechanism. We find that typical photon luminosities in the v=1, J=1→0 SiO line are ~1 x 10^{43} s^{-1} for Miras, ~4 x 10^{43} s^{-1} for OH/IR stars, and ~2 x 10^{46} s^{-1} for the supergiants VY CMa and VX Sgr. The results for Miras and OH/IR stars, both of which have similar L_*'s, suggest a possible correlation between L_{SiO} and $\dot{M}$. The results for the supergiants indicate that L_{SiO} is more sensitive to L_* than $\dot{M}$, however. A correlation between SiO flux and stellar bolometric flux has been previously found (Cahn 1981).

Our observations of the v=0, J=1→0 SiO line produced several new detections. Most of the studies of v=0 SiO have been in the higher rotational transitions such as the J=2→1 at 86 GHz. In these transitions the emission is generally thermal or quasi-thermal with broad, parabolic line profiles. In the 43 GHz v=0 line, we observed several Mira variables (*e.g.*, o Ceti, χ Cyg, and R Cas) with strong, narrow emission features, indicating maser emission. A distinct change in the excitation between v=0, J=1→0 and v=0, J=2→1 clearly occurs.

Our observations were made with a single channel receiver equipped with a linearly polarized feed. Clark and co-workers have demonstrated that Miras such as R Leo often exhibit strong linear polarization in their v=1, J=2→1 SiO spectra (*e.g.*, Clark *et al.* 1984). We observed R Leo with the feed polarization angle in two orthogonal positions and checked for changes in the relative strengths of spectral features. Linear polarization >50% was observed in some R Leo v=1, J=1→0 features.

Finally, we observed and mapped the v=0, J=1→0 emission from the carbon-rich object IRC +10216. The emission angular extent of SiO in that source is comparable to our 22" beam. In comparison, the extents of HC_3N and HC_5N, recently observed with the 100m telescope at wavelengths near 1.3 cm, are >40". No significant difference in the linewidths of the three species is observed, which indicates that the SiO has reached terminal flow velocity. The spatial distribution of SiO relative to other species is consistent with the idea that it condenses onto grains as the gas drops below a certain temperature.

The authors thank N. Keen, A. Korn, and others in the MPIfR Millimeter Laboratory for the development of the cooled mixer receiver used in this study. L. E. S. was an Alexander von Humboldt-Stiftung Senior Scientist Awardee when these observations were made. P. R. J. thanks the U. S. National Radio Astronomy Observatory for travel support to attend this conference.

REFERENCES

Cahn, J. H. 1981, *A. J.*, 86, 1935.
Clark, F. O., Troland, T. H., Pepper, G. H., and Johnson, D. R. 1984, *Ap. J.*, 276, 572.
Jewell, P. R., Batrla, W., Walmsley, C. M., and Wilson, T. L. 1984, *Astr. Ap.*, 130, L1.

MASS LOSS MECHANISMS FOR COOL, LOW-GRAVITY STARS

Thomas E. Holzer and Keith B. MacGregor
High Altitude Observatory
National Center for Atmospheric Research[1]
P. O. Box 3000
Boulder, CO 80307

ABSTRACT. In this paper we explore a number of physical effects relevant to mass loss from cool, low-gravity stars. It is our primary intention to provide a basis for understanding the various mechanisms proposed to explain such quasi-steady and non-steady mass loss. We begin by considering a few general implications of the requirements of mass, momentum, and energy balance and continue by examining in some detail four specific mass loss mechanisms. Emphasis is placed on the basic physical constraints imposed uniformly on all these mechanisms by inferences drawn from observations of a broad range of stars. We suggest that these physical constraints, rather than the techniques used to observe particular objects, should serve as the dominant factor in organizing our thinking about massive, low-speed, late-type stellar winds.

1. INTRODUCTION

There has long been evidence that cool, low-gravity stars can lose mass at very large rates--so large, in fact, that the mass loss can have a significant effect on stellar evolution (e.g., Adams and MacCormack 1935; Deutsch 1956, 1960; Weymann 1962a,c; and reviews by Reimers 1975; Goldberg 1979, 1984; Hagen 1980; Castor 1981; Hartmann 1981; Cassinelli and MacGregor 1984). These massive stellar winds, in contrast to those associated with early-type stars, generally appear to be characterized by an asymptotic flow speed (u_∞) that is considerably less than the stellar surface gravitational escape speed (v_{go}) and by a ratio of these speeds (u_∞ / v_{go}) that decreases with decreasing escape speed (v_{go}) (Deutsch 1956, 1960; Weymann 1962a; Reimers 1977). Such massive, low-speed winds are not usually detected in late-type stars with surface gravities greater than ~1% of that of the sun, and the presence of transition region and coronal lines in the spectra of these higher gravity stars has led to the inference that they possess hot, tenuous winds, like the solar wind, in which $u_\infty \approx v_{go}$ (e.g., Dupree 1975; Linsky and Haisch 1979; Dupree and Hartmann 1980; Stencel and

[1] The National Center for Atmospheric Research is sponsored by the National Science Foundation.

M. Morris and B. Zuckerman (eds.), Mass Loss from Red Giants, 229–255.

Mullan 1980). There is, however, a small group of intermediate-gravity stars that exhibit both high-temperature spectral lines and evidence for massive winds (e.g., Hartmann, Dupree, and Raymond 1981; Linsky 1982).

Cool, massive late-type stellar winds arise in a broad range of stellar atmospheres and thus are amenable to study through a broad range of observational approaches: studies of spectral line profiles in the visible and near *UV* resonance lines of neutral and singly ionized metals, in $H\alpha$, and in the infrared vibration-rotation bands of *CO*; studies of infrared dust emission in the 10 μm silicate band and the far infrared continuum, of thermal millimeter wave radiation in rotational lines of molecules, of thermal centimeter wave radiation from partially ionized chromospheres, and of radio-frequency maser emission (e.g., Goldberg 1984). Although it may be tempting to classify these winds according to the remote sensing techniques through which they are observed, we will follow the approach of Goldberg (1984) in resisting this temptation and will initially consider all the massive winds from cool, low-gravity stars to fall in a single category. This category can be defined by the large mass loss rates and low asymptotic flow speeds that characterize these winds and sharply distinguish them from the hot, tenuous solar wind and from massive winds of early-type stars. Such a categorization emphasizes the similar dynamical constraints placed on all the mechanisms considered as candidates for driving any of these cool, massive, low-speed winds.

In discussing these mass loss mechanisms, our emphasis will be on steady (and quasi-steady), spherically symmetric wind descriptions, because such descriptions allow us to concentrate on a set of relatively simple physical concepts. Understanding these concepts should help us to develop the physical intuition necessary to think effectively about the inherently complex spatial and temporal structure likely to characterize late-type stellar winds. Our discussion begins with some general comments on mass, momentum, and energy balance in stellar winds, and it continues with the development of wind descriptions that illustrate four mass loss mechanisms and emphasize physical principles rather than mathematical formalism. The order of development of the four wind descriptions (thermally driven winds, wave-driven winds, radiation-driven winds, shock-driven winds) is intended to facilitate an understanding of them, not to imply any indication of their relative importance. Following the discussion of these four mass loss mechanisms, we consider briefly some other (less well developed) ideas about driving mass loss and conclude with a few comments about the possible roles played by the various mechanisms in driving massive, low-speed winds from low-gravity, late-type stars.

2. MASS, MOMENTUM, AND ENERGY BALANCE: GENERAL CONSIDERATIONS

The equations describing mass, momentum, and energy balance in a steady, radial, spherically symmetric stellar wind flow can be written in the form

$$\rho u r^2 = f = constant \tag{1}$$

$$u \frac{du}{dr} = - \frac{1}{\rho} \frac{dp}{dr} - \frac{GM_*}{r^2} + D \;, \qquad \mathbf{(2)}$$

$$\frac{1}{\gamma - 1} \rho u \frac{d}{dr} \ln \left(\frac{p}{\rho^\gamma} \right) = - \frac{1}{r^2} \frac{d}{dr} (q r^2) + Q \;, \qquad (3)$$

where ρ, u, p, and q are the mass density, flow speed, thermal pressure, and heat flux density, γ is the ratio of specific heats, ρD and Q are the net volume rates at which momentum and heat are added to the flow, and G, M_*, and $4\pi f = -\dot{M}_*$ are the gravitation constant, stellar mass, and stellar mass loss rate. Energy balance can also be described by an equation of the form

$$f \left(\frac{1}{2} u^2 + \frac{\gamma}{\gamma - 1} \frac{p}{\rho} - \frac{GM_*}{r} \right) + q r^2$$
$$= F = F_o + \int_{r_o}^{r} dr' \; (f D + Q r^2) \;, \qquad (4)$$

which is derived by combining equations (2) and (3) and integrating. Here, F is the energy flux associated with the advection of flow energy, enthalpy, and gravitational potential energy and with the conduction of heat; in the absence of net energy addition (i.e., for $D = Q = O$), F is a second constant of the flow (the first being the mass flux, f).

There are two general constraints that can be placed on a stellar wind driving mechanism by making use of equations (1) and (4), along with certain observational inferences. These constraints involve the outward energy flux density at the stellar surface that is required to drive the wind and the location in the wind where energy must be added to either the internal energy or flow energy of the expanding atmosphere. Let us first consider the energy flux required to drive the wind.

2.1. Wind energy flux

For simplicity we assume that our expanding stellar atmosphere is gravitationally bound and very subsonic at the atmospheric base ($r = r_o \approx R_*$) and is highly supersonic at large distances from the star ($r = r_\infty$), where the wind has achieved its asymptotic flow speed. If we define the squares of the adiabatic sound speed and gravitational escape speed by

$$v_s^{\,2} = \gamma \, v_t^{\,2} = \gamma \, p / \rho \;, \qquad (5)$$

$$v_g^{\,2} = 2 G M_* / r \;, \qquad (6)$$

our assumptions can be expressed as inequalities:

$$v_{go}^2 >> u_o^2 + 2\gamma v_{to}^2 / (\gamma-1) , \tag{7}$$

$$u_\infty^2 >> 2\gamma v_{t\infty}^2 / (\gamma-1) , \tag{8}$$

$$u_\infty^2 >> v_{g\infty}^2 . \tag{9}$$

Now, defining the energy-per-unit-mass of the flow by

$$E = \frac{1}{2} u^2 + \frac{\gamma}{\gamma-1} v_t^2 - \frac{1}{2} v_g^2 , \tag{10}$$

we find that $E_o \approx -v_{go}^2/2$, $E_\infty \approx u_\infty^2/2$, and (4) becomes

$$F_{do} = q_o r_o^2 + \int_{r_o}^{r_\infty} dr\,(fD + Qr^2) \tag{11a}$$

$$\approx \frac{1}{2} f\,(u_\infty^2 + v_{go}^2) . \tag{11b}$$

The atmospheric base energy flux required to drive the wind (F_{do}) is taken to include the atmospheric base heat flux density (assuming $q_\infty r_\infty^2 << q_o r_o^2$) and the additional energy flux emanating from the atmospheric base that is eventually deposited in the wind through either heating (Q) or direct acceleration (D).

The information contained in equation (11) can be simply stated in words: the energy flux required to drive a stellar wind is that necessary to lift the mass transported by the wind out of the stellar gravitational field and to accelerate the wind to its asymptotic flow speed. In the massive winds from early-type stars, for which $u_\infty^2 >> v_{go}^2$, virtually all of the driving energy of the wind goes into accelerating the flow to its asymptotic speed. In solar-type winds, for which $u_\infty^2 \approx v_{go}^2$, comparable parts of the wind's driving energy go into lifting the expanding atmosphere out of the stellar gravitational field and into accelerating it to the asymptotic flow speed. *In the massive winds from low-gravity, late-type stars, for which $u_\infty^2 << v_{go}^2$, virtually all of the driving energy of the wind goes into lifting the expanding atmosphere out of the gravitational field.*

Assuming $r_o \approx R_*$, the energy flux density emanating from the atmospheric base that is required to drive a stellar wind can be written in the form [cf. equation (11)]

$$\frac{F_{do}}{r_o^{\,2}} \approx 3.3 \times 10^3 \left(\frac{-\dot{M}_*}{10^{-7}} \right) \left(\frac{M_*}{M_\odot} \right) \cdot \left(\frac{400\, R_\odot}{R_*} \right)^3 \cdot \left(1 + \frac{u_\infty^{\,2}}{v_{go}^{\,2}} \right) erg\ cm^{-2} s^{-1} , \quad \mathbf{(12)}$$

where the mass loss rate $(-\dot{M}_*)$ is expressed in solar masses per year. For the solar wind $(-\dot{M}_\odot \approx 2 \times 10^{-14}\, M_\odot\, y^{-1})$ equation (12) gives an energy flux density requirement of about $10^5\ erg\ cm^{-2} s^{-1}$; the value of $5 \times 10^5\ erg\ cm^{-2} s^{-1}$ normally quoted for high-speed solar wind streams results from the observational inference that the solar sources of these streams (i.e., solar coronal holes) occupy only about 20% of the area that would be occupied by these sources if the streams expanded spherically symmetrically. For a K5 supergiant with $-\dot{M}_* \approx 10^{-7}\, M_\odot y^{-1}$, $M_* \approx 16 M_\odot$, and $R_* \approx 400 R_\odot$, the energy flux density requirement of equation (12) is about $6 \times 10^4\ erg\ cm^{-2}\ s^{-1}$, and for a Mira with $-\dot{M}_* \approx 2 \times 10^{-6}\, M_\odot\, y^{-1}$, $M_* = M_\odot$, and $R_* = 400 R_\odot$, it is only about $5 \times 10^4\ erg\ cm^{-2}\ s^{-1}$. The reason the stellar surface energy flux densities required to drive massive winds from late-type, low-gravity stars are rather modest in comparison with that required to drive the low-mass-loss-rate solar wind is that the radii of the stars associated with massive winds are so large that the gravitational field out of which the wind must be lifted is relatively weak $(\propto R_*^{-1})$ and the stellar surface area from which the energy flux is supplied is relatively large $(\propto R_*^{\,2})$. Specifically, although $\dot{M}_* / \dot{M}_\odot \approx 10^7$, this is more than made up for by the fact that $(R_* / R_\odot)^3 \approx 6 \times 10^7$ [cf. equation (12)]. It is important to remember, however, that much of the energy flux emanating from the stellar base in a mechanical form (e.g., as small or large amplitude waves) can be dissipated and radiated away in the massive winds, without taking part in driving the atmospheric expansion (see, particularly, sections 3, 4, and 6). Equation (12), therefore, really represents only a lower limit to the required energy flux from the atmospheric base, and knowledge of the characteristics of the mass loss mechanism and the expanding atmosphere is necessary to determine the actual energy flux requirement.

2.2. Wind momentum flux

In some special, but important, cases it is useful to consider requirements placed on the mass loss mechanism by the wind momentum flux.[2] Specifically, when the wind is driven purely by radiation pressure above some atmospheric base level, $r = r_o$ (not necessarily, $r_o = R_*$), momentum balance can be described by

$$\frac{1}{r^2} \frac{d}{dr} \left[\rho u^2 r^2 + (1 - \eta)\, L / 4\pi c \right] = - \frac{\rho G M_*}{r^2} = - \frac{\rho v_{go}^{\,2} r_o}{2 r^2} , \quad (13)$$

where $\eta\,(r)$ measures the fraction of the asymptotic radiative momentum flux (L/c) that has been received by the gas at a level r in the wind. This fraction can exceed

[2] The wind momentum flux refers here to a scalar quantity: viz., $4\pi\rho u^2 r^2$, which corresponds to L/c for the stellar radiation field, where L is the stellar luminosity. The net vector momentum flux of a spherically sym-

unity: for example, in a scattering atmosphere in which there is multiple scattering of some photons (and all photons scatter at least once) $\eta_\infty > 1$. Integration of equation (13) from the atmospheric base to the region of asymptotic flow (denoted by the subscript ∞) yields

$$-\dot{M}u_\infty = \frac{\eta_\infty L}{c}\left(1 + \frac{v_{go}^2}{u_\infty^2}\,\frac{u_\infty}{2<u>}\right)^{-1}, \tag{14}$$

where the weighted average is defined by $<u>^{-1} = \int_{r_o}^{r_\infty} dr\ r_o/ur^2$. In the limiting case defined by $v_{go}^2 << 2<u>u_\infty$, equation (14) reduces to

$$-\dot{M}u_\infty \approx \eta_\infty L/c\ . \tag{15}$$

This limiting case is applicable to many radiatively driven winds from early-type stars, in which the flow accelerates rapidly ($<u> \gtrsim u_\infty/2$) and $v_{go}^2 << u_\infty^2$ (with $r_o = R_*$). It may also have some applicability to massive, late-type stellar winds in which the stellar atmosphere is extended by waves and the wind is accelerated to its asymptotic speed by the stellar radiation field. In this case, it is necessary that the radiation not interact strongly with the atmosphere close to the star, so that $r_o >> R_*$ (but $u_o^2 << u_\infty^2$) and $v_{go}^2 << u_\infty^2$; it is also necesssary that the (non-radiative) mechanism extending the atmosphere in $r < r_o$ is of negligible importance in $r > r_o$. The possibility that both of these requirements are met in some late-type stellar winds is considered when we discuss radiatively driven winds and shock-driven winds in sections 5 and 6.

In general, however, one must avoid applying equation (15) to the massive winds from low-gravity, late-type starts without carefully considering the assumption on which it is based.

2.3. Wind mass flux

If we can discover how the mass flux in the stellar wind is determined it will become possible to say something about where energy must be added to the wind in order to produce its observationally inferred mass flux and asymptotic flow speed. Consider first a thermally driven wind in which the temperature is constant: equations (1) and (2), with $D = 0$, yield

$$(u^2 - v_t^2)\,\frac{1}{u}\,\frac{du}{dr} = \frac{2}{r}\,(v_t^2 - \frac{1}{4}\,v_g^2)\,, \tag{16}$$

where $v_t^2 = p/\rho =$ constant. Equation (16) can be integrated from the coronal base to

metric wind or radiation field is, of course, zero.

the sonic[3] point in the flow ($r_s/r_o = v_{go}^2/4v_t^2$: e.g., Parker 1958) to obtain

$$\frac{u_o}{v_t} = \left(\frac{v_{go}^2}{4\,v_t^2} \right)^2 exp \left(- \frac{v_{go}^2}{2v_t^2} + \frac{3}{2} + \frac{u_o^2}{2v_t^2} \right), \tag{17}$$

where $-\dot{M}_* = 4\pi f \propto u_o$ [cf. equation (1)]. If the stellar atmosphere is strongly gravitationally bound at its base (i.e., $v_{go}^2 >> 4v_t^2$), then $u_o^2 << v_{go}^2$, the temperature dependence of u_o is dominated by the first term in the argument of the exponential in equation (17), and the wind mass flux increases exponentially with increasing atmospheric temperature. This strong dependence of mass flux on temperature results from the substantial extension of the atmosphere (and the consequent increase in atmospheric density at the sonic point in the flow) when the temperature is increased in a subsonic flow in which the pressure scale height is small in comparison with the size of the subsonic region. The dependence of the mass flux on atmospheric temperature becomes much weaker when the atmosphere is not strongly gravitationally bound ($v_{go}^2 \gtrsim 4v_t^2$), because in this case the pressure scale height is comparable to the size of the region of subsonic flow, and the atmospheric extension produced by an increase in temperature results in a rather small change in density at the sonic point of the flow.

In considering the wind mass flux, we have so far discussed only the region of subsonic flow, because in the isothermal wind relevant information from the region of supersonic flow cannot propagate upstream (i.e., as a compressive wave) and thus cannot reach the base of the atmosphere, to which it must penetrate in order to have an influence on the mass flux. If the sonic speed is defined as the speed at which compressive waves propagate, then in all the winds we consider below this general principle will continue to hold: viz., the wind mass flux is determined by effects occurring in the region of subsonic flow. This principle can, however, be violated. For example, in the radiatively driven winds of early-type stars, information concerning changes in the radiation field in the region of supersonic flow propagate upstream at the speed of light. Such effects will not, however, concern us in our consideration of the radiation-driving of dusty winds (see section 5).

From the preceding discussion of the mass flux in an isothermal wind, we can conclude that adding heat to the region of subsonic flow of a thermally driven wind produces an increase in mass flux through an increase in the average temperature of the subsonic region (e.g., Leer and Holzer 1979). A similar conclusion regarding the addition of momentum to the subsonic flow can be drawn from the generalization of equation (17) to the case where $D \neq 0$:

$$\frac{u_o}{v_t} = \left(\frac{v_{go}^2}{4\,v_t^2} \right)^2 exp \left[- \frac{v_{go}^2}{2v_t^2} \cdot \left(1 - \frac{2}{v_{go}^2} \int_{r_o}^{r_s} drD \right) + \frac{3}{2} + \frac{u_o^2}{2v_t^2} \right]. \tag{18}$$

[3] Sonic here refers to the isothermal sound speed, rather than the adiabatic sound speed defined in equation (5). The former has the subscript t in this paper and the latter has the subscript s.

Thus the addition of energy (by either heating or direct acceleration) to the region of subsonic flow increases the wind mass flux, but energy addition to the supersonic region has no effect on the mass flux (e.g., Leer and Holzer 1980).

2.4. Asymptotic flow speed

From equation (11), we can write the asymptotic flow speed of a highly supersonic wind in the form

$$u_\infty^2 \approx 2F_{do}/f - v_{go}^2 , \tag{19}$$

where F_{do}/f represents the energy-per-unit-mass added to the flow above the atmospheric base by the energy flux (over and above the atmospheric base enthalpy flux) responsible for driving the wind. The mass flux, f, is controlled by the fraction of this energy flux, F_{do}, that is deposited in the subsonic flow, so the asymptotic flow speed is sensitive not only to the magnitude of F_{do}, but also to the region where this energy flux is added to the flow. If equation (4) is evaluated at the sonic point of the flow, rather than at the atmospheric base, equation (11) becomes

$$F_{ds} \approx \frac{1}{2} f \left[u_\infty^2 + v_{gs}^2 - 2\left(\frac{\gamma+1}{\gamma-1}\right) v_{ts}^2 \right] . \tag{20}$$

In general, for the wind descriptions considered below, $v_{go}^2 >> | v_{gs}^2 - 2(\gamma+1)\, v_{ts}^2/(\gamma-1) |$ and $u_\infty^2 << v_{go}^2$, so $F_{do} >> F_{ds}$, and we conclude that all but a small fraction of the energy flux driving the wind is added to the flow in the subsonic region.

The constraints placed on the mass loss mechanism by observational inferences of the mass flux and asymptotic flow speed are thus rather severe: the driving energy flux must be of such a form and magnitude that when deposited largely in the region of subsonic flow it drives the observed wind mass flux, but the small fraction of this energy flux added to the supersonic flow must be just large enough to accelerate the wind to its low asymptotic flow speed. The exact nature of these constraints depends on the mass loss mechanism and stellar atmosphere under consideration, as will become clear below.

3. THERMALLY DRIVEN WINDS

When we refer to a stellar wind as thermally driven we mean that the only important outward force exerted on the expanding atmosphere is the thermal pressure gradient force: viz., $D \approx 0$ in equation (2) (e.g., Parker 1958, 1963, 1964a,b; Weymann 1960, 1962a,b). The viability of suggestions that the massive winds from late-type stars are thermally driven (e.g., Hearn 1975; Mullan 1978; Watanabe 1981) can be tested by using a simple approach in which the general nature of the atmospheric temperature structure required for thermal driving is determined, and the detectable radiation from an atmosphere with such a temperature structure is compared with the observed radiative signatures of the winds purported to be thermally driven. This approach, which is pursued below, leads us to the conclusions that thermally driven winds would produce much larger radiative energy fluxes than are observed (especially in the UV and EUV) and that the mechanical energy flux necessary to maintain the required atmospheric temperature structure is so large that its concomitant non-thermal force would dominate the thermal pressure gradient force over a significant fraction of the wind acceleration region. These two points are most easily illustrated by considering a thermally driven wind model in which the atmospheric temperature is held constant in the region where the flow speed is less than the thermal speed.

In section 2 [cf. equations (16) and (17)] it was shown that the flow speed in an isothermal wind depends only on v_t^2 and v_{go}^2 and thus only on the atmospheric temperature (T) and the stellar mass (M_*) and radius (R_*). Specification of the stellar mass loss rate ($-\dot{M}_*$) allows us to determine the radial profile of the atmospheric density through use of the mass conservation equation (1), along with equation (16), and thus to calculate the energy radiated from the atmosphere through application of a reasonable radiative loss function, which we take to be

$$Q_r = - n_e \, n_h \, \Phi(T) \, , \tag{21}$$

$$\begin{aligned} \Phi(T) = 10^{-22} \Big\{ & -0.3 \; exp \; [-100 \; (\; logT \; - 4.09 \;)^2 \;] \\ & + 2.5 \; exp \; [- 30 \; (\; logT \; - 4.3)^2 \;] \\ & + 4.5 \; exp \; [- 16 \; (\; logT \; - 4.9 \;)^2 \;] \\ & + 4.5 \; exp \; [- 16 \; (\; logT \; - 5.35 \;)^2 \;] \\ & + 2 \; exp \; [- 4 \; (\; logT \; - 6.1 \;)^2 \;] \Big\} \, . \end{aligned} \tag{22}$$

Equation (22) is a modified form of the function used by Athay (1981) and provides a reasonable fit to the loss function of Summers and McWhirter (1979) for $T > 1.5 \times 10^4 K$, to that of Raymond (in Rosner, Tucker, and Vaiana 1978) for $1.5 \times 10^4 > T > 10^4 K$, and to that used by Hartmann and MacGregor (1980) for $T < 10^4 K$. The electron density (n_e) and total hydrogen density (n_h) are estimated by using the solar chromosphere model of Vernazza, Avrett, and Loeser (1973), and helium is assumed to compose 10% (by number) of the atmosphere. These qualitative estimates allow us to approximate (adequately for our purposes) the radiative energy flux arising from the isothermal wind by [cf. equation (4)]

$$F_{ro} = - \int_{r_o}^{\infty} dr' \;\; Q_r \; r^2 \, . \tag{23}$$

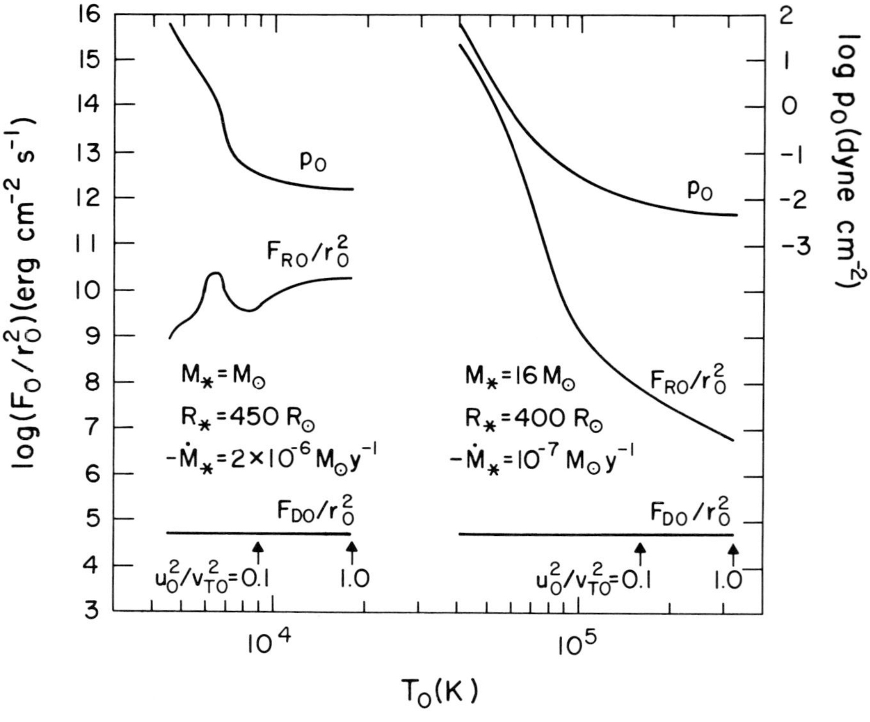

Figure 1. Radiative energy flux density normalized to the atmospheric base $(F_{ro}/r_o^{\,2})$ emitted by isothermal, thermally driven winds versus wind temperature (T_o). Each value of T_o corresponds to a complete wind model and characterizes both the wind dynamics and the spectral nature of the emitted radiation. Also shown are the atmospheric base pressure (p_o) required to supply the specified stellar mass loss rate $(-\dot{M}_*)$ in a wind of temperature T_o and the energy flux density $(F_{do}/r_o^{\,2})$ required to drive such a wind. Two stars are considered: a typical Mira, with $M_* = M_\odot$ and $R_* = 450\,R_\odot$, and a typical K5 supergiant, with $M_* = 16\,M_\odot$ and $R_* = 400\,R_\odot$. For each star, the wind temperatures (T_o) corresponding to $u_o^{\,2}/v_{to}^2 = 0.1$ and 1.0 are marked (u_o is the flow speed at the base of the atmosphere and v_{to} is the thermal speed characteristic of the wind).

With the use of equations (1), (16), and (21)-(23), we can now determine the radiative properties of massive, thermally driven winds from low-gravity stars for which the stellar radius (R_*), mass (M_*), mass loss rate $(-\dot{M}_*)$, and atmospheric temperature (T) are specified. Figure 1 shows the radiative energy flux density normalized to the atmospheric base $(F_{ro}/r_o^{\,2})$ for a range of atmospheric temperatures of two typical late-type, low-gravity stars: a Mira with $M_* = M_\odot$, $R_* = 450R_\odot$, and $-\dot{M}_* = 2 \times 10^{-6}\,M_\odot\,y^{-1}$; a K5 supergiant with $M_* = 16M_\odot$, $R_* = 400R_\odot$, and

$-\dot{M}_* = 10^{-7} M_\odot\, y^{-1}$. Also shown in Figure 1 are the atmospheric base energy flux density required to drive the wind (F_{do}/r_o^2 : cf. equation (11) with $u_\infty^2/v_{go}^2 = 0.1$) and the atmospheric base pressure (p_o). The rapid decrease of the pressure with increasing wind temperature results from the exponential increase of the wind speed at the atmospheric base with increasing temperature and the resultant exponential decrease of the density required to supply the specified mass loss rate [cf. section 2.3: particularly equation (17)].[4] The temperature range shown covers a range of isothermal winds from that in which the flow speed equals the thermal speed at the atmospheric base ($u_o = v_{to}$)[4] to those with base pressures comparable to or exceeding the expected photospheric pressure.

It is clear that all the winds included in Figure 1 have exceedingly large radiative energy flux densities--orders of magnitude larger than the solar radiative flux densities at comparable temperatures. Virtually all of this radiation is emitted from a region extending a few hundredths of a stellar radius (for the lower temperatures in each stellar atmosphere) to a few tenths of a stellar radius (for the higher temperatures) above the stellar base. It is, of course, the extent of this radiating region, which is orders of magnitude larger than the extent of the comparable region in the sun, that accounts for the magnitude of the radiative energy fluxes. Such large radiative fluxes would be readily detectable, as they are often comparable to the stellar luminosity and always at a temperature much higher than that characterizing the bulk of the stellar radiation. But these fluxes are not observed, and this is the first argument against the massive winds from cool, low-gravity stars being thermally driven.

The second argument relates to the mechanical energy flux required to supply this radiative energy flux and thus maintain the relatively high atmospheric temperature required by a thermally driven wind. As a general rule, one can represent a mechanical energy flux density (F_m/r^2) as the product of the characteristic energy density (ϵ_m) and a characteristic speed (v_m) at which the energy is transported :

$$F_m \approx \epsilon_m v_m r^2 . \tag{24}$$

Such a mechanical energy flux not only heats the atmosphere as it is dissipated, but also exerts an outward force on the atmosphere, which is analogous to the thermal pressure gradient force and results from the outward decline of energy density (ϵ_m):

$$\rho D_m \approx -\frac{d\epsilon_m}{dr} \approx \epsilon_m \left\{ \frac{1}{\lambda_m} + \frac{2}{r} \right\} , \tag{25}$$

where λ_m is the characteristic length over which the mechanical energy flux is dissipated, $r/2$ reflects the spherical expansion of a radially propagating mechanical

[4] The exponential increase of u_o and decrease of p_o with increasing T_o occurs only when the sonic point of the flow lies well above the atmospheric base and $u_o^2 << v_{to}^2$. The relatively weak dependence of u_o and p_o on T_o when $u_o^2 \gtrsim 0.1\, v_{to}^2$ also marks a transition to winds that are not properly called thermally driven, because of nonthermal forces necessarily exerted on the flow below the nominal atmospheric base (r_o).

energy flux, and D_m is the force per unit mass of equation (2).[5] For the wind models of Figure 1, the dissipation length (λ_m) is comparable to the pressure scale height, and if we apply the reasonable[5] constraint that $v_m \lesssim 10 v_{to}$, it follows that $\rho D_m >> - dp/dr$ over a sufficiently large region that the atmosphere will be much more extended than suggested by wind models including only thermal driving. The thermally driven wind assumption is thus violated.

If there is one point to remember from the preceding discussion, it is that thermal driving is a very inefficient mechanism for cool, massive winds from low-gravity stars, simply because virtually all the mechanical energy flux required to maintain the high wind temperature is radiated away, and only a tiny fraction of this energy flux is actually converted into that which drives the wind. This point and the other conclusions we have drawn are not dependent on the isothermal assumption characterizing Figure 1: thermally driven wind models, for example, that are characterized by constant flow speeds and monotonically declining temperatures ($T \propto r^{-1}$) produce qualitatively the same results. It has, however, been suggested by Watanabe (1981) that there exists a very special type of atmospheric temperature structure which provides a way of getting around the aforementioned difficulties characteristic of thermally driven winds. Before leaving this subject we will examine the two primary reasons why Watanabe's idea does not work.

Watanabe (1981) has pointed out that if the dissipation length (λ_m) characterizing the mechanical energy flux is very small ($\lambda_m / R_* \approx 10^{-5}$), the atmospheric temperature can rise to a high level and fall back to its radiative equilibrium value in such a short distance (near the atmospheric base) that an appropriately small radiative energy flux will be emitted at the elevated temperatures. Provided the peak temperature is high enough that $v_{tp}^2 \gtrsim v_{go}^2/2$ (the subscript p refers to the temperature peak), a thermally driven wind can be produced. The asymptotic flow speed of such a wind will be $u_\infty^2 \approx 2v_{tp}^2 - v_{go}^2$, and because $u_\infty^2 \lesssim 0.2 v_{go}^2$ for the winds under consideration, it follows that $0.5 \lesssim v_{tp}^2/v_{go}^2 \lesssim 0.6$. This constraint means that both the mechanical energy flux and its dissipation length must be extremely finely tuned to produce a low-speed, massive, thermally driven wind. Such fine tuning seems unlikely to occur over a wide range of stars, so such a mechanism is unlikely to have very general applicability (cf. section 4).

A more fundamental difficulty with Watanabe's (1981) wind models results from the fact that the required radial structures of the atmospheric temperature and flow speed are inconsistent with the assumption that the winds are thermally driven [i.e., the assumption that $D \approx O$ in equation (2)]. To illustrate this difficulty, let us consider the case (representative of Watanabe's models) in which the temperature increases over a fraction of a stellar radius from T_1 to its maximum value of T_2 where ($T_2 >> T_1$), and let us assume that the flow speed is comparable to or just larger than the thermal speed ($u_2 \approx v_{t2}$) at the temperature peak.[6] Using equation (5), we

[5] Equations (24) and (25) are intended only as qualitative indicators of the effects of a mechanical energy flux on an ambient atmosphere. Appropriate forms of equations (24) and (25) for the specific mechanical energy transport mechanisms of weak acoustic shock waves and intermediate hydromagnetic (Alfven) waves are given later in this section and in section 4.

[6] Note that the apparent disparity between this case and Watanabe's model shown in his Figure 3 results from the mislabelling of the flow speed scale in that figure. The flow speeds appropriate to his temperature profile are actually a factor of 10 smaller than indicated in the figure, and the mass loss rate is more than a factor of 20 smaller than indicated in his Table 1.

can integrate equations (1) and (2) over the region of this temperature increase to obtain

$$\rho_1 u_1 = \rho_2 u_2 \tag{26}$$

$$\rho_1 u_1^2 + \rho_1 v_{t1}^2 = \rho_2 u_2^2 + \rho_2 v_{t2}^2 \ . \tag{27}$$

Applying the conditions $u_2 \approx v_{t2}$ and $v_{t2}^2 >> v_{t1}^2$ (i.e., $T_2 >> T_1$), equations (26) and (27) can be combined to yield

$$u_1 \approx v_{t1}^2 / 2v_{t2} \ . \tag{28}$$

Now, integration of the energy balance equation (4) over the same region leads to

$$\frac{1}{2} u_1^2 + \frac{5}{2} v_{t1}^2 + \frac{\Delta F_m}{f} \gtrsim \frac{1}{2} u_2^2 + \frac{5}{2} v_{t2}^2 \ , \tag{29}$$

where the effects of radiative cooling, which can be minimized by narrowing the region over which the temperature increases, are accounted for by the inequality. Here, ΔF_m refers to that portion of the mechanical energy flux that is dissipated as heat in the region of integration. Making use of equation (28) and the conditions $v_{t2}^2 >> v_{t1}^2$ and $u_2 \approx v_{t2}$, equation (29) yields

$$\Delta F_m / r^2 \gtrsim 3\rho_2 v_{t2}^3 \tag{30}$$

The dissipation of such a large mechanical energy flux inevitably involves the exertion of some nonthermal force on the gas (i.e., $D > O$), and equation (30) allows us to estimate the magnitude of this force, which must appear in the momentum balance equation (27). If the mechanical energy flux is transported by weak acoustic shock waves, the associated nonthermal force is (e.g., Hartmann and MacGregor 1980)

$$\rho D \approx \frac{d}{dr} \left(\frac{F_m}{v_t r^2} \right) , \tag{31}$$

so the correction term in equation (27) is greater than $3\rho_2 v_{t2}^2$, which is substantially larger than any of the other terms in the equation. Thus, if weak acoustic shocks transport the mechanical energy flux required by Watanabe's postulated temperature profiles, acceleration of the wind to supersonic flow is *dominated* by a nonthermal force (i.e., $D > O$), and the wind cannot be considered to be thermally driven. To avoid this pitfall, the mechanical energy flux must be transported at a characteristic speed $v_m >> v_{t2}$. We are not aware of such a transport mechanism that is also consistent with energy dissipation over such a narrow region as is required in Watanabe's (1981) wind models.

4. WAVE-DRIVEN WINDS

If the thermal pressure gradient force alone cannot drive massive winds from cool, low-gravity stars, we must turn our attention to other possible driving forces. One such force, as noted above, arises from the transport of a mechanical energy flux through the stellar atmosphere. Let us first consider energy transport by small amplitude waves, including weak shock waves (strong shock waves driven by stellar pulsations are discussed in section 6). In the presence of a magnetic field, small amplitude waves may be either compressive (e.g., acoustic waves) or non-compressive (e.g., Alfvén waves).[7] The compressive waves tend to steepen quite rapidly into (weak) shocks and are dissipated within a few pressure scale-heights of the atmospheric base. Thus, although these waves can extend the atmosphere near its base, they are unable to lift a massive wind out of the stellar gravitational field (e.g., Hartmann and MacGregor 1980). Because Alfvén waves are non-compressive, they are much less readily damped and thus are able to extend a stellar atmosphere over a much larger radial distance and can lift a massive stellar wind out of the stellar gravitational field (e.g., Belcher and Olbert 1975; Hartmann and MacGregor 1980; Haisch, Linsky, and Basri 1980; Holzer, Flå, and Leer 1983).

To gain a feeling for how Alfvén waves can drive a stellar wind, let us first consider the particularly simple case of undamped Alfvén waves in a stellar atmosphere (Belcher and Olbert 1975). Alfvén waves exert a force on a stellar atmosphere because the variation of the wave energy density in space leads to a wave pressure gradient force, which is analogous to the thermal pressure gradient force. The wave energy density is

$$\epsilon = <\delta B^2>/4\pi = \rho <\delta v^2> . \tag{32}$$

where $<\delta B^2>$ and $<\delta v^2>$ are the mean-square magnetic field and velocity field of the wave. (Because the Alfvén wave is a transverse wave that propagates parallel to the background magnetic field, B, the wave magnetic and velocity fields are perpendicular to the background magnetic field, parallel to each other, and 90° out of phase with each other.) The energy flux density transported by an outward propagating Alfvén wave in spherically symmetric wind with radial magnetic field is

$$F_w / r^2 = \epsilon \left(v_a + \frac{3}{2} u \right) , \tag{33}$$

where the Alfvén speed, v_a, is defined by

[7] In a plasma where $v_a{}^2 > v_s{}^2$ ($v_a{}^2 = B^2/4\pi\rho$, the Alfvén speed), the hydromagnetic wave modes are the slow (acoustic), intermediate (Alfvén), and fast (magnetoacoustic). The fast-mode wave has characteristics of both the slow and intermediate modes and will not be considered specifically here.

$$v_a^{\,2} = B^2/4\pi\rho \ . \tag{34}$$

Thus the total energy per unit mass available to the flow, if all the Alfven wave energy flux can be converted to flow energy, is

$$F_w/f \;=\; <\delta v^2> (v_a/u \;+\; 3/2) \ . \tag{35}$$

When the Alfven waves are undamped (i.e., there is no heating of the wind by wave dissipation) this energy conversion must be accomplished entirely through the action of the wave pressure gradient force, which enters the wind momentum balance equation (2) in the form

$$D \;=\; -\frac{1}{\rho}\,\frac{d}{dr}\left(\frac{<\delta B^2>}{8\pi}\right) \ . \tag{36}$$

If the Alfvén waves are to remain undamped, their amplitude must be small (large-amplitude Alfvén waves are compressive), and in the region of subsonic[8] flow this requirement implies that $u^2 << v_a^{\,2}$: i.e., the subsonic flow is highly sub-Alfvénic. In the sub-Alfvénic region, only a small fraction of the Alfvén wave energy flux is transferred to the flow, so equations (33) and (34) yield

$$<\delta B^2> \;\approx\; <\delta B_o^{\,2}> (\rho/\rho_o)^{\frac{1}{2}} \ . \tag{37}$$

The sonic[8] point in a flow dominated by the Alfvén-wave pressure gradient force is thus found from equations (2), (36), and (37) to occur near $1.75R_\odot$, and the mass loss rate associated with such a flow is (e.g., Holzer et al. 1983)

$$-\dot{M}_* \;\approx\; 1.8 \times 10^{-13} \left(\frac{F_{wo}/r_o^{\,2}}{10^6\, B_o}\right)^2 \left(\frac{R_*}{R_\odot}\right)^{\frac{7}{2}} \left(\frac{M_\odot}{M_*}\right)^{3/2} M_\odot y^{-1} \ , \tag{38}$$

where $F_{wo}/r_o^{\,2}$ is the Alfvén wave energy flux density (in $erg\ cm^{-2}\ s^{-1}$) emanating from the base of the atmosphere, and B_o is the atmospheric base magnetic field (in *gauss*). Because only a small fraction of the Alfvén wave energy flux is deposited in the region of subsonic flow, most of it must be transferred to the supersonic wind, and the asymptotic flow speed is thus quite large (cf. section 2.4). From equations (19) and (35), together with the requirements that the sonic point occur near $1.75R_\odot$ and that $<\delta v^2> \;\approx\; v_g^{\,2}$ at the sonic point, we find that

[8] Sonic here refers to the speed of a compressive wave in the presence of Alfvén waves, which modify the compressibility of the medium. The square of this modified sound speed is $v_s^{\,2} + <\delta v^2>/4$ (e.g., Holzer et al. 1983).

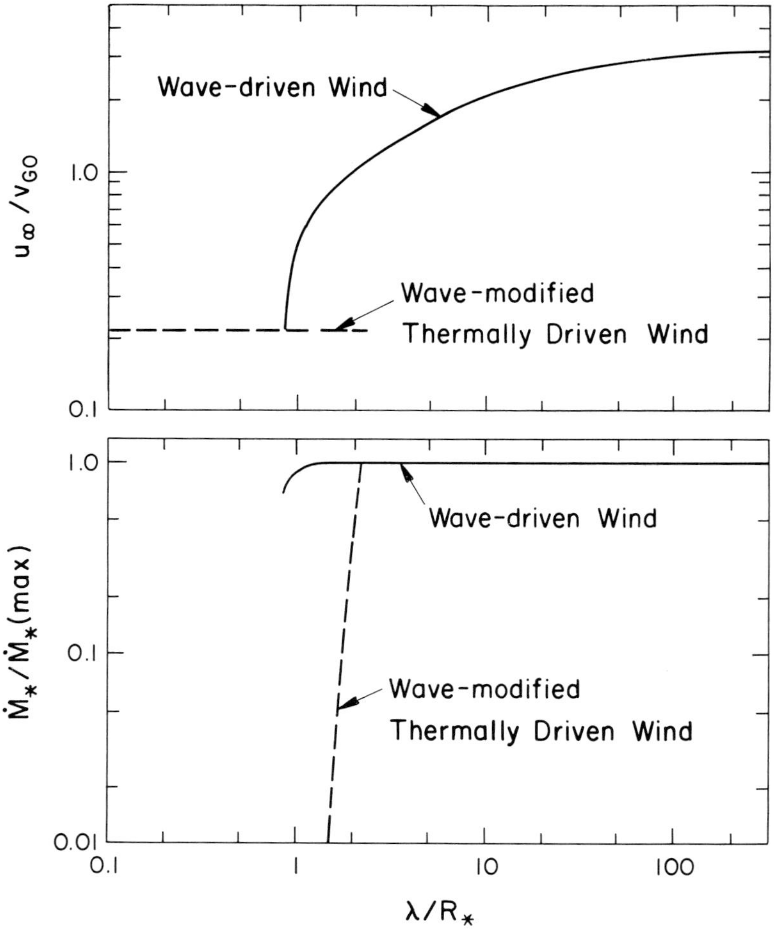

Figure 2. Asymptotic flow speed (u_∞) and mass loss rate ($-\dot{M}_*$) as functions of the constant damping length (λ) of Alfvén waves in wave driven winds from our typical K5 supergiant: $M_* = 16 M_\odot$, $R_* = 400 R_\odot$, $T = 10^4 K$ (isothermal wind), $B_o = 10$ *gauss*, $F_{wo}/r_o^{\,2} = 3.36 \times 10^6$ $erg\ cm^{-2} s^{-1}$, $\rho_o = 1.07 \times 10^{-13}$ $g\ cm^{-3}$. The maximum mass loss rate occurs for $\lambda \rightarrow \infty$ and is $-\dot{M}_*$ (max) $= 5 \times 10^{-7}$ $M_\odot\ y^{-1}$. The wave-driven wind becomes sonic near 1.8 R_*, and both its mass flux and asymptotic flow speed are determined by the Alfvén wave pressure gradient force. The wave-modified thermally driven wind becomes sonic near 31 R_*, and its mass flux is determined by the wave force, while its asymptotic flow speed is determined by the thermal pressure gradient force. For damping lengths less than about 0.85 R_* the waves cannot lift the wave-driven wind out of the stellar gravitational field, and for damping lengths greater than about 2.25 R_*, the atmosphere is so dominated by waves that the thermally driven wind does not exist. The wave-modified thermally driven wind is probably an artifact of the isothermal assumption and is unlikely to exist in a real stellar atmosphere. (This figure is based on results of Holzer et al. 1983.)

$$u_\infty^2 \approx v_{go}^2 \left[\frac{8}{7} \left(\frac{v_a}{u} \right)_s + \frac{5}{7} \right], \qquad (39)$$

where the subscript s refers to the sonic point of the flow.

Evidently, the mass loss rates of the typical Mira and K5 supergiant winds of section 3 (cf. Fig. 1) can be driven by undamped Alfvén wave energy flux densities of $7.6 \times 10^4 B_o$ and $1.7 \times 10^5 B_o$ $erg\ cm^{-2} s^{-1}$ at the atmospheric base, but such wave-driven winds are characterized by $u_\infty^2 >> v_{go}^2$ rather than by the observationally inferred $u_\infty^2 << v_{go}^2$. Hartmann and MacGregor (1980) pointed out that if the Alfvén waves are damped over a characteristic length of about a stellar radius very little energy is deposited in the region of supersonic flow, and the asymptotic flow speed is substantially reduced. Most of the wave energy flux in such a wind model is transformed into thermal energy of the expanding atmosphere and is radiated away. The wave damping thus provides an extended warm atmosphere (which is an observed characteristic of many of the winds under consideration: e.g., Goldberg 1984), a large mass loss rate (because the damping has little effect on the wave energy flux in the region of subsonic flow), and a low asymptotic flow speed (because the wave energy flux is severely reduced in the region of supersonic flow). The range of damping lengths that can provide these effects is, however, quite small, as is illustrated in Figure 2 for the typical K5 supergiant of section 3.

The question arises as to what mechanism(s) can provide a damping length in this narrow range for a large number of stars (assuming the Alfvén wave driving mechanism is responsible for many of the massive winds from late-type stars). Hartmann and MacGregor (1980) have noted that, for a reasonable wave period, ion-frictional damping of the Alfvén waves provides an appropriate damping length for our typical K5 supergiant. Yet for a reasonable frequency *spectrum* of Alfvén waves, it seems unlikely that an appropriate damping length will be achieved and the basic question of how a stellar atmosphere might finely tune the characteristic Alfvén wave damping length (perhaps, through some self-regulating mechanism) remains open (e.g., Holzer et al. 1983). This difficulty with the Alfvén wave driving mechanism is one that is encountered when considering almost any other driving mechanism: as noted in section 2.4, the driving energy flux must be of such a form that most of the energy addition to the flow occurs in the subsonic region and that the small amount of energy added to the flow in the supersonic region is just the right amount to accelerate the wind to its low asymptotic flow speed.

The physical process whereby this difficulty can be overcome has not yet been discovered, but the possibility of such a discovery still exists. Of course, if the damping length difficulty is resolved, then one must ask whether Alfvén wave energy flux densities of some $10^5\ B_o$ $erg\ cm^{-2}\ s^{-1}$ are to be expected at the surfaces of cool, low-gravity stars (note that to avoid strong nonlinear damping of the waves in the subsonic region it is required that $B_o \gtrsim 10$ *gauss*). Unfortunately, our understanding of the nature of stellar magnetic fields and of the generation of atmospheric turbulence seems inadequate to resolve this last question at present.

5. RADIATION-DRIVEN WINDS

The electromagnetic radiation field of a star provides a reservoir of momentum that can be tapped to drive a stellar wind, if a suitable mechanism exists for coupling the radiation field to the gas in the stellar atmosphere. In hot stars the radiative flux is substantial in the ultraviolet portion of the spectrum, where the atmosphere exhibits many strong resonance lines, and this good spectral match between the stellar radiation and the atmospheric opacity provides the coupling necessary for radiative driving of massive stellar winds. The spectral match is not so favorable in cool stars, where the stellar radiation field peaks in the red or near-infrared, and the strong resonance lines of the atoms and ions expected to be prevalent in the wind occur in the visible and ultraviolet. Both molecules (e.g., Fix and Alexander 1974; Maciel 1976, 1977) and dust (e.g., Gehrz and Woolf 1971; Gilman 1972; Salpeter 1974; Kwok 1975; Goldreich and Scoville 1976; Lucy 1976; Menietti and Fix 1978; Philips 1979; Woodrow and Auman 1982; Tielens 1983; Jura 1984) have been suggested as possible sources of opacity that can couple the stellar radiation field to the wind. Quantitative evidence that molecules can provide a sufficiently strong coupling mechanism has yet to be presented. Although the same statement can be made for dust, the idea that radiation pressure on dust grains can drive the observationally inferred massive winds from cool, low-gravity stars has become quite popular in the last several years. Despite this popularity, little attention has been paid to some basic dynamical and energetic constraints on the dust mechanism, so we will focus our attention on these constraints. Because of this focus, we will not give a broad review of the theoretical work in this area; interested readers are referred to the recent, excellent review of Goldberg (1984).

The force per unit mass associated with radiative driving of a stellar wind is usually written in the form [cf. equation (2)]

$$D = \frac{GM_*}{r^2}\,\Gamma(r) = \frac{v_g^{\,2}}{2r}\,\Gamma(r)\,, \tag{40}$$

where Γ is the ratio of the magnitude of the local radiative force and the local gravitational force. If the r^{-2} factor in equation (40) is associated with the radial variation of the radiative momentum flux density, then Γ can be considered a measure of the local strength of momentum coupling between the radiation field and the wind. In the dust-driven wind models considered here, the dust and gas are always sufficiently strongly coupled that all radiative momentum coupled to the dust can be assumed to be frictionally coupled to the gas, which accounts for most of the stellar wind mass. Thus, the radial variation of Γ is determined by such effects as grain condensation and destruction, spectral redistribution of the emergent stellar radiation through interaction with the dust, and radial variation of the dust density through atmospheric expansion. As is shown below, the accurate determination of $\Gamma(r)$ is the single most important task confronting the builder of dust-driven wind models, but it is so difficult a task that it has yet to be satisfactorily completed for any stellar wind purported to be driven through radiation pressure on dust grains.

Let us consider some representative forms of $\Gamma(r)$ and examine the nature of the associated radiation-driven winds. First, we assume that dust grains are formed over a very narrow radial region located at $r = r_c$, so that $\Gamma(r)$ has the form of a step function, with $\Gamma = 0$ in $r < r_c$ and $\Gamma \gtrsim 1$ at $r = r_c$ (the variation of Γ in $r > r_c$ will be considered below). It is clear from equations (1), (2), and (40) that a radiation-driven wind characterized by such a Γ will make its transition from subsonic to supersonic flow in the grain formation region (i.e., at $r = r_c$), so the stellar mass loss rate is

$$-\dot{M}_* \approx 4\pi \, \rho_c \, v_{tc} \, r_c^{\,2} \, . \tag{41}$$

Because the wind is taken to be radiation-driven, and no significant radiative force is exerted on the atmosphere in $r < r_c$, the mass density ρ_c and thermal speed v_{tc} are approximately those characterizing a hydrostatic atmosphere in radiative equilibrium in $r < r_c$. The pressure scale-height in such an atmosphere is quite small near the stellar surface, so the grain formation region must be located rather near the stellar surface if a substantial mass loss rate is to be effected. The asymptotic flow speed of this wind can be calculated using equation (11), if we note that

$$\int_{r_o}^{r_c} dr \; f D = 0 \, , \tag{42a}$$

$$\int_{r_c}^{\infty} dr \; f D = f \int_{r_c}^{\infty} dr \; \Gamma \, \frac{v_g^{\,2}}{2r} \approx \frac{1}{2} \, v_{go}^{\,2} \left(\frac{r_o}{r_c} \right) \left(\frac{\Gamma_c}{1+\beta_c} \right) , \tag{42b}$$

where $\beta_c = \langle -d \ln \Gamma / d \ln r \rangle$ is an average value over the region just beyond $r = r_c$. Equations (11) and (42) can be combined to yield

$$u_\infty^{\,2} \approx v_{go}^{\,2} \left(\frac{r_o}{r_c} \right) \left(\frac{\Gamma_c}{1+\beta_c} - 1 \right) . \tag{43}$$

If we are to ensure first that the wind carries a substantial mass flux (i.e., r_c / r_o not too large), second that the mass flux is lifted out of the gravitational field (i.e., $u_\infty^{\,2} > 0$), and third that the asymptotic flow speed is small (i.e., $u_\infty^{\,2} << v_{go}^{\,2}$), then it is clear that r_c, β_c, and Γ_c must be very finely tuned. The necessity of such fine tuning of model parameters is precisely the difficulty encountered (cf. section 4) when one proposes that wave-driven wind models can describe a broad range of massive winds from cool, low-gravity stars. We must remember that when the basic parameters of a stellar wind model need to be finely tuned to produce results that are in *qualitative* agreement with fundamental observational inferences (viz., large mass loss rate, low asymptotic flow speed), the wind model cannot be considered to have general applicability until the means whereby a star and its atmosphere can produce such fine tuning are discovered.

One might ask whether this difficulty disappears when physical effects that produce different forms of $\Gamma(r)$ are considered. For example, the constraint that the sonic point in the flow must occur relatively near the stellar surface (i.e., r_c/r_o not too large) and thus that $\Gamma \gtrsim 1$ relatively near the star can be eased if the radiative force near the star is large enough (but still $\Gamma < 1$) to extend the atmosphere substantially. Such atmospheric extension might result from gradual grain formation over a broad radial region near the star or a gradual increase in grain opacity with radial distance (e.g., "clean" grains formed near the star gradually become "dirty": Draine 1981). Similarly, the variation of the radiative force in the region of supersonic flow (measured by β_c above) is sensitive to a variety of processes determining grain formation and destruction, grain opacity, and the spectral redistribution of the radiation field. Yet, whatever the general form the radiative force (Γ) takes, analysis similar to that used in obtaining equations (41) - (43) shows that the parameters specifying $\Gamma(r)$ must be finely tuned if a wind with large mass flux and low asymptotic flow speed is to be driven by the radiative force alone.

Clearly, the grain formation problem is central to the determination of the radial structure of the radiative force $\Gamma(r)$. Although there have been extensive calculations of grain formation in the atmospheres of cool, low-gravity stars (e.g., Kwok 1975; Deguchi 1980; Draine 1981; Lefevre, Bergeat, and Daniel 1982; Woodrow and Auman 1982), questions have been raised about the consistency of such calculations with experimental studies (e.g., Nuth and Donn 1984). Until our knowledge of grain formation in stellar atmospheres is substantially advanced (presumably through a combined theoretical, experimental, and observational effort), it seems that our understanding of the role of radiation pressure on dust grains in driving massive, low speed stellar winds will be severely limited.

Before leaving this subject, let us briefly consider an idea that is currently thought to support the contention that some late-type stellar winds are radiatively driven. This idea (e.g., Castor 1981; Knapp et al. 1982) is that if the fraction of the stellar radiative momentum flux transferred to grains ($\eta_\infty L/c$: cf. section 2.2) can be shown observationally to equal the stellar wind momentum flux ($\dot{M}u_\infty$) over a broad range of stars, then one can infer that these winds are radiatively driven. As indicated in section 2.2, however, such an observational result would show (assuming single scattering of photons) that the radiation field has provided only a small fraction ($\approx u_\infty^2/v_{go}^2$) of the energy required to drive the wind: i.e., to lift the mass flux out of the stellar gravitational field and provide it with its asymptotic flow speed. It would thus be more reasonable to infer from such observations that some non-radiative mechanism does most of the work involved in driving the wind by extending the stellar atmosphere to a large radial distance (where $v_g^2 < u_\infty^2$), that grains form at this distance, causing the radiative force to dominate, and that the radiative force supplies the last, small bit of energy required to drive the wind to its asymptotic flow speed. This sort of process is discussed in the next section in the context of winds driven by pulsation-generated shocks, together with radiation pressure on dust grains.

6. SHOCK-DRIVEN WINDS

Many of the stars whose atmospheres are sufficiently dusty to provide strong coupling between the stellar radiation field and the atmosphere also exhibit regular pulsations, which can produce large-amplitude shock waves propagating outward through the atmosphere. It has been suggested, particularly for the Mira stars, that the heat and momentum added to the stellar atmosphere by such pulsation-driven shocks can play an important role in driving stellar mass loss (e.g., Willson and Hill 1979; Wood 1979, 1981; Willson 1981; Willson and Pierce 1981; Jones, Ney, and Stein 1981; Jura 1984). Although winds driven by large-amplitude shock waves are inherently non-steady, it is useful to view the *average* effect of quasi-periodic shocks as a steady process, in order to gain some feeling for the nature of the resulting winds and the relationship of the shock-driving mechanism to the other mass loss mechanisms we have discussed. When interpreting observations, however, one must consider not only the average characteristics of shock-driven mass loss, but also the shell structure associated with post-shock compressions, which should be most prevalent relatively near the star.

When a shock front passes through a fluid element in a stellar atmosphere, that fluid element experiences a rapid compressive heating and outward acceleration associated with the strong inward-directed pressure gradient characterizing the shock front. Behind the shock front the fluid element experiences a gradual expansive cooling and deceleration (i.e., an inward acceleration) associated with the modest (outward-directed) pressure gradient characterizing the back side of the shock wave. The strong heating and acceleration arising from passage of the shock front occurs over a relatively short time period (determined by the thickness and speed of the shock front), while the gradual cooling and deceleration on the back side of the shock wave occurs over a relatively long period (approximately the stellar pulsation period, τ). In the outflowing stellar wind, the extended cooling and deceleration of the fluid element are not sufficient to overcome the initial heating and acceleration in the shock front, and the net *average* effect of the entire shock wave is an acceleration and heating approximately described by [cf. equations (2) and (3)]

$$D_s \approx \delta u / \tau , \tag{44}$$

$$Q_s \approx \rho (\delta u)^2/\tau , \tag{45}$$

where δu is half the velocity jump across the shock front, and τ is the stellar pulsation period. The net heating in the stellar atmosphere must, however, also include the effect of radiative cooling, so the source term in equation (3) is given by equations (21) and (45):

$$Q = Q_s + Q_r + H_r \approx \rho (\delta u)^2/\tau - n_e n_h \Phi(T) + H_r . \tag{46}$$

Here we have added a radiative heating term, H_r, to the radiative cooling term, Q_r, given by equation (21). Calculations of the effects of pulsation-driven shock waves on a stellar atmosphere have been carried out for the two limiting cases in which radiative cooling and heating are (1) assumed negligible (adiabatic shocks: viz., $Q_s >> -Q_r$, and $Q_s >> H_r$), and (2) assumed completely dominant (isothermal shocks: viz., $Q_s << -Q_r$, and $-Q_r \approx H_r$). We summarize the results of the calculations of Wood (1979) by averaging them over time and presenting the results in

Figure 3. Wood's (1979) calculations were carried out for a Mira of mass $M_* = M_\odot$, radius $R_* = 450\ R_\odot$, and pulsation period $\tau = 373$ days, for which the observationally inferred mass loss rate is about $-\dot{M}_* \approx 2 \times 10^{-6}\ M_\odot\ y^{-1}$. The dotted curves in Figure 3 represent the radial temperature and density profiles characterizing the radiative-equilibrium hydrostatic atmosphere that Wood used as an initial condition in his calculations. Although none of the quasi-steady winds exhibits a density profile like that of the initial atmosphere, all the winds except that driven by adiabatic shocks (solid curves in Figure 3) exhibit the radiative equilibrium temperature profile. Wood (1979) considered four types of wind: a wind driven by adiabatic shocks (solid curves), a wind driven by isothermal shocks (dashed curves), a wind driven by radiation pressure (dash-double-dot curves), and a wind driven by both isothermal shocks and radiation pressure (dash-dot curves), with the same radiative force function $\Gamma(r)$ (cf. section 5) as that of the purely radiatively driven wind.

The most obvious characteristics of the wind driven by adiabatic shocks are its elevated temperature and its large flow speed at the base of the atmosphere, which corresponds to a mass loss rate of the order $-\dot{M}_* \approx 2 \times 10^{-2}\ M_\odot\ y^{-1}$ (about four orders of magnitude larger than the observationally inferred rate). This wind is driven primarily by the thermal pressure gradient force arising from shock heating [cf. equations (2), (3), and (45)]: the force arising from shock acceleration [cf. equations (2) and (44)] provides less than 20% of the driving force. Such a wind, of course, is plagued with the same difficulties as the purely thermally driven winds of section 3--namely, an unobserved high-temperature radiative signature and an extremely large radiative cooling rate. Near the atmospheric base, the radiative cooling rate is several orders of magnitude larger than the average shock heating rate in equation (46), so the adiabatic shock assumption (i.e., $-Q_r << Q_s$) is clearly violated. This strong radiative cooling is the reason that Wood, as well as Willson and her colleagues, considered the other extreme limit of isothermal shocks, in which shock heating is ignored.

Although the time-averaged parameters displayed in Figure 3 and used in equations (44) - (46) provide a relatively accurate and quite useful description of a wind driven by adiabatic shocks, they obscure one important aspect of a wind driven by isothermal shocks: viz., the process of shock coalescence. It is well known to solar wind researchers (e.g., Gosling, Hundhausen, and Bame 1976) that quasi-periodic shocks associated with corotating, high-speed solar wind streams coalesce as they are transported through interplanetary space and evolve from the original smaller scale shock structures with a characteristic period of a few days into larger scale structures with a characteristic period on the order of a month. This process of shock coalescence is not particularly important in the adiabatic-shock-driven wind of Figure 3, because the wind has already approached its asymptotic supersonic state, in which the mass loss rate and asymptotic flow speed are largely determined, before any coalescence occurs (the same is true of the wind with radiation pressure effects described below). In the wind driven by isothermal shocks, however, the implicit radiative damping of the shocks is so strong and the resultant decline of δu (cf. Figure 3) and D_s [cf. equations (2) and (44)] is so rapid that the wind remains subsonic out to the region where shock coalescence becomes significant.[9] Yet, the extension of

[9] There is the important additional effect of the inward movement of the coalescence region in this case, which occurs because of the minimal contribution of advection to the transport of shocks through circumstellar space, in

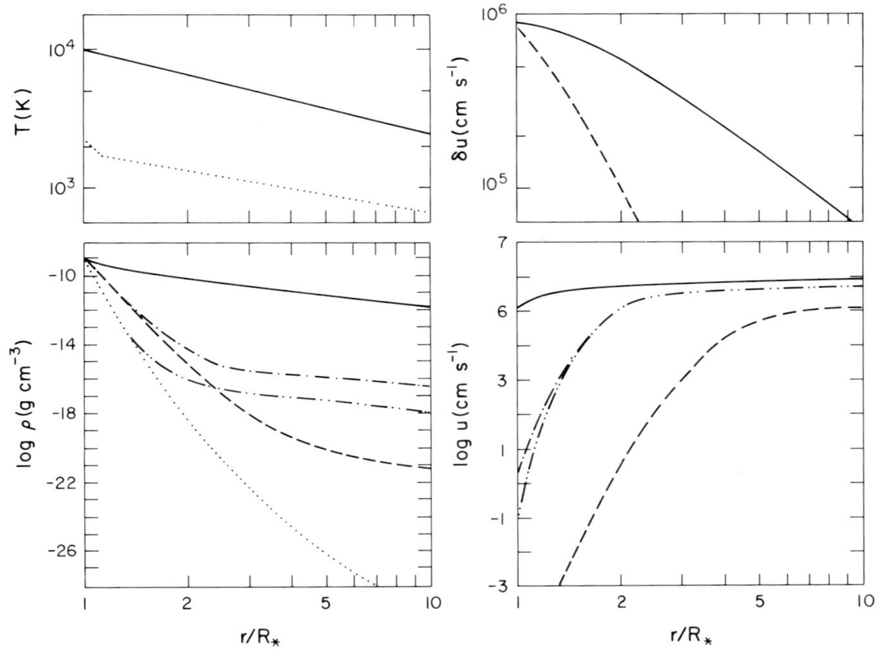

Figure 3. Time averaged radial profiles of atmospheric temperature (T), mass density (ρ), flow speed (u), and shock velocity amplitude (δu = half the shock front velocity jump) deduced from the shock-driven wind results of Wood (1979). Profiles of a hydrostatic, radiative equilibrium atmosphere (dotted curves), a wind driven by adiabatic shocks (solid curves), a wind driven by isothermal shocks (dashed curves), a wind driven by radiation pressure (dash-double-dot curves), and a wind driven by a combination of isothermal shocks and radiation pressure (dash-dot curves) are shown. The radiative equilibrium temperature profile characterizes all the winds except the one driven by adiabatic shocks. Wood's (1979) calculations were carried out for a Mira variable with $M_* = M_\odot$, $R_* = 450 R_\odot$, and pulsation period $\tau = 373$ *days*; the pressure at the stellar surface was taken to be $p_o = 10^2\ erg\ cm^{-3}$.

the atmosphere out to the coalescence region is sufficiently modest (see the density profiles of Figure 3) that the ejection of mass by the (long-period) coalesced shocks can produce a mass loss rate of only about $-\dot{M}_* \approx 10^{-12}\ M_\odot\ y^{-1}$, some six orders of magnitude less than the observationally inferred mass loss rate.

By making use of the temperature, density and shock velocity amplitude (δu) profiles of Figure 3, we find that for isothermal shocks the shock heating (implied by δu) dominates radiative cooling [cf. equation (46)], and the isothermal assumption (i.e., $Q_s << Q_r$) is violated. Evidently, the proper description of shock-driven winds lies somewhere between the adiabatic and isothermal descriptions and requires a full

contrast to the dominance of advective shock transport in the adiabatic case.

treatment of energy balance (as both Wood and Willson realize), but the development of such a description is considerably more difficult than the already quite difficult task of developing isothermal and adiabatic descriptions. A possible first step in the direction of the full treatment of energy balance might involve the use of a polytropic description of the jump conditions across the shock front. Although, the adiabatic index ($\gamma = 5/3$) would be used to describe the bulk of the flow, it would be replaced for the shock fronts by a polytropic index that could take on values less than 5/3 (but greater than 1, which corresponds to an isothermal shock). Estimates of average values of Q_r and Q_s [cf. equation (46)] could then indicate what range of polytropic indices can produce reasonably realistic (in terms of energy balance) winds and what the mass loss rates and asymptotic flow speeds of these winds are. Such an approach should, at least, help to clarify the nature of the role that pulsation-driven shock waves might play in driving massive winds from cool, low-gravity stars.

Wood (1979) carried out another study, again summarized in Figure 3, which gives us a feeling for the possible interplay between two (or more) mass loss mechanisms operating in a single stellar atmosphere. Wood chose, somewhat arbitrarily, a radial profile of a radiation force acting on the atmosphere (cf. $\Gamma(r)$ in section 5) that produced an asymptotic flow speed of about 5 $km\ s^{-1}$ when imposed on his initial hydrostatic, radiative equilibrium atmosphere. Then he added the effect of isothermal shocks and found that the mass loss rate was increased by some two orders of magnitude, owing to the extension of inner atmosphere (in $r \lesssim 2R_*$) through the outward force applied by the shocks (cf. $\rho(r)$ in Figure 3). In this model, the force due to the isothermal shocks dominates near the star ($r \lesssim 2R_*$), and the radiation pressure force dominates beyond ($r \gtrsim 2R_*$), so each mass loss mechanism is responsible for supplying about half the energy required to drive the resulting wind [note that $v_g^{\,2}(2R_*) = v_{go}^{\,2}/2$, and recall the implications of equations (4) and (11)]. Thus, although the radiation field has supplied far more energy to this wind than that associated with the wind momentum flux ($-\dot{M}_* u_\infty$: cf. sections 2 and 5), it can be considered no more than an equal partner (with the isothermal shocks) in driving the wind.

7. SUMMARY AND CONCLUSIONS

We have considered four mechanisms for driving massive winds from late-type stars. Although these mechanisms have been discussed in the context of steady (or quasi-steady), radial, spherically symmetric flow, their basic physical character remains essentially the same when the mechanisms are invoked in models of asymmetric, non-steady mass loss. If we can understand these mechanisms in the relatively simple context, we are thus provided with a sound basis for understanding their possible roles in the more complex context of a realistic stellar atmosphere.

There are, of course, other mass loss mechanisms that we have not specifically considered (e.g., Goldberg 1984). For example, Schwarzschild (1975) has suggested that large-scale convection might trigger sporadic ejection of massive clouds, and Mullan (1981) and Pneuman (1983) have proposed that magnetic-field-line reconnection might eject blobs of mass and magnetic fields from a stellar atmosphere. In the former case, no specific ejection process has been identified. In the latter case, as well as in the case of related proposals by Mullan (1980, 1982), there has not been sufficient quantitative analysis of the mechanisms to allow one to evaluate their viability or, in most cases, to understand fully their physical basis.

The general discussion of mass, momentum, and energy balance in section 2 raised a number of points (reiterated in sections 4-6) that are relevant to all mass loss mechanisms. In particular, it was noted that the low asymptotic flow speed ($u_\infty^2 << v_{go}^2$) observationally inferred to be a characteristic of all the massive winds from cool, low-gravity stars tells us several things about the nature of the mechanism(s) driving the winds. First, most of the energy added to the atmosphere in driving the wind goes into lifting the wind out the stellar gravitational field, and only a small fraction is reflected in the energy associated with the asymptotic flow. This fact implies, for example, that when one considers the origin of the stellar wind momentum flux ($-\dot{M}_* u_\infty$), one is dealing with only a small fraction [viz., $u_\infty^2/(u_\infty^2 + v_{go}^2)$] of the energy required to drive the wind, and one can thus expect to learn relatively little about the overall mass loss mechanism(s) responsible for driving the wind. Second, a large fraction of the energy required to drive the wind must be added in the region of subsonic flow, and the small fraction added to the supersonic flow is rather tightly constrained by the required low asymptotic flow speed. This constraint generally implies that the parameters of a wind model must be finely tuned if the model is to produce results consistent with the most basic observational inferences, and the requirement of such fine tuning calls into question the applicability of the model to a broad range of stars, unless some sort of atmospheric self-regulation mechanism that provides a physical basis for the fine tuning can be found. Third, much of the energy added to the wind must come in the form of momentum addition rather than heat addition. This requirement results from the fact that the thermal pressure gradient force, which is enhanced by heat addition, is a very inefficient driving mechanism for cool, dense winds, because of the enormous capacity of such winds for cooling through emission of electromagnetic radiation.

As illustrated specifically in section 6, more than one mass loss mechanism may play a significant role in driving a massive wind from a cool, low-gravity star, and different combinations of mechanisms may be dominant in different types of these winds. Eventually, quantitative descriptions of mass loss will have to be developed in the context of asymmetric, highly non-steady winds, as one can clearly infer from detailed observations of a star like α-Orionis (e.g., Goldberg 1979, 1984). Yet, whatever type of wind model is being considered, one should always try to remember the lessons that can be learned from the simple, steady, spherically symmetric descriptions considered above. These lessons should be equally useful for those interpreting observations and those constructing theoretical models.

ACKNOWLEDGEMENTS. We are grateful to Egil Leer for useful discussions, to Tom Bogdan for reviewing this paper, and to Lorraine Hori for preparing the camera-ready manuscript.

REFERENCES

Adams, W.S. and MacCormack, E. 1935, *Ap. J.,* **81,** 119.

Athay, R.G. 1981, *Ap. J.,* **249,** 340.

Belcher, J.W. and Olbert, S. 1975, *Ap. J.,* **200,** 369.

Cassinelli, J.P. and MacGregor, K.B. 1984, in *Physics of the Sun,* ed. P.A. Sturrock, T.E. Holzer, D. Mihalas, and R.K. Ulrich (Chicago: University of Chicago Press), in press.

Castor, J.J. 1981, in *Physical Process in Red Giants,* ed. I. Iben, Jr. and A. Renzini (Dordrecht: Reidel), p. 285.

Deguchi, S. 1980, *Ap. J.,* **236,** 567.

Deutsch, A.J. 1956, *Ap. J.,* **123,** 210.

________. 1960, in *Stars and Stellar Systems VI,* ed. J.L. Greenstein (Chicago: University of Chicago Press), p. 543.

Draine, B.J. 1981, in *Physical Processes in Red Giants,* ed. I. Iben and A. Renzini (Dordrecht: Reidel), p. 317.

Dupree, A.K. 1975, *Ap. J. (Letters),* **200,** L27.

Dupree, A.K. and Hartmann, L. 1980, in *IAU Colloquium 51, Stellar Turbulence,* ed. D.F. Gray and J.L. Linsky (New York: Springer-Verlag), p. 279.

Fix, J.D. and Alexander, D.R. 1974, *Ap. J. (Letters),* **188,** L91.

Gehrz, R.D. and Woolf, N.J. 1971, *Ap. J.,* **165,** 285.

Gilman, R.C. 1972, *Ap. J.,* **178,** 423.

Goldberg, L. 1979, *Quart. J. R. A. S.,* **20,** 361.

________. 1984, to appear in *CNRS-NASA Monograph Series on Nonthermal Phenomena in the Stellar Atmosphere* (Washington, D.C.: NASA), Ch. 7.

Goldreich, P. and Scoville, N. 1976, *Ap. J.,* **205,** 144.

Gosling, J.T., Hundhausen, A.J., and Bame, S.J. 1976, *J. Geophys. Res.,* **81,** 2111.

Hagen, W. 1980, in *Cool Stars, Stellar Systems, and the Sun,* ed. A.K. Dupree (Cambridge: Smithsonian Astrophysical Observatory Special Report 389), p. 143.

Haisch, B.M., Linsky, J.L., and Basri, G.S. 1980, *Ap. J.,* **235,** 519.

Hartmann, L. 1981, in *Solar Phenomena in Stars and Stellar Systems,* ed. R.M. Bonnet and A.K. Dupree (Dordrecht: Reidel), p. 33.

Hartmann, L., Dupree, A.K., and Raymond, J.C. 1981, *Ap. J.,* **246,** 193.

Hartmann, L. and K. B. MacGregor 1980, *Ap. J.,* **242,** 260.

Hearn, A.G. 1975, *Astron. Ap.,* **40,** 355.

Holzer, T.E., Flå, T., and Leer, E. 1983, *Ap. J.,* **275,** 808.

Jones, T.W., Ney, E.P., and Stein, W.A. 1981, *Ap. J.,* **250,** 324.

Jura, M. 1984, *Ap. J.,* **282,** 200.

Knapp, G.R., Phillips, T.G., Leighton, R.B., Lo, K.Y., Wannier, P.G., Wooten, H.A., and Huggins, P.J. 1982, *Ap. J.,* **252,** 616.

Kwok, S. 1975, *Ap. J.,* **198,** 583.

Leer, E. and Holzer, T.E. 1979, *Solar Phys.,* **63,** 143.

________. 1980, *J. Geophys. Res.,* **85,** 4681.

Lefevre, J., Bergeat, J., and Daniel, J.-Y. 1982, *Astron. Ap.,* **114,** 341.

Linsky, J.L. 1982, in *Advances in Ultraviolet Astronomy: Four Years of IUE Research,* ed. Y. Kondo, J.M. Mead, and R.D. Chapman (NASA Conf. Pub. 2238), p. 17.

Linsky, J.L. and Haisch, B.M. 1979, *Ap. J. (Letters)*, **229,** 127.
Lucy, L.B. 1976, *Ap. J.*, **205,** 482.
Maciel, W.J. 1976, *Astron. Ap.*, **48,** 27.
_______. 1977, *Astron. Ap.*, **57,** 273.
Menietti, J.D. and Fix, J.D. 1978, *Ap. J.*, **224,** 961.
Mullan, D.J. 1978, *Ap. J.*, **226,** 151.
_______. 1980, in *Cool Stars, Stellar Systems and the Sun*, ed. A.K. Dupree (Cambridge: Smithsonian Astrophysical Observatory Special Report 389), p. 189.
_______. 1981, in *Physical Processes in Red Giants*, ed. I. Iben and A. Renzini (Dordrecht: Reidel), p. 355.
_______. 1982, *Astr. Ap.*, **108,** 279.
Nuth, J.A. and Donn, B. 1984, *Bull. Am. Astron. Soc.*, **16,** 445.
Parker, E.N. 1958, *Ap. J.*, **128,** 664.
_______. 1963, *Interplantary Dynamical Processes* (New York: Interscience).
_______. 1964a, *Ap. J.*, **139,** 72.
_______. 1964b, *Ap. J.*, **139,** 93.
Philips, J.P. 1979, *Astron. Ap.*, **71,** 115.
Pneuman, G.W. 1983, *Adv. Space Res.*, **2,** 279.
Reimers, D. 1975, in *Problems in Stellar Atmospheres and Envelopes*, ed. B. Baschek, W.H. Kegel, and G. Traving (Berlin: Springer), p. 229.
_______. 1977, *Astr. Ap.*, **57,** 395.
Rosner, R., Tucker, W.H., and Vaiana, G.S. 1978, *Ap. J.*, **220,** 643.
Salpeter, E.E. 1974, *Ap. J.*, **193,** 585.
Schwarzschild, M. 1975, *Ap. J.*, **195,** 137.
Stencel, R.E. and Mullan, D.J. 1980, *Ap. J.*, **238,** 221.
Summers, H.P. and McWhirter, R.W.P. 1979, *J. Phys. B: Atom. Molec. Phys.*, **12,** 2387.
Tielens, A.G.G.M. 1983, *Ap. J.*, **271,** 702.
Vernazza, J.E., Avrett, E.H., and Loeser, R. 1973, *Ap. J.*, **184,** 605.
Watanabe, T. 1981, *Publ. Astron. Soc. Japan*, **33,** 679.
Weymann, R. 1960, *Ap. J.*, **132,** 380.
_______. 1962a, *Ann. Rev. Astron. Ap.*, **1,** 97.
_______. 1962b, *Ap. J.*, **136,** 476.
_______. 1962c, *Ap. J.*, **136,** 844.
Willson, L.A. 1981, in *Physical Processes in Red Giants*, ed. I. Iben and A. Renzini (Dordrecht: Reidel), p. 225.
Willson, L.A. and Hill, S.J. 1979, *Ap. J.*, **228,** 854.
Willson, L.A. and Pierce, J.N. 1981, in *Cool Stars, Stellar Systems, and the Sun*, ed. M.S. Giampapa and L. Golub (Cambridge: Smithsonian Astrophysical Observatory Special Report 392), Vol. I, p. 147.
Wood, P.R. 1979, *Ap. J.*, **227** 220.
_______. 1981, in *Physical Process in Red Giants*, ed. I. Iben and A. Renzini (Dordrecht: Reidel), p. 205.
Woodrow, J.E.J. and Auman, J.R. 1982, *Ap. J.*, **257,** 247.

PULSATION, MASS LOSS AND GRAIN FORMATION IN COOL GIANTS

M.J.Drinkwater and P.R.Wood
Mt. Stromlo and Siding Spring Observatories
Australian National University

ABSTRACT. The effect of pulsation on the outer layers of cool, luminous giant stars has been studied. Models typical of Mira variables and pulsating IRC and OH/IR sources are considered. Pulsation is assumed to be in the fundamental mode (first overtone pulsation in Mira variables was considered in an earlier paper - Wood 1979). The pulsation can not by itself produce significant mass loss rates. Radiation pressure acting on grains in the atmosphere of the static red giant models is also incapable of producing significant mass loss rates. Pulsation enhances the mass loss rate due to a grain-driven stellar wind by a factor of $\sim 10^6$ but, in the present models, the resultant mass loss rates are still small compared to observed values.

I. INTRODUCTION

No satisfactory explanation for large mass loss rates from red giants has yet been produced (eg. Castor 1981). One property of all cool red giants is that they vary to some extent, and the stars with the largest mass loss rates (Mira variables, OH/IR and other infra-red sources) all exhibit well-developed pulsations. This indicates that pulsation may well be implicated in the production of large mass loss rates. Here we present some results of an examination of the role played by pulsation in the mass loss process.

The interaction of pulsation and mass loss has been investigated previously by Wood (1979) and Willson and Hill (1979). The models calculated by Wood (1979) examined the properties of a typical Mira variable with P= 375 days under the assumption that the star was a first overtone pulsator (Wood 1982 discusses the arguments for first overtone pulsation in Mira variables). Here we examine the dynamical properties of the atmospheres of cool red giants pulsating in the fundamental mode. Because of the higher gravities of the fundamental mode pulsators (compared to first overtone pulsators), it was hoped that high shock velocities and hence high mass loss rates would result. However, it turns out that instead of one powerful shock per cycle developing in a fundamental mode pulsator, several weaker shocks develop (this will be discussed elsewhere). We explore the properties of several models with P=373, 600, and 1200 days; these periods cover the range spanned by the Mira variables, IRC sources, and OH/IR stars (eg. Jones *et al.* 1983).

M. Morris and B. Zuckerman (eds.), Mass Loss from Red Giants, 257–260.

The OH/IR stars are almost certainly fundamental mode pulsators (Jones *et al.* 1983, but see de Jong 1983), and Willson (1982) argues that the Mira variables are also fundamental mode pulsators.

The numerical method used here is almost identical to that given in Wood (1979). Basically, we vary the pressure periodically at the base of the stellar atmosphere to simulate the pulsation of the underlying star. An Eulerian code is used to follow the propagation of pressure and shock waves out through the atmosphere, the temperature of which decreases with radius but is constant in time.

II. RESULTS

The properties of the models studied are shown in Table 1. Masses, luminosities and periods were chosen to approximate observed values for Mira variables, IRC sources and OH/IR stars, based on results in Wood, Bessell and Fox (1983) and Jones *et al.*(1983). Radii (and then T_{eff} values) were derived by adopting a pulsation constant $Q = 0.1$, which is appropriate for fundamental mode pulsation (Fox and Wood 1982).

The behaviour of the present models as regards mass loss is very similar to the behaviour of the first overtone models studied by Wood (1979). Basically, the first shock propagates rapidly down the steep density gradient of the initial hydrostatic atmosphere (Figure 1). Material infalling after the passage of this shock impedes the progress of subsequent shocks. These shocks coalesce and gradually build up enough momentum to drive another shock away from the star. This process of coalescing of shocks from many cycles to drive occasional shocks away from the star is repeated continually (see Figure 1). The shells of gas driven away from the star by coalesced shocks may explain the observational result (Bernat 1981; Ridgway 1981) that circumstellar absorption lines indicate the presence around mass-losing stars of multiple discrete expanding shells.

Once the first two or three shocks have passed the outer boundary, the mass loss rates in models 1, 2 and 3 are much smaller than observed values. This is due to the very low density in the outer layers. Average mass loss rates after many periods are given for models 2 and 3 in Table 1.

For models 2 and 3, separate calculations were run in which grains were allowed to form in the cool atmospheric material, and then the effect of radiation pressure on these grains was studied when the

TABLE 1
Model Properties

Model	Type	$M/M_\odot$	$L/L_\odot$	P(days)	$R/R_\odot$	T_{eff}	$\langle\dot{M}_1\rangle$	$\dot{M}_2$	$\langle\dot{M}_3\rangle$
1	Mira	1	7050	373	240	3400	–	7×10^{-28}	–
2	IRC	1	11200	600	330	3340	1×10^{-20}	4×10^{-21}	3×10^{-15}
3	OH/IR	1	17700	1200	524	2970	8×10^{-13}	4×10^{-15}	2×10^{-10}

Notes. $\langle\dot{M}_1\rangle$ is the mass loss rate ($M_\odot$/yr) produced by pulsation alone.
$\dot{M}_2$ is the mass loss rate produced by radiation pressure on grains.
$\langle\dot{M}_3\rangle$ is the mass loss rate produced by the combination of pulsation and radiation pressure on grains.

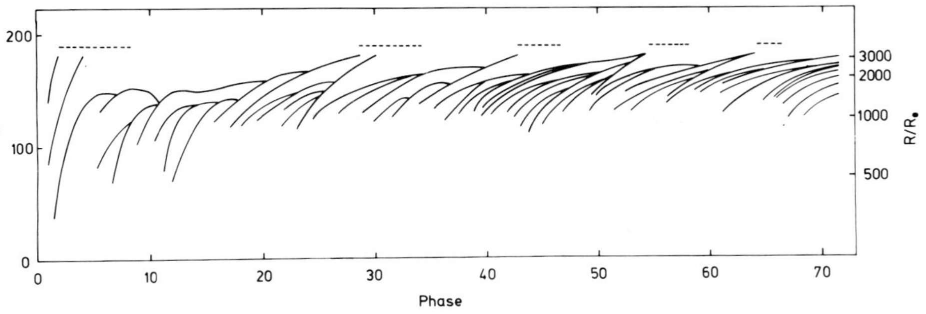

Figure 1. The position (zone number J) of various shock fronts in the atmosphere of model 2 as a function of phase of the pulsation cycle. Many shocks coalesce to drive occasional large shocks through the outer boundary. The dotted lines indicate these events. Note that the shock positions have only been plotted in the outer regions of the atmosphere where they move slowly.

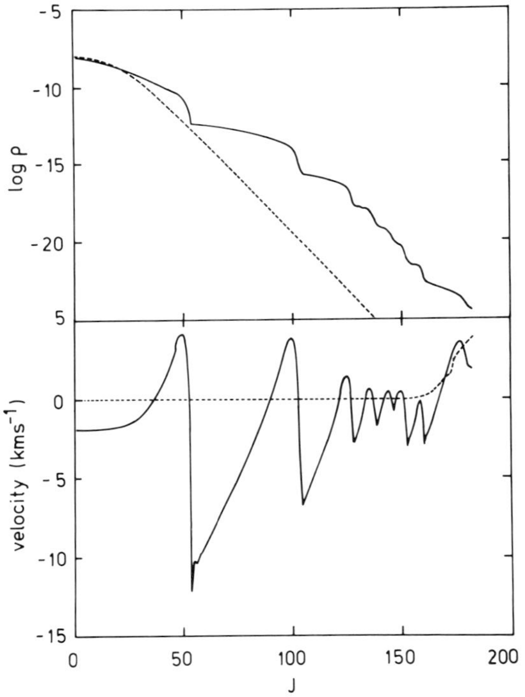

Figure 2. Density and velocity profiles through the atmosphere of model 2, when the action of radiation pressure on grains is considered. The dotted lines indicate the density and flow velocity in the absence of any pulsation.

gas and grains were dynamically coupled. An oxygen-rich atmosphere (solar abundance) was assumed, and the grain opacity was obtained from the tables of Alexander, Johnson and Rypma (1983). The results of these calculations were similar in all models: we will now discuss model 2 as a typical case.

With a constant pressure at the inner boundary a flow developed with a velocity at R= 3345 $R_\odot$ of 3.8 kms^{-1} (Figure 2). Extrapolating to larger radii using the conservation equations for adiabatic flow gives a velocity at R = ∞ of 12 kms^{-1}. The mass loss rate was 4×10^{-21} $M_\odot yr^{-1}$ (Note that this mass loss rate scales essentially linearly with the adopted pressure in the photosphere.) When the atmosphere was pulsated, its lower parts became extended, causing a higher density (by a factor of $\sim10^6$) at the base of the region where grains form and radiation pressure acting on the grains drives the mass flow. The density enhancement in the grain forming layers caused a proportionally large enhancement of the mass loss rate to ≈ 3×10^{-15} $M_\odot yr^{-1}$. Such a mass loss rate is still far below the observed mass loss rates (≈10^{-5} $M_\odot yr^{-1}$) for IRC sources with P≈600 days (Knapp *et al.*1982)

III. SUMMARY

The influence of stellar pulsation on the atmospheres of cool red giants has been studied. Pulsation by itself does not seem capable of generating a significant rate of mass loss in models for Mira variables, IRC sources or OH/IR stars. Mass loss driven by radiation pressure acting on grains is greatly enhanced by pulsation but the rate produced in these simple models still falls short of the very large mass loss rates observed in the OH/IR stars.

The models studied here treat the radial temperature variation in a very simple manner. A better treatment of the temperature structure may allow grains to form closer to the star, in which case the factor of $\sim10^6$ enhancement in mass loss rate produced by pulsation may lead to significant mass loss rates.

REFERENCES

Alexander,D.R., Johnson,H.R. and Rypma,R.L. 1983, Ap.J.,**272**,773.
Bernat,A.P. 1981, Ap.J.,**246**,184.
Castor,J.I. 1981, in "Physical Processes in Red Giants", eds. I.Iben and A.Renzini (Reidel:Dordrecht), p.285.
de Jong,T. 1983, Ap.J.,**274**,252.
Fox,M.W. and Wood,P.R. 1982, Ap.J.,**259**,198.
Jones,T.J., Hyland,A.R., Wood,P.R. and Gatley,I. 1983, Ap.J.,**273**,669.
Knapp,G.R., Phillips,T.G., Leighton,R.B., Lo,K.Y., Wannier,P.G. and Wootten,H.A. 1982, Ap.J.,**252**,616.
Ridgway,S. 1981, in "Physical Processes in Red Giants", eds. I.Iben and A.Renzini (Reidel:Dordrecht), p.305.
Willson,L.A. and Hill,S.J. 1979, Ap.J.,**228**,854.
Willson,L.A. 1982, in "Pulsation in Classical and Cataclysmic Variable Stars", eds. J.P.Cox and C.J.Hansen (Boulder:JILA reprint), p.269.
Wood,P.R. 1979, Ap.J.,**227**,220.
Wood,P.R. 1982, in "Pulsation in Classical and Cataclysmic Variable Stars", eds. J.P.Cox and C.J.Hansen (Boulder:JILA reprint), p.284.
Wood,P.R., Bessell,M.S. and Fox,M.W. 1983, Ap.J.,**272**,99.

EFFECTS OF ROTATION AND CONVECTION ON MASS-LOSS FROM RED GIANTS

S.R. Sreenivasan and W.J.F. Wilson
Department of Physics, The University of Calgary
Calgary, Alberta, Canada
T2N 1N4

1. INTRODUCTION

Scalo (1981) has drawn attention to the presence of peculiar red giants (or p giants as he abbreviates them) which show abundance anomalies and more recently Luck, Bond and Lambert (1984) have pointed out the presence of high latitude F supergiants which are found at heights of 1 - 5 kpc above the Galactic disk.

We demonstrate in this paper that both of these phenomena can be understood as a consequence of the combined effect of rotation and convection in red giants suffering mass loss.

2. THE OBSERVATIONS

2.1 High Latitude F Supergiants

An apparent puzzle concerning these high latitude F supergiants is that their luminosities are indicative of massive population I stars and their height above the Galactic plane is indicative of ages characteristic of lower mass population II stars. Recently Luck, Bond and Lambert (1984) reached the conclusion that they are "low mass population II stars masquerading as young massive objects" because they are rather metal-poor.

2.2 p Giants

These are late type luminous stars exhibiting peculiarities in their spectra. They show anomalous abundances of C and/or O and also heavy elements. Small $^{12}C/^{13}C$ ratios, as well as deficiencies in hydrogen and enhancements of Li, are present. Furthermore, p giants are found amongst pop I as well as pop II objects. The suspicion is that these anomalies are caused by some form of mixing between the interior and the surface. Since p giants are found among warm as well as cool stars, one requires a mechanism that operates continuously and in a general fashion. More details of the observational features are summarized in the comprehensive review by Scalo (1981).

M. Morris and B. Zuckerman (eds.), Mass Loss from Red Giants, 261–263.

3. THEORY

3.1 Effects of Rotation, Convection and Mass-Loss

We have recently examined the consequences of spinning massive stars losing mass. (Sreenivasan and Wilson 1978a, b; 1982; 1984a, b, c, d; Sreenivasan 1980 for a review). We have shown that a spinning model has an extended convective core and that the differential rotation between the core and surface induces shear turbulence caused by well-known hydrodynamic instabilities. This in turn generates the heating of the outer layers and a stellar wind. The stellar wind spins the surface down and exacerbates the turbulence in turn. The enlarged core increases the amount of nuclear fuel available and extends the main sequence lifetime of the models. When the models reach the red giant phase they generate winds analogous to the solar wind due to the acoustic energy flux liberated in the outer convection zone.

3.2 The Models, Computing Procedure and Input Physics

We have evolved 2 and 3 $M_{\odot}$ models of pop I composition (X=0.70;Y=0.27) and considered ZAMS initial rotational speeds of 50-100 km/s. The breakup speed for these models is 500 and 540 km/s respectively. The rest of the input physics is as described in our papers referred in the previous subsection.

We find that when the ZAMS initial rotational speed is at the lower end of the range the red giant branch ascends high enough to produce models which have the same luminosity as those of massive stars of $\sim$ 10 $M_{\odot}$.

When the initial rotational speed is slightly greater or we assume the same efficiency of generating nonthermal energy dissipation as the sun then the models are not as luminous along the red giant branch but display an increased amount of mixing and hence abundance anomalies.

The assumption of population II composition would not change our conclusions drastically.

4. CONCLUSIONS

We conclude that the interaction of rotation and convection adequately explains both the characteristics of p giants and those of high latitude F supergiants. Although other mechanisms might be found to explain these two seemingly different classes of objects, the simple mechanism we have described here resolves both the puzzles satisfactorily. Fuller details will be submitted for publication elsewhere.

The work reported here was supported in part by the Natural Sciences and Engineering Research Council of Canada and The University of Calgary.

REFERENCES

Luck, R.E., Bond, H.E., and Lambert, D.L. 1983, Pub.A.S.P., 95, 413.
Luck, R.E., Bond, H.E., and Lambert, D.L. 1984, Bull.A.A.S., 16, 490.
Scalo, J.M. 1981, in Physical Processes in Red Giants, ed. I. Iben Jr., and A. Renzini, (Reidel: Dordrecht, Holland), 77-114.
Sreenivasan, S.R 1980, in Highlights in Astronomy, 5, ed. P.A. Wayman, (Reidel: Dordrecht, Holland), 601-613.
Sreenivasan, S.R., and Wilson, W.J.F. 1978a, Astron. Ap., 70, 755-761.
Sreenivasan, S.R., and Wilson, W.J.F. 1978b, in Stellar Pulsation Instabilities, ed. D. Fischel, J.W. Lesh and W.M. Sparks, (NASA TM 80625: Greenbelt, MD), 363-380.
Sreenivasan, S.R., and Wilson, W.J.F. 1982, Ap.J., 254, 287-296.
Sreenivasan, S.R., and Wilson, W.J.F. 1984a, Ap.J., submitted.
Sreenivasan, S.R.., and Wilson, W.J.F. 1984b, in Origin of Non-Radiative Momentum and Heating in Hot Stars, ed. A.B. Underhill (NASA TM: Greenbelt, MD), in press.
Sreenivasan, S.R., and Wilson, W.J.F. 1984c, Ap.J., submitted.
Sreenivasan, S.R., and Wilson, W.J.F. 1984d, Ap.J., submitted.

HYDROGEN DEFICIENCY AND MASS LOSS FROM AGB STARS

Gordon C. Augason
Space Science Division
NASA-Ames Research Center
Moffett Field, CA 94035

The mechanism for mass loss from stars is not understood, but theoretical studies predict that it should accelerate once stars on the AGB begin thermal pulsations (TP) (Shoenberner, 1979). We have discovered that the stars on this part of the AGB are apparently hydrogen deficient which may result from the mass loss predicted by theory. Recent observations of non-Mira AGB stars (Goebel and Johnson, 1984, Augason et al, 1985) show that these stars have less flux at 1.6 microns (the H^- opacity minimum) than do models with solar abundance. We interpret this missing flux as being due to a lack of H^- resulting from a significant hydrogen deficiency. Johnson et al (1985) have computed opacity sampled models of cool, non-Mira carbon stars where hydrogen is replaced with helium and they find that it is necessary to reduce the hydrogen abundance of the models below 30% of the solar value to match the observations. Hydrogen deficiency is also supported by the nondetection of H_2 in: non-Mira stars (Tsuji, 1983), S stars (Johnson, 1983) and carbon stars (Johnson et al, 1983).

If hydrogen is truly missing it must either have been burned in the interior of the star accompanied by extensive mixing or the hydrogen rich outer layers of the star were removed by a mass loss process exposing hydrogen depleted material. We consider the possibility that the observed apparent hydrogen deficiency has resulted from a mass loss process. Using the infrared flux method of Blackwell and Shallis (1977) to determine effective temperatures and the period-luminosity relationship of Fox and Wood (1982), we have located 7 recently observed S stars (Augason et al, 1985) on the HR diagram. The S stars are found to fall into two groups: A warm group (apparently just beginning TP (Iben, 1982)) with a mild hydrogen deficiency and a cool group (higher up the AGB and undergoing TP) with a strong hydrogen deficiency. Iben and Renzini (1983) estimate that it should take between 10^4 and 10^6 years, depending on the mass, for a star to move up to the AGB from where TP pulsations begin to the region where TP cease and planetary nebulae (PN) form. If 70% of the hydrogen rich envelope of .2 $M_\odot$ (Iben and Renzini, 1983) is removed from a star in 1×10^5 years, the mass loss rate would be 1.4×10^{-6} $M_\odot$ yr^{-1}. This mass loss rate is

M. Morris and B. Zuckerman (eds.), Mass Loss from Red Giants, 265–267.

high compared to the Reimers (1975) rate of 6.2 x$10^{-8} M_\odot$ yr^{-1} but it is comparable to that observed for similar stars (Spergel et al, 1983).

It is believed that the AGB stars are precursors of Mira stars and PN. If AGB stars are truly hydrogen deficient then Miras and PN should be hydrogen deficient. Molecular hydrogen has been observed in Miras (Johnson et al, 1983). However, the amount of hydrogen is considerably less than that predicted by molecular equilibrium computations. Apparently the Miras have no more hydrogen in their atmospheres than do their progenitors undergoing TP on the upper part of the AGB. The detection of H_2 is apparently only possible because of their greatly extended atmospheres and increased column densities.

Currently there are two models for the formation of PN: the "superwind" model and the "interacting" wind model (Kwok, 1982). In the "superwind" model, a stellar wind produces a high mass loss rate towards the end of the AGB which is responsible for removing matter from the star and creating the PN. In the "interacting" wind model a high velocity wind at the very end of the AGB compresses the circumstellar material, previously ejected from the star, and creates the PN. In either case a PN should reflect the composition of the material which has been ejected from the AGB star up to ~10^4 years (the escape time for the stellar system) prior to the PN formation. Zuckerman and Aller (1985) find no evidence for hydrogen deficiency in the PN they have observed. Hydrogen deficiency should be accompanied by an increase in helium. If PN are composed of material half of which is hydrogen rich, ejected from the envelope of an AGB star during early TP and half is hydrogen depleted by conversion to helium and ejected from the envelope of an AGB star during later TP (Mira phase), the following He/H ratios are obtained: 70% depletion, He/H = 0.29; 50% depletion, He/H = 0.22 and 30% depletion, He/H = 0.16. Kaler (1979) found the He/H ratio for most of the PN he observed to be between 0.10 and 0.12 with a few PN having He/H between 0.12 and 0.24. If an envelope of an AGB star becomes hydrogen deficient during the last 10% of TP, the following He/H ratios are found in the PN: 70% depletion, He/H = 0.13; 50 depletion, He/H = 0.12 and 30% depletion, He/H = 0.11. While these He/H ratios are in reasonable agreement with the PN observations, they require that hydrogen depletion not occur until almost the end of the Mira phase. Our observations indicate that TP-AGB stars become hydrogen deficient before the Mira phase. It is difficult to understand TP-AGB stars we have observed in terms of current theories about Miras and PN. It is possible that these apparently hydrogen deficient stars have arrived at their position on the AGB by a yet unknown evolutionary path. However, it is difficult to accept this conclusion because of the number and types of stars that we have found there. Also, if these stars are not progenitors of PN, which stars are?

REFERENCES

Augason, G.C., Bregman, J.D., Johnson, H.R. and Witteborn, F.C. 1985, Ap.J., Submitted.

Blackwell, D.E. and Shallis, M.J. 1977, M.N.R.A.S., 180, 177.

Fox, M.W. and Wood, P.R. 1982, Ap.J., 259, 198.

Goebel, J.H. and Johnson, H.R. 1984, Ap.J. (Letters), 284, L39.

Iben, I. 1982, Ap.J., 260, 821.

Iben, I. and Renzini, A. 1983, Ann. Rev. Astr. Ap., 21, 271.

Johnson, H.R. 1983, Private Communication.

Johnson, H.R., Bernat, A.P. and Krupp, 1980, Ap.J., 206, 201.

Johnson, H.R., Goebel, J.H., Goorvitch, D., and Ridgeway, S.T. 1983, Ap.J. (Letters), 270, L63.

Johnson, H.R., Alexander, D.R., Bower, C.D., Lemke, D.A., Luttermoser, D.G., Petrakis, J.P., Reinhart, M.D., Welch, K.A., and Goebel, J.H. 1985, Ap.J., In Press.

Kaler, J.B. 1979, Ap.J., 228, 163.

Kwok, S. 1982, Ap.J., 258, 280.

Reimers, D. 1975, "Problems in Stellar Atmospheres and Envelopes", eds, Baschek B., Kegel W.H., and Traving G., Springer Verlag., Berlin.

Shoenberner, D. 1979, Astr. Ap., 79, 108.

Spergel, D.N., Giullani, J.L., and Knapp, G.R. 1983, Ap.J., 275, 330.

Tsuji, J. 1983, Astr. Ap., 122, 314.

Zuckerman, B. and Aller, L.H. 1985, Ap.J., Submitted.

THE PHYSICAL AND CHEMICAL STRUCTURE OF CIRCUMSTELLAR ENVELOPES

Alain Omont
Groupe d'Astrophysique, Université de Grenoble
B.P. 68
38402 Saint Martin d'Hères Cedex
France

ABSTRACT. The physical and chemical structure of massive circumstellar envelopes is reviewed in relation with infrared and millimeter observations, with a particular emphasis on the carbon-rich shell IRC+10216. The spatial distribution of the physical parameters (temperature, density, velocity field, mass-loss, ionization, etc.) and their relations are discussed. The different chemical processes responsible for the observed molecular abundances are examined: freezing-out of chemical equilibrium, radical and ion reactions, grain processes, photochemistry, etc.

1. INTRODUCTION

The observational knowledge and the analysis of the physical structure of circumstellar envelopes of late-type stars is of prime importance for the understanding of their evolution and basic processes such as dynamics, wind origin and dust formation. Similarly, detailed studies of their chemical composition are indispensable to derive element and isotopic abundances which are direct witnesses of nucleosynthesis and of dredge-up mechanisms; and they can also afford important information on topics such as UV radiation field, ionization, grain processes, etc.

The structure of envelopes of red giants is indeed discussed at length in most of the papers of these proceedings, in particular as inferred from the observations (see reviews by Goldberg, Linsky, Beckers, Wannier, Beckwith, Werner, Morris, Bowers and Knapp).

Accordingly, the purpose of the present paper is first to summarize the general characterictics of the global structure of the envelopes and of the corresponding physical processes; and then to discuss their chemical composition and our present understanding of its origin. We will emphasize mostly the intermediate and external layers of the envelopes because most of our detailed information is related to them through infrared or millimetric spectroscopy. For the same reason we will systematically refer to the most studied envelope IRC+10216: it is one of the most massive and relatively nearby (probably 200-300 pc)

M. Morris and B. Zuckerman (eds.), Mass Loss from Red Giants, 269–289.

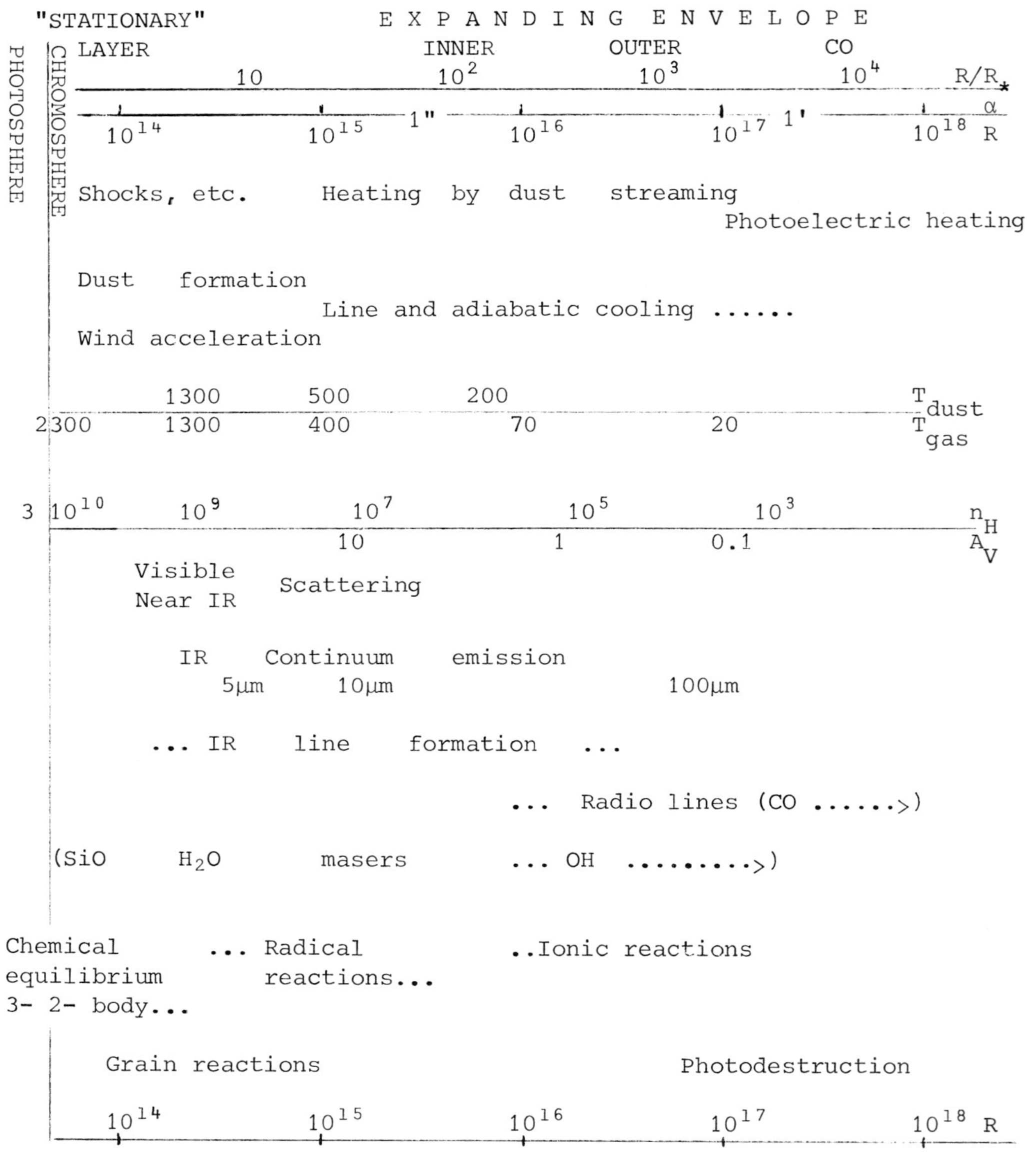

Figure 1. Structure of a massive envelope (IRC+10216). Assumed parameters : d = 200 pc, $R_* = 1000\ R_\odot = 7\ 10^{13}$ cm, $\dot{M} = 4\ 10^{-5}\ M_\odot$/yr. In the outer envelope : $n_H = 5\ 10^{37}\ r^{-2}\ cm^{-3}$,

$A_V = 2.5\ 10^{-22}\ N_H$. The range of maser emission is indicated for similar O-rich envelopes, it is obviously irrelevant for C-rich shells such as IRC+10216.

which allows some angularly resolved infrared and millimetric studies and insures the best sensitivity for infrared and millimetric spectroscopy. As it is a carbon shell, its chemistry is particularly rich and 20 molecules have already been detected there.

Besides the other papers of these proceedings, several recent reviews have discussed similar or related topics (see e.g. Zuckerman 1980 and Wannier 1984). In particular, the very recent review by Glassgold and Huggins (1984, hereafter referred to as GH) is entirely devoted to the chemistry of circumstellar envelopes. Accordingly, we will considerably shorten the corresponding topics and we will systematically refer the reader to their paper for complementary developments and points of view.

2. GLOBAL STRUCTURE

2.1. IRC+10216

A general picture of the envelope of IRC+10216 is given in Figure 1. It is supplemented by the reproduction in Figure 2 of the figure of Keady, Hall and Ridgway (1984) representing the quantitative estimates of the different physical parameters derived from their model of this envelope. Of course, the results of this model should not be taken too literally; however it is probably one of the best representations of the intermediate envelope based on infrared observations of continuum and CO lines (see also Goldreich and Scoville 1976, Kwan and Hill 1977, Tielens 1983). The significance of the values of the different parameters as well as the implied physical processes will be discussed in the next section.

Figure 1 displays the regions where some of these physical processes take place, as well as where the different observed radiations (lines and continuum) are emitted, and the different important chemical processes are at work in the different parts of the envelope. The location of these different processes, as well as the variation of the physical parameters, allows one to delineate the following regions:

- Just above the photosphere, a somewhat "stationary" layer: This is a rather confused and certainly complicated region. It is stationary in that sense that the massive wind of the more external envelope is not yet developed. But it can include chromospheric activity (see e.g. Linsky in these proceedings), shock waves, and complex motions with infalling matter as attested, for instance, by infrared CO lines (Keady et al. 1984) or by the interpretation of SiO masers in oxygen-rich envelopes (Langer and Watson 1984). This region terminates with wind acceleration which is far from being understood (see e.g. Holzer in these proceedings), and is certainly closely related to dust formation, the location and the mechanism of which are also very uncertain. It is possible as advocated by Keady et al. 1984, that this acceleration occurs in different steps associated with the formation of different types of grains.

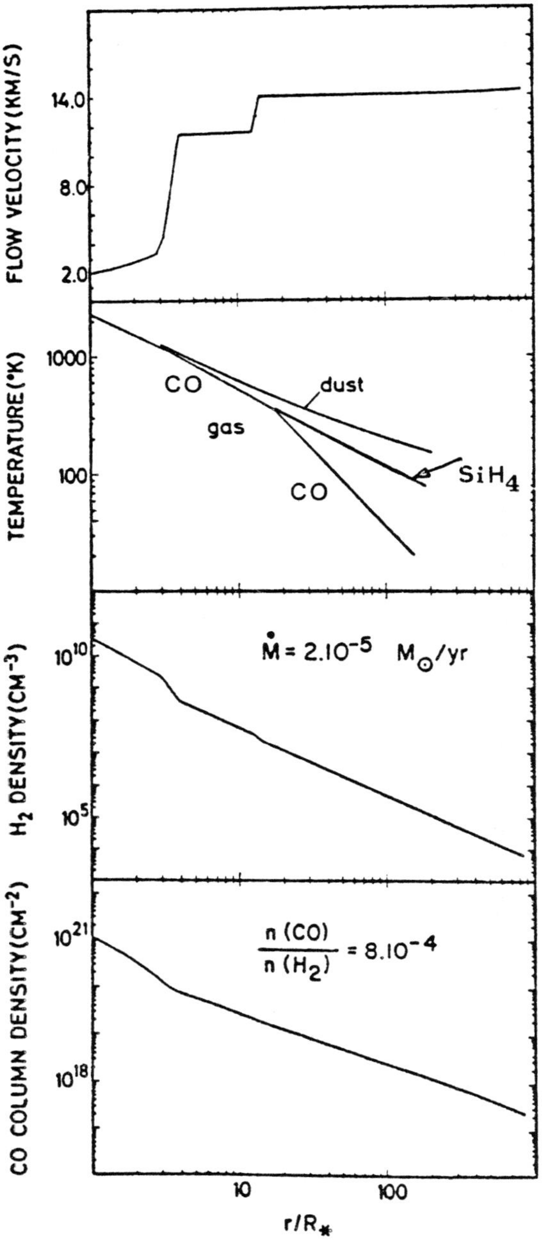

Figure 2. Gas parameters of IRC+10216 in the model of Keady, Hall and, Ridgway (1984). $R_* \sim 7\ 10^{13}$ cm. For $R > 20\ R_*$, the temperature of the gas derived from IR CO lines is probably not relevant, and one should prefer the estimation derived from SiH_4 lines (Keady and Ridgway 1984) which is in good agreement with the theoretical model of Kwan and Linke (1982).

- In the expanding envelope itself which in IRC+10216 ranges from 10^{15} to 10^{18} cm, and where the velocity is close to the terminal velocity, ~10–20 km/sec, one can distinguish, with some arbitrariness, on the basis of the spectral range of the observations and of the penetration of galactic UV radiation:

i) A warm molecular shell (T ~ 100–500 K, R $\lesssim 10^{16}$ cm) where most IR lines are formed, but its angular size ($\lesssim$ 3") is too small to allow any radio study with single dishes. The chemistry is only moderately active there because of relatively small time constants, freezing of reactions between closed shell molecules, and complete shielding from external UV. However, radical reactions are still fast and grain processes are potentially important.

ii) A cold molecular shell (T $\lesssim$ 100 K, R $\lesssim 10^{17}$ cm) with an angular extent (~ 1') comparable to typical beam sizes of 10-m millimetric radiotelescopes and with physical conditions similar to those in molecular clouds. Accordingly, single dish radio observations probe this region as a whole with maximum weight on external layers, while millimetric interferometry already allows one to study its internal part and should even yield some information on the external layers of the IR line region. On the other hand, this "millimetric" region is almost inaccessible to infrared studies because of relatively small column densities. It is also responsible for most of the submillimeter continuum emission (see Werner in these proceedings, Sopka et al. 1984). All chemical reactions are too slow to be important, except ionic reactions; but photochemistry is very active in its external layers. UV radiation creates there different radicals and some ions from stable abundant molecules such as C_2H_2, HCN, (and H_2O in O-rich shells), etc., and finally destroys any molecule except H_2 and CO.

iii) An extended CO shell (up to R ~ 10^{18} cm) because CO is much more stable to photodissociation than other molecules, which probably implies line photodissociation and self-shielding of CO (Morris and Jura 1983).

- Finally, it is not impossible that the envelope of IRC+10216 is extended somewhat further than R ~ 10^{18} cm by a sparse halo ($n < 10^2$ cm^{-3}) composed mainly of H_2 and atoms and ions of the other elements.

2.2 Other envelopes

The same global pattern shows in most of the envelopes. Changes in most of the main characteristics of the object can introduce important quantitative variations in the physical parameters and in the extent of the different regions, but do not alter the overall structure:

- Oxygen-rich shells have a different chemical composition in both the gas and the dust. This introduces important modifications in the cooling function and in dust emissivity and formation. Significant differences in wind formation and in the temperature structure of the shell can result (see e.g., Tielens 1983). Of course, their molecular lines are completely different from those of carbon shells (except for CO), and they present strong maser emission of OH in the partially photodissociated molecular shell, and of SiO and H_2O mostly in the different layers above the photosphere.

- Changes in the luminosity directly affect the dust temperature and radiation pressure, and hence dust formation, wind development, gas temperature, etc.

- IRC+10216 has one of the highest known mass loss rates, $\dot{M}$. Density in the expanding envelope directly scales with $\dot{M}$, as does the shielding from external UV. Accordingly, less intense winds do not only imply less dense shells, but also a much smaller extent of the CO shell and of the cold molecular shell (e.g. CIT 6).

However, there is a point where important modifications in $\dot{M}$ and/or the properties of the central star can fundamentally alter the overall structure of the envelope. If the shell is not completely opaque to chromospheric UV, photodissociation and photoionization extend to the whole shell; the number of molecules is drastically reduced and the ionization considerably increased (a typical example is α Ori, see e.g. GH, Clegg et al. 1983). Atomic and ionic lines are then the best tools for studying the envelope: absorption lines or chromospheric emission lines for probing the internal layers (see Linsky and Goldberg in these proceedings) and scattered lines (KI, etc.) in the very external shell.

As discussed in Section 3.1., most of the massive external envelopes are spherically symmetric. However, a lot of them are clearly asymmetric in their internal parts (e.g. IRC+10216), and some of them, mostly pre-planetary nebulae, have a well defined bipolar structure in the whole envelope : they then combine regions completely penetrated by UV radiation in the polar directions, and a massive molecular ring well shielded from UV (with a particular structure of the photodissociation region as discussed by Jura 1983b): a typical example is CRL2688.

3. PHYSICAL CONDITIONS

3.1 Geometrical and velocity structure

As already mentioned, most of the envelopes present in their external layers a very good spherical symmetry. This was verified by different techniques for more than 25 objects: millimetric CO mapping with single telescopes, millimetric interferometry in HCN and CO lines, VLA maps of 1612 MHz OH masers and heterodyne 10 μm interferometry (see Wannier 1984 for the references and a more complete discussion).

However, there is a whole class of bipolar objects with highly flattened shells (e.g. CRL2688, CRL618, NGC7027, VY CMa, etc.) (see Cohen and Masson in these proceedings, Morris 1981). An important fraction of them is probably in an advanced stage of evolution towards a planetary nebula.

Perhaps more intriguing is the fact that there are several examples, including IRC+10216, where envelopes which are highly spherically symmetric in their external layers exhibit strong anisotropies in their central parts ($R \lesssim 10^{15}$ cm in IRC+10216).

The origin of these asymmetries is not well understood and could involve close binary companions as discussed for example by Morris (1981) and Bowers et al. (1983).

Most of the terminal expansion velocities fall in the range 10-20 km/sec with some notable exceptions (see e.g. Knapp et al. 1982, Knapp and Morris 1984, Bowers et al. 1983). A velocity close to this terminal velocity is probably reached deep inside the envelope, soon after full condensation of dust (see e.g. Keady et al. 1984 and Figure 2). As discussed by Holzer in these proceedings, the exact mechanism of wind development is still unknown; however, the observed values of $\dot{M}$ and L are compatible with acceleration by radiation pressure.

Radiation pressure on grains insures that they stream through the gas at velocities of a few km/sec in the external envelope (Goldreich and Scoville 1976) (1-2 km/sec in IRC+10216, Kwan and Hill 1977). This is the main heating mechanism of the gas (see Section 3.4.).

Velocity dispersion is relatively modest in the external layers ($\lesssim$ 1 km/sec in IRC+10216, Henkel, Matthews and Morris 1983, Morris et al. 1984), but it can reach several km/sec in OH stars. As discussed for instance by Wannier and Alcock in these proceedings, the velocity structure of the internal layers of the envelope is extremely complicated with possibilities of blob ejection, infalling matter, shocks, multi-shell ejections, multi-step acceleration (as proposed by Keady et al. 1984, see Figure 2), etc. IR spectroscopy provides detailed information on this velocity structure in IRC+10216, but its interpretation implies elaborate radiative transfer modelling (see Keady et al. 1984 and Wannier in these proceedings).

3.2. Mass loss rate and density

Mass loss rates $\dot{M}$ range from 10^{-7} to 10^{-4} $M_\odot$/yr. As discussed for instance by Bowers et al. 1983, there are numerous methods of estimation of $\dot{M}$: depth of the 10 μm silicate absorption feature, model fitting to the IR excess, momentum conservation, infrared CO emission, millimetric CO emission, outer radius of 1612 MHz OH maser emission, etc. The two latter methods proved very fruitful in recent years in determining $\dot{M}$ for tens of envelopes. They are relatively reliable, but probably no better than a factor ~ 2-3. In both cases the derivation of $\dot{M}$ involves some modelling and the assumption of relatively empirical relations.

The case of CO was discussed by Kwan and Hill 1977, Kwan and Linke 1982, Knapp et al. 1982, Knapp and Morris 1984, and Morris in these proceedings. At first glance the antenna temperature T_A^* directly depends on the density n and hence $\dot{M}$ (Eq. 3.1) because of the collisional excitation of CO. However, T_A^* also depends on the kinetic temperature T_K, and on the CO abundance because the extent of the region where LTE is achieved by collisions depends on the opacity of the line. T_K depends on the luminosity of the star and possibly on external UV (see below). Finally, for a given T_A^* , $\dot{M}$ scales approximately as the square of the distance, d.

For OH masers, the modelling is also difficult (see Goldreich and Scoville 1976). However, there is a good empirical proportionality between $\dot{M}$ and R_{OH} the extent of OH emission (Bowers et al. 1983, Bowers these proceedings), which can be understood since the shielding of OH against photodissociation is directly proportional to $\dot{M}$ (Huggins and Glassgold 1982). Furthermore, R_{OH} can be determined by VLA mapping up to distances of several kpc, and even further by phase lag measurements between the near and far OH peaks of variable sources (see e.g. Bowers and Morris 1984 and references therein). This relation between the extent of the non-photodissociated shell and $\dot{M}$ is obviously general and could also be used to determine $\dot{M}$ from observations of other molecules (see e.g. Huggins et al. 1984, Glassgold et al. 1985).

For spherical symmetry, there is an obvious relation between local mass-loss $\dot{M}$, expansion velocity v_e and density n

$$\dot{M}(r) = 4\pi r^2 \, v_e(r) \, n(r) \, \bar{\mu} \qquad (3.1)$$

where $\bar{\mu}$ is the mean particle mass ; or

$$n_H = \frac{C}{r^2} = \frac{3\ 10^{37}}{r^2} \; \frac{\dot{M}_H}{10^{-5} M_\odot/yr} \; \frac{10 \text{ km/sec}}{v_e} \qquad (3.2)$$

where n_H is the total number of H nuclei per cm^3 and $\dot{M}_H$ is the hydrogen mass loss rate (for $n_{He}/n_H = x$; $\dot{M} = (1+4x)\dot{M}_H$ and for hydrogen entirely in molecular form $n = (0.5 + x)n_H$).

There is good evidence for an approximately constant mass loss rate in the expanding envelope of IRC+10216 inferred from an approximate r^{-2} variation of the density. However, an increase of $\dot{M}$ by a factor ~ 2-3 in the external layers has been advocated from CO (Kwan and Linke 1982) and HCN (Bieging et al. 1984) observations. However, the interpretation of these measurements is difficult and, for CO, depends on the temperature model.

In the region of wind acceleration, the density decreases faster than r^{-2} for constant $\dot{M}$ as seen in Figure 2. In the still inner regions above the photosphere, modelling of the atmosphere is difficult and rather arbitrary.

3.3. Influence of dust processes

The properties and the distribution of dust in circumstellar shells and their relations to infrared emission have been discussed at length (see e.g. Zuckerman 1980, Beckers, Wannier, Beckwith, and Werner in these proceedings, Keady et al. 1984, and references therein). We will not reproduce these discussions here and we will give only some general comments, in particular on grain processes important for the structure of the gas envelope.

The amount of infrared data on late-type shells is considerable and increases rapidly. Accordingly, there is significant progress in our knowledge of dust composition and emissivity, but a lot of points are still obscure and debated, in particular the detailed composition of dust. It is clear that very much is expected in the future in that field from IR photometry and spectroscopy on numerous objects from ground-based observations, from IRAS results or from future space facilities (ISO, SIRTF, etc.).

The total amount of dust is known in a number of objects with relatively good accuracy (see e.g. Werner in these proceedings), and it can be compared to the mass of gas. The ratio dust/gas is typically comparable to the one in the interstellar medium, ~ 1%, with possibly slightly smaller values in IRC+10216 (1/300 following Werner in these proceedings). The amount of dust is obviously important for UV shielding; however, little is known about the grain size distribution, except that the UV opacity can be approximately inferred from the extent of photodissociated regions (Jura 1983a).

The distribution of dust and its temperature have been inferred with relatively good precision in a few objects from modelling and from high angular resolution IR observations (interferometry, lunar occultation, etc.). A central question, fundamental for the structure of the envelope, is the location and conditions of dust condensation. This is a very difficult problem because the basic physics of nucleation still require some further clarification and modelling of the actual conditions of circumstellar shells, taking into account the exact chemical composition of the gas, non-equilibrium effects (see e.g. Salpeter 1974) and interaction with radiation, in particular with the possibility of the inverse greenhouse effect (McCabe et al. 1979) which prevents the evaporation of good IR radiators such as SiC. The formation of dust is thus entangled with infrared radiative transfer, and also with the dynamics of the envelope: radiation pressure on dust is not only the most likely process for wind acceleration, but dynamical instabilities can favor density and temperature inhomogeneities achieving locally the conditions of dust formation.

As usual, both the energetics of the envelope and the dust temperature T_D are dominated by radiation. The modelling of T_D is thus relatively straightforward <u>if</u> the dust properties are known (see e.g. Keady et al. 1984 and Figure 2).

3.4. Gas temperature

The temperature of the gas T_K has been modelled by Goldreich and Scoville (1976), Kwan and Hill (1977), Kwan and Linke (1982), and Tielens (1983). In the high density regions, especially where the wind is not fully accelerated, there is a strong coupling between T_K and T_D which are nearly equal (Figure 2). In the expanding envelope, collisions with the grains are still the main heating process of the gas, but mainly because of the streaming of the grains pushed by radiation pressure through the gas, and not by the difference between T_D and T_K which are thus unrelated. Efficient cooling by adiabatic expansion and mainly by molecular lines (CO in C-rich envelopes, H_2O, OH and CO in O-rich ones) induces $T_K \ll T_D$ (Figure 2). In this region the variation of T_K can be approximated by an $r^{-\alpha}$ law with $\alpha \sim 0.7$ (Goldreich and Scoville 1976, Kwan and Hill 1977, Kwan and Linke 1982).

In the outer shell T_K can decrease to a few tens of Kelvins as seen in Figure 2. However, T_K increases again in the very external layers penetrated by galactic UV, as in the boundary of interstellar clouds because of photoelectric heating and of possible heating by photodestruction of abundant species. In IRC+10216 this effect could be important for $r \gtrsim 0.5$-$1\ 10^{17}$ cm as mentioned by Kwan and Linke (1982). However, it has not yet been discussed in detail.

Kwan and Hill (1977) and Kwan and Linke (1982) have shown that their model predictions of T_K are in reasonable agreement with radio observations of CO. However, Wannier and Sahai (1982, 1984) have found some indications that T_K is significantly higher than the mentioned models in the region of formation of IR CO lines. It should be noted that, besides CO, it is difficult to find a good "thermometer" for these shells: because of the very large IR intensities, most of the rotational levels of molecules are not at LTE, but their populations are mainly determined by the absorption and emission of the photons in the IR lines (Morris 1975). The only exceptions in the conditions of IRC+10216 are: i) when high saturation of IR lines makes the rotational collision rates faster than the IR absorption rates: this is likely the case for CO lines if the gradient in velocity is not too large; ii) when selection rules prevent rotational excitation through absorption of IR photons (e.g. ortho NH_3, see Betz et al. 1979).

4. OBSERVED MOLECULAR ABUNDANCES

4.1. IRC+10216

A list of observed molecules in IRC+10216 is given in Table I, where we have reproduced the estimated abundances of GH with slight modifications and additions (see also McCabe et al. 1979, Lafont et al. 1982,hereafter referred to as LLO, Glassgold et al. 1985). As mentioned by GH, uncertainties on these values are large and can reach an order of magnitude in the worst cases.

In an ideal IR absorption observation, one measures the optical thickness τ of the molecular absorption layer in front of the continuum IR source in different rovibration lines. One thus derives the column density in each rotational level and hence the total column density of the observed molecule. However, in practice considerable difficulty may arise : saturation can prevent the determination of τ; the line strengths of convenient lines, those that avoid saturation, atmospheric absorption and too weak continuum intensity, are often very uncertain; as in stellar atmospheres the regions of continuum emission and of line formation are not separated and complicated effects of radiative transfer (such as PCyg profiles) can occur requiring elaborate modelling. The case of the important molecule C_2H_2 (observable only in the infrared) is exemplary: although it is one of the best studied molecules in the laboratory, the strength of the weak observed band has been revised by an order of magnitude from new measurements (Rinsland et al. 1982), implying a diminution of the estimated abundance by the same factor.

TABLE I. Observed abundances in IRC+10216 adapted from Table I of GH (see also LLO). $\bar{X}(i)$ is the mean abundance of i, defined in the text. All abundances are uncertain by factors $\gtrsim 3$. $X(CO) \cong 3\times10^{-4}$. Significant limits for $\bar{X}(i)/X(CO)$ for some unobserved species are respectively: HCO^+, 4×10^{-6}, HCl and AlCl, 2×10^{-5}, and H_2S, 2×10^{-4}.

$\bar{X}(i)/X(CO)$	Species				$\bar{X}(i)$
10^{-1}					
	C_2H_2				10^{-5}
10^{-2}	HCN				
					10^{-6}
				CN	
10^{-3}	NH_3				
			HC_3N		
	CH_4, C_2H_4	CS, SiS	HC_5N	C_2H, C_4H	10^{-7}
				C_3N, SiC_2	
10^{-4}			HC_7N		
		SiO			
	HNC		$HC_{11}N$		10^{-8}
				C_3H	
10^{-5}		SiH_4			
	CH_3CN				
					10^{-9}

On the other hand, radio lines are observed in emission in the outer shell. The angular sizes of radio beams (~ 1') are such that they integrate the emission over regions of size comparable to the whole envelope and, hence, very inhomogeneous. For optically thick radio lines radiative transfer modelling is also necessary. It implies a good knowledge of the rotational excitation (by IR vibrational lines for molecules other than CO) and it can be complicated by hyperfine structure for molecules such as HCN, CN, etc. In the case of optically thin lines, the radio signal is essentially proportional to the total number Ω_{iu} of molecules in the upper level of the transition in the antenna beam. Accordingly, observations of all excited lines allow, in principle, a direct determination of the total number Ω_i of a given species i in the envelope. However, this requires observations at different frequencies implying different antennas and/or different beam sizes with the usual calibration uncertainties, so that obtaining a complete and homogeneous set of data implies a considerable effort. Uncertainties in the source distance d and, possibly, in the molecular dipole moment must be added. Finally the ambiguity in the definition of the abundance should be emphasized: as mentioned, the basic quantity is Ω_{iu}/d^2; it is often given in terms of beam averaged column density $\bar{N}_{iu}$ which avoids the uncertainty in d:

$$\Omega_{iu} = \pi r_B^2 \bar{N}_{iu} \tag{4.1.}$$

where r_B is the beam radius, with a similar relation for the mean total column density $\bar{N}_i$ in the beam. It is obvious that $\bar{N}_i$ is completely different from the radial column density N_{iIR} measured in the infrared. For an r^{-2} distribution

$$n_i = f_i \; r^{-2} \tag{4.2.}$$

one has $$N_{iIR} = \int_{r_c}^{\infty} n_i dr = f_i/r_c \tag{4.3.}$$

and $$\bar{N}_i = 4 \; f_i/r_B \tag{4.4.}$$

where r_c is approximately the (line dependent) size of the continuum source, typically ~ a few 10^{15} cm, while r_B ~ 10^{17} cm.

When f_i is constant, the abundance $X_i(\equiv X(i))$ is also constant and $X_i = f_i/C$ (Eq. 3.2.). However, if f_i varies with r, the mean abundance $\bar{X}_i$ is beam dependent:

$$\bar{X}_i = \bar{N}_i r_B/4C = \Omega_i/(4\pi r_B C)$$

In all cases, abundance values bear the additional uncertainty on C and hence on $\dot{M}$.

The rich variety of molecules observed in IRC+10216 is naturally divided into the four columns of Table I (see GH): i) Simple stable molecules (plus HNC), naturally expected in a carbon-rich environment. ii) Molecules of relatively depleted heavy elements Si and S, which is not surprising, but the abundance of silicated molecules is unexpectedly high. iii) A whole series of acetylenic chains which are the most complex interstellar molecules observed, and are apparently the most common path towards complexity as attested to by their presence in dark clouds. Their origin is probably the most challenging problem of the chemistry of IRC+10216. iv) A corresponding collection of very reactive radicals which could naturally be formed by photochemistry.

It is worth recalling that several of the unusual species of Table I have been discovered in IRC+10216. Some of them are not observed elsewhere. And most others have abundances much larger than in typical interstellar clouds. Besides the expected absence of oxygenated molecules (except CO and SiO), the absence of ions is noteworthy (see Section 6.2.). Several significant upper limits for the abundances of undetected molecules are also reported in Table I.

4.2. Other envelopes

The status of molecular observations in other envelopes has been reviewed in detail by GH. We will not reproduce it here, adding just a few comments on prominent objects:

There is yet no clear example of another shell that is really similar to IRC+10216. The relatively well studied and nearby envelope CIT6 is much less massive.

As mentioned, preplanetary nebula objects are of particular interest and their chemistry is relatively rich. CRL2688 has probably the richest chemistry with salient peculiarities such as an abundance of atomic carbon comparable to that of CO (Beichman et al. 1983) and an abundance of NH_3 much larger than in IRC+10216 (Nguyen-Q-Rieu et al. 1984).

Oxygen rich envelopes have a poorer chemistry and abundance determinations are sparse, except for CO, which is of course very abundant. Maser emission from H_2O and OH also implies large abundances ($X_i \sim 10^{-5}$–10^{-4}). SiO is also very abundant in the very internal layers where its maser emission takes place.

A particularly interesting object is the dense shell OH231.8+4.2, the only one where gaseous SH_2 has been detected (Ukita and Morris 1983), and one of the few where solid ice has been identified.

5. CHEMISTRY IN THE INNER SHELL

5.1. General considerations and time constants

As already mentioned, circumstellar chemistry has been recently reviewed in detail by GH, we therefore intend in the present and next sections to summarize its general features with some further developments (grain processes, ion chemistry, etc.) relying mainly on the works of LLO and Glassgold, Lucas and Omont (1985; hereafter referred to as GLO).

Compared to interstellar cloud chemistry, the main characteristics of circumstellar shells are: i) the very wide variety of physical conditions from the photosphere to the external layers similar to diffuse clouds; ii) the very short time constants and the fast outside transportation of the molecular gas; iii) a carbon rich environment in carbon envelopes. The chemical structure of the envelopes is thus essentially determined by the physical conditions and by the relative values of the different rates and time constants. The most important characteristic time is just the dynamical time $t_{dyn} = r/v$. Its inverse has to be compared to the different rates of the main chemical processes:

- Chemical reactions with a species of abundance X, that we write:

$$R_{ch} = A\, e^{-T_B/T}\, X\, n_H \qquad (5.1.)$$

where typical values of A are 10^{-11} cm^3 sec^{-1} for reactions with neutrals, and 10^{-9} cm^3 sec^{-1} for ionic reactions. T_B represents the activation energy with typical values: $T_B = 0$ for ionic reactions, $0 \leqslant T_B \lesssim 10^4$K for radical reactions, $T_B \gtrsim 10^4$ for reactions between closed shell molecules.

-Photodestruction by the shielded galactic UV radiation, that, following GH, we write

$$\Gamma_{pd} = G\ e^{-d/r}. \tag{5.2.}$$

Here G is the photodestruction rate of the considered molecule in the ambient galactic UV field with typical values ranging from 10^{-11} to 10^{-9} sec^{-1}, and d is a characteristic length for shielding by dust with typical values in the range (10^{16}-10^{17}) ($\dot{M}_H/10^{-5}$ $M_\odot$/yr) cm.

- Rate of collisions between grains and molecules

$$\tau_{gr}^{-1} = k_{gr} n_H \tag{5.3.}$$

with $k_{gr} \sim 10^{-16} - 10^{-17}$ cm^3 sec^{-1}.

Figure 3 displays the variations with r of τ_{dyn}^{-1} and of characteristic examples of R_{ch}, Γ_{pd} and τ_{gr}^{-1} in IRC+10216 with the model of Figure 2. The comparison of these quantities with τ_{dyn}^{-1} shows that one can clearly distinguish between the chemical processes important in the inner dense shells where chemical reactions and grain processes can be faster than τ_{dyn}^{-1}, and the ones to be considered (photodissociation and possibly ionic reactions) in the external envelope. The boundary between these two regions is rather clearly delineated at r ~ 10^{16}cm for dense shells such as IRC+10216.

5.2. Freezing out and radical reactions in internal shells

In the very dense layers in the photosphere and above, chemical equilibrium is probably a good approximation if effects of UV radiation are negligible (see Clegg et al. 1983 for a discussion of situations such as α Ori where this condition is not met). The computation of molecular abundances is then relatively easy and involves only the abundances of the elements and the thermochemical data (ΔG_f) of the different molecules (see Tsuji 1964, for photospheres, McCabe et al. 1979 and LLO for models of internal envelopes, and the review by GH).

Because of the overwhelming abundance of hydrogen, the boundary of the region of full chemical equilibrium probably coincides with the point where the rates of three-body reactions with two hydrogens equals τ_{dyn}^{-1}. More precisely, the critical process is the freezing-out of three-body formation of H_2 at r_3, which occurs for $n_H \sim 10^{11}$ cm^{-3}. This process and the resulting abundance of atomic hydrogen have been discussed in detail by Glassgold and Huggins (1983) and by GH. The frozen-out value of H/H_2 critically depends on the physical conditions in the upper stellar atmosphere and is always relatively high, $\gtrsim 10^{-3}$.

It should be noted however that one could consider other mechanisms for H_2 formation: LLO have pointed out that catalytic radiative association with heavy molecules could be relevant, but would be at most of marginal importance; grain processes are highly uncertain and will be discussed in the next section.

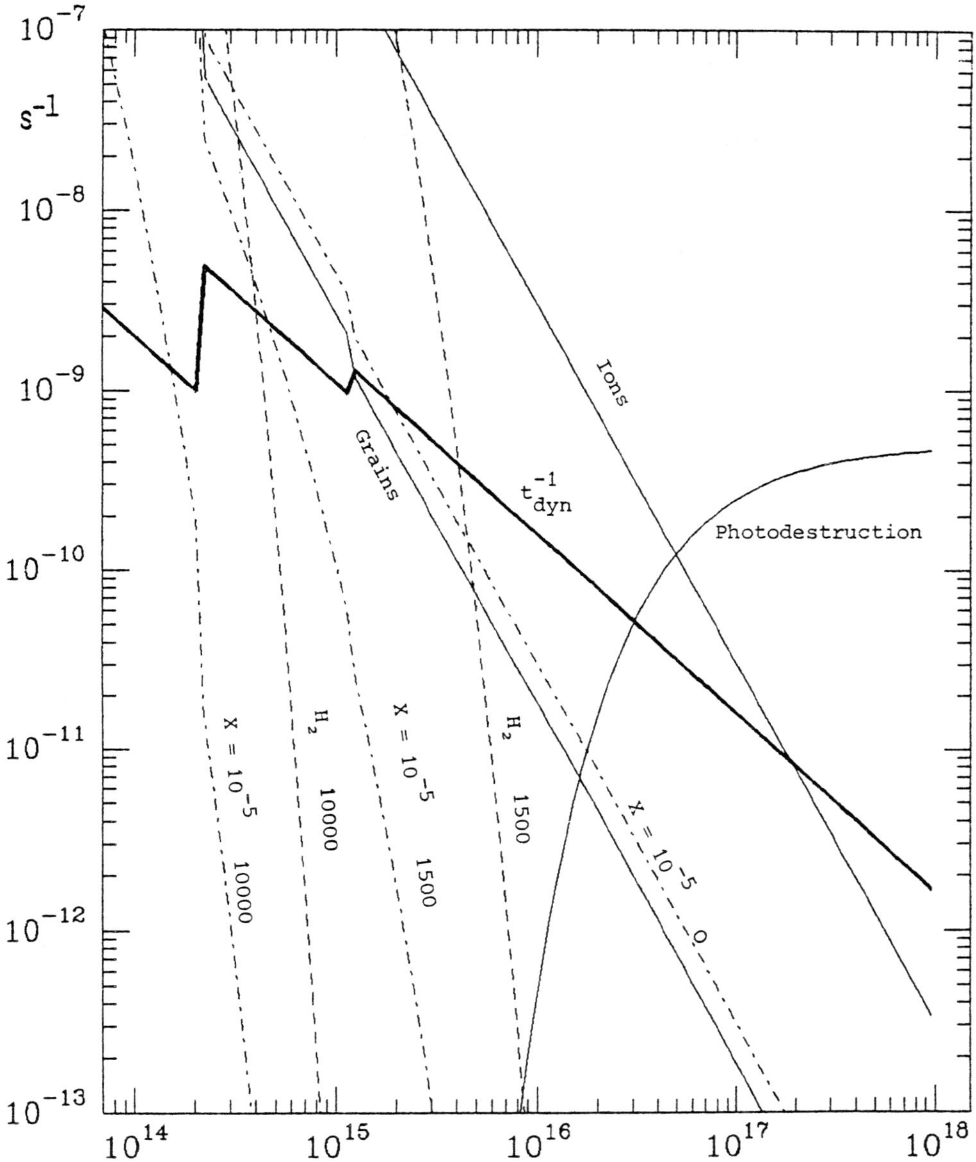

Figure 3. Time constants and reaction rates (sec^{-1}) in the envelope of IRC+10216 with the model of Figure 2. The assumed typical rates are the following: photodissociation $5\ 10^{-10}\ \exp(-7\ 10^{16}/r)$; grains $6\ 10^{-17}\ n_H$; ions $10^{-9}\ X n_H$ with $X = 10^{-5}$; reactions with H_2 $5\ 10^{-12}\ n_H\ \exp(-T/T_B)$ with $T_B = 1500$ and 10000 K; reactions with other abundant species (e.g. C_2H_2) $10^{-11}\ X n_H\ \exp(-T/T_B)$ with $X = 10^{-5}$ and $T_B = 0$, 1500 and 10000 K.

In the region above r_3, two-body reactions between two closed shell species rapidly freeze out because of their high activation energies. This is also the case with very exothermic reactions such as most collisional dissociations. Gas phase chemistry is then dominated by radical reactions. The most important are by far those involving H such as

$$AH + H \rightleftarrows A + H_2 \qquad (5.4.)$$

(e.g. A = OH, O, C_2H, CN, etc.). The ratio A/AH is then at equilibrium with H/H_2,

$$A/AH \sim (H/H_2) \exp [-(D_{AH} - D_{H_2})/kT] \qquad (5.5.)$$

Accordingly, all radicals with a binding energy to H, D_{AH} larger than D_{H_2} = 4.5 eV (e.g. CN and C_2H) have a very rapidly decreasing abundance when T decreases.

Reactions 5.4 have typical activation energies in the range 1000-3000 K. Accordingly, they freeze out for temperatures lower than a few hundred K (Figure 3). However, other radical reactions with abundant species such as

$$C_2H + C_2H_2 \rightarrow C_4H_2 + H \qquad (5.6.)$$

have lower activation energy, and can proceed a little further out.

Thus, it is clear that the assumption of freezing out of thermodynamical chemical equilibrium is completely irrelevant for radical and subordinated species. The evaluation of their abundance requires a complete solution of the time dependent network of chemical reactions (see e.g. Goldreich and Scoville 1976, Scalo and Slavsky 1980) with the appropriate reaction rates, which are often highly uncertain (see e.g. LLO).

Such reactions could process a non-negligible fraction of C_2H_2 into heavy molecules, but it is impossible to assess their real importance in the absence of a precise knowledge of the thermochemical data and of the reaction rates (see LLO).

In conclusion, abundances at the boundary of this internal envelope ($r \sim 10^{16}$cm) reflect the freezing out of processes in the interior, but at various depths. Every species which can be affected by radical reactions has an abundance which is very different from that of LTE chemical equilibrium achieved in the very internal layers. The only species for which the abundances remain frozen out from LTE chemical equilibrium are the very abundant and very stable molecules which contain a major fraction of a given element: CO, C_2H_2, N_2, HCN, etc. Their abundances yield nearly no information on chemical processes, but only element abundances in the gas. The abundances of all other species are potentially dependent on radical reactions with H being particularly important. It should also be kept in mind that dissociation by UV radiation or by shocks in the very internal layers and grain processes can play a fundamental role in the determination of the abundances of many species.

5.3. Grain processes

Grains contain a major fraction of refractory elements, and, in fact, at equilibrium they should deplete nearly completely the gas in the heavy molecules observed in IRC+10216 (except CO). Accordingly, a major but difficult goal of circumstellar chemistry is to understand qualitatively and quantatively the relatively mild depletions observed. However, grain processes are still poorly known, and it is difficult to go much further than to write a list of open questions:

-Condensation: as already mentioned, conditions are far from the conventional homogeneous nucleation theory: very complex composition of the gas, the possibility of several condensation steps, radiative processes, hydrogenation of most of the carbon, etc. Accordingly, it is very difficult to foresee what is the composition of the gas in the condensation region and just above it.

-Accretion onto grains: As seen in Figure 3, in IRC+10216 the point where $\tau_{gr} \sim \tau_{dyn}$ is well outside the region of grain formation. Accordingly, in such massive shells every gas molecule or atom hits a grain a large number of times on its way to the external layers of the envelope. This number N_{CG} is at least ~10 in IRC+10216, but it is very difficult to evaluate because it depends critically on the exact density in the region of dust formation and on the acceleration processes. Accordingly, if the sticking coefficient S is not negligible, gas depletion of the corresponding molecules and growth of dust grains proceed up to relatively external regions where $\tau_{gr} S^{-1} \sim \tau_{dyn}$. The order of magnitude of this effect is in good agreement with the amount of ice observed in the very massive and cold O-rich envelope of OH 231.8+4.2 as discussed by Jura and Morris (1984) (accretion of H_2O requires low dust temperatures which in most envelopes corresponds to large radii and thus to densities too small to allow efficient accretion). This effect could also contribute to the observed depletion of heavy elements (Si, S, Al...) in the gas of IRC+10216 (Jura and Morris 1984). However, it should be noted that it stops rather early ($r \sim 3\ 10^{15}$ cm), and also that, as N_{CG} is probably not that large, the fraction of the corresponding atoms in the gas, $\sim (S\ N_{CG})^{-1}$ is never vanishingly small.

-Chemical reactions on grains. At the dust temperatures of interest, molecules do not absorb for an appreciably length of time on grain surfaces (except the very special case of H_2O mentioned above). Accordingly, molecules impinging on grains either only bounce or suffer a chemical reaction with species already chemically bound to the grain. Again the results are probably different for the radicals where the relatively modest activation energies are likely to be more easily overcome than for closed shell molecules such as C_2H_2 and HCN. Grains are then probably a sink for radicals (LLO) and in particular for H which can lead to the formation either of hydrides with atoms of the grain (CH_4, C_2H_4, SiH_4, etc.), or of H_2. But the yields of these processes are rather uncertain (see LLO, Draine 1979, Balooch and Olander 1975).

-Grain collisions. The streaming velocity v_G through the gas of a grain of radius a varies roughly as $a^{1/2}$. Accordingly, the relative velocities of grains are of the same order of magnitude as v_G. Grain collisions in envelopes have been advocated by Bierman and Harwitt (1980) for accounting for the grain size distribution through shattering. However, this process probably requires $v_G \gtrsim 10$ km/sec, much larger than in dense envelopes such as IRC+10216 where $v_G > \sim 1$ km/sec. Even in that case the relative kinetic energy is large and can reach $\sim 0.1\ v_5^2$ eV per atom for the smaller grains, i.e., $\sim 10^{-2}$ of its binding energy (v_5 is the relative velocity in km/sec). Hence, it is possible that a small fraction of this energy is used to eject a significant number of atoms or radicals (such as C_n or SiC_n) into the gas (LLO).

6. PHOTOCHEMISTRY IN THE OUTER SHELL

6.1. Photodissociation

As reviewed in detail by GH to which we refer the reader, photodissociation by the ambient UV galactic radiation limits the radial extent of the molecules and is probably the major source of radicals in dense envelopes. There exist now numerous detailed models of this photodissociation: e.g. Goldreich and Scoville (1976), Scalo and Slavsky (1980), GH, LLO, Huggins et al. (1984), Glassgold et al. (1985). The main uncertainties of these models come from: i) the photodestruction rates, which are not easily measured in the laboratory (see e.g. Glassgold et al. (1985) for an illustration of these difficulties) and ii) the unknown UV properties of dust in circumstellar shells which control the penetration of UV radiation. On the other hand, it is possible to infer some information on UV dust properties from the extent of the observed molecules and of their photodissociation products. (Jura 1983a).

The main observational evidence of this photodissociation is the large extent of OH maser emission (Goldreich and Scoville 1976). In IRC+10216 there is evidence of both a very large extent of CN (Wootten et al. 1982), and of a limited extent of HCN (Olofsson et al. 1982, Bieging et al. 1984). Accordingly, it seems likely that at least most of the observed CN is created by photodissociation of HCN. However, for other radicals such as C_2H the discussion may be more difficult (Huggins et al. 1984), but the question should be soon clarified with progress in millimeter observations (especially better angular resolution) and in infrared spectroscopy.

6.2. Ionic reactions

For some species such as C_2H_2, photodestruction can produce ions such as $C_2H_2^+$ with branching ratios of several tenths. Furthermore, the successive photodestruction chain of every carbon containing molecule terminates with the production of C^+ (e.g. $C_2H_2 \rightarrow C_2H \rightarrow C_2 \rightarrow C^+$ or $CO \rightarrow C \rightarrow C^+$). The production of these ions (except CO photodissociation) occurs in regions ($r \lesssim 10^{17}$ cm) where the density is still large enough to allow reactions that form carbon chains of the type,

$$C_2H_2^+ + C_2H_2 \rightarrow C_4H_2^+ + H_2 \qquad (6.1.)$$

$$C^+ + C_2H_2 \rightarrow C_3H^+ + H \qquad (6.2.)$$

Formation of C_3H from the second reaction has been modelled by Nejad et al. (1984) to explain the observation of C_3H in IRC10216. However, they assume CO as the most important source of C^+ which is probably irrelevant because of self-shielding of CO. GH proposed that reactions of the type of Eq. (6.1.) could be an important mode of formation of long carbon chains. Both points have been considered in detail by Glassgold et al. (1985) with the following conclusions: i) it is easy to account for the observed amount of C_3H with reactions (6.2.) and ii) it is possible to form by reaction (6.1.) sufficient chains with four heavy atoms to be comparable to the observed amount of C_4H. It is not impossible that such processes could account for all the heavy species but this is not warranted because of unknown reaction rates and would require rather favorable conditions.

In the internal layers where UV radiation does not penetrate, the major source of ions is probably ionization of H_2 by cosmic rays as in interstellar clouds. This process has been also discussed by Glassgold et al. (1985). Almost every H_2 ionization is successively processed into H_3^+, then into HCO^+ and then into $C_2H_3^+$. The case of HCO^+ is of particular interest because it has been actively searched for in IRC+10216 (Johansson et al. 1984). In the inner layers its density should be roughly constant, with abundances not much smaller than in dense clouds. However, it is very efficiently destroyed by the electrons coming from the photoionization of C_2H_2. Accordingly, the main reason for its non-observation is probably that the whole mass of the inner shells, where it is present, is much smaller than typical interstellar sources such as TMC1, and also that the line width is much larger in IRC+10216.

ACKNOWLEDGEMENTS

The author wishes to thank A.E. Glassgold for providing the manuscript of his review before publication and for many fruitful discussions.

Note added in proof. In a recent preprint submitted to the Astrophysical Journal, D.B. Slavsky and J.M. Scalo present a very complete modelling of the chemistry of oxygen-rich circumstellar shells.

REFERENCES

Balooch, M. and Olander, D.R. 1975, J. Chem. Phys. 63, 4772.

Beichman, C.A., Keene, J., Phillips, T.G., Huggins, P.J., Wootten, H.A., Masson, C., and Frerking, M.A. 1983, Ap.J., 273, 633.

Betz, A.L., Mc Laren, R.A. and Spears, D.L. 1979, Ap. J. (Letters), 229, L97.

Bieging, J.H., Chapman, B. and Welch, W.J. 1984, Ap.J. in press.

Bierman, P. and Harwitt, M. 1980, Ap.J., (Letters) 241, L105.

Bowers, P.F., Johnston, K.J. and Spencer, J.H., 1983, Ap.J. 274, 733.

Bowers, P.F., and Morris, M. 1984, Ap.J., 276, 646.

Clegg, R.E.S., Van Ijzendoorn, L.J. and Allamendola, L.J. 1983, Mon. Not. Roy. Astr. Soc., 203, 125.

Draine, B.T. 1979, Ap.J., 230, 106.

Glassgold, A.E. and Huggins, P.J. 1983, Mon. Not. Roy. Astr. Soc., 203 517.

(GH) Glassgold, A.E. and Huggins, P.J. 1984, in The M, S and C Stars, eds H.R. Johnson and R. Querci, to be published by NASA and CNRS.

Glassgold, A.E., Lucas, R. and Omont, A., 1985, Astr. Ap. submitted.

Goldreich, P. and Scoville, N.Z. 1976, Ap.J., 205, 144.

Henkel, C. Matthews, H.E. and Morris, M. 1983, Ap.J., 267, 184.

Huggins, P.J. and Glassgold, A.E. 1982, Ap.J., 252, 201.

Huggins, P.J., Glassgold, A.E. and Morris, M. 1984, Ap.J., 279, 284.

Johansson, L.E.B., Anderson, C., Elder, J., Friberg, P., Hjalmarson, A., Höglund, B., Irvine, W.M., Olofsson, H. and Rydbeck, G. 1984, Astr. Ap., 130, 227.

Jura, M. 1983a, Ap.J. 267, 647.

Jura, M. 1983b, Ap.J.,275, 683.

Jura, M. and Morris, M. 1985, Ap.J., in press.

Keady, J.J., Hall, D.N.B., and Ridgway, S.T. 1985, Ap.J., in press,

Keady, J.J., and Ridgway, S.T. 1984, private communication.

Knapp, G.R., Phillips, T.G., Leighton, R.B., Lo, K.Y., Wannier, P.G., Wooten, H.A. and Huggins, P.J., 1982, Ap.J., 266, 701.

Knapp, G.R. and Morris, M. 1985, Ap.J., in press.

Kwan, J. and Hill, F. 1977, Ap.J., 215, 781.

Kwan, J. and Linke, R.A. 1982, Ap.J., 254, 587.

(LLO) Lafont, S., Lucas, R. and Omont, A. 1982, Astr. Ap., 106, 201.

Langer, S.H. and Watson, W.D. 1984, Ap.J., 284, 751.

McCabe, E.M., Smith, R.C. and Clegg, R.E.S., 1979, Nature, 281, 263.

Morris, M. 1975, Ap.J., 197, 603.

Morris, M. 1981, Ap.J., 249, 572.

Morris, M., Lucas, R. and Omont, A. 1985, Astr. Ap., 142, 107.

Nejad, L.A.M., Millar, T.J. and Freeman, A., 1984, Astr. Ap., 134, 129.

Nguyen-Q-Rieu, Graham, D. and Bujarrabal, V. 1984, Astr. Ap., 138, L5.

Olofsson, M., Johansson, L.E.B., Hjalmarson, A. and Nguyen-Q-Rieu 1982, Astr. Ap., 107, 128.

Rinsland, C.P., Baldacci, A. and Rao, K.N. 1982, Ap.J. Suppl., 49, 487.

Salpeter, E.E. 1974, Ap.J., 193, 579 and 585.

Scalo, J. and Slavsky, D.B. 1980, Ap.J., (Letters), 239, L73.

Sopka, R.J., Hildebrand, R.H., Jaffe, D.T., Gatley, I., Roelig, T., Werner, M.W., Jura, M. and Zuckerman, B. 1985, Ap.J., in press.
Tielens, A.G.G.M. 1983, Ap.J., 271, 702.
Tsuji, T. 1964, Ann. Tokyo Astr. Obs., 9, 1.
Tsuji, T. 1973, Astr. Ap., 23, 411.
Ukita, N. and Morris, M. 1983, Astr. Ap., 121, 15.
Wannier, P.G. 1984, in Cool Stars, Stellar Systems, and the Sun. Springer-Verlag, New York, p.296.
Wannier, P.G. and Sahai, R. 1982, Bull. A.A.S., 14, 894.
Wannier, P.G. and Sahai, R. 1984, in preparation.
Wooten, A., Lichten, S.M., Sahai, R. and Wannier, P.G. 1982, Ap.J., 257, 151.
Zuckerman, B. 1980, Ann. Rev. Astr. Ap., 18, 263.

BIPOLAR NEBULAE AND MASS LOSS FROM RED GIANT STARS

Martin Cohen
NASA-Ames Research Center
Mailstop 245-6, Moffett Field
CA 94035, U. S. A.
and
Radio Astronomy Laboratory
University of California
Berkeley, CA 94720, U.S.A.

ABSTRACT. Observations of several bipolar nebulae are used to learn something of the nature of mass loss from the probable red-giant progenitors of these nebulae. Phenomena discussed are: probable fluorescent emission bands in the optical and infrared in HD 44179 and GL 2688's optical molecular emissions; newly discovered very-high velocity knots along the axis of OH 0739-14 which reveal evidence for mass ejections of $\pm$300 km/s from the M9 III star embedded in this nebula; the bipolar structure of three extreme carbon stars, and the evidence for periodic mass ejection in IRC+30219, also at high speed (about 80 km/s); the curious cool TiO-rich region above Parsamian 13 which may represent the very recent shedding of photospheric material from a cool, oxygen-rich giant. Several general key questions about bipolar nebulae, that relate to the process of mass loss from their progenitor stars, are raised.

1. INTRODUCTION

The organizers of this meeting have given me licence to talk about some interesting aspects of bipolar and/or protoplanetary nebulae. I'd like to talk about relatively few individual nebulae but to focus upon some very specific aspects of these. The common theme will be to use bipolar nebulae to learn something of the physics, chemistry, or geometry of mass loss from their red giant predecessors.
A table of these curiosities follows.

TABLE I. Bipolar nebulae to be discussed.

HD 44179	The Red Rectangle Nebula
GL 2688	The Cygnus Egg Nebula
OH 0739-14	An OH/IR M9 III star on which Mark Morris is resident pundit

M. Morris and B. Zuckerman (eds.), Mass Loss from Red Giants, 291 -303.

TABLE I (continued).

IRC+30219=GL 1403, GL 2699, IRC+10216=GL 1381	Three "extreme carbon stars"
Parsamian 13	An M-star that has extremely recently shed its outer layers

What I'd like to do, as we discuss these nebulae, is to pose some relatively simple questions to which I cannot produce simple answers. Perhaps somebody in the audience can. Each question bears upon the unique aspects of specific nebulae and has relevance to the process of mass loss in red giants or the origins of these bipolar nebulae. In some cases one can suggest specific and definitive experiments that might help to resolve these mysteries.

2. INDIVIDUAL OBJECTS

2.1 HD 44179 and GL 2688

A few years ago strange molecular bands were found in emission in the optical spectrum from the vicinity of HD 44179 (Schmidt, Cohen and Margon 1980). These take the form of both a very broad emission feature and a set of well-defined, resolved sharp features (Fig. 1).

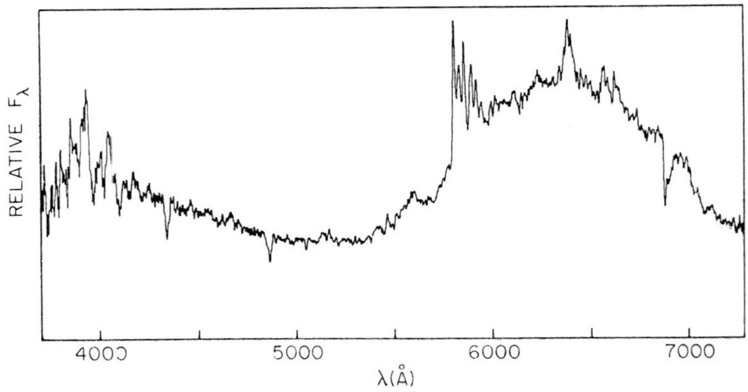

Figure 1. Optical spectrum of the nebulosity near HD 44179. Note both the broad emission feature between 5500 and 7250 Å and the sharp unidentified features near 5800 Å in emission.

Nebular spectropolarimetry indicates that these emissions originate strictly in the associated biconical nebula. There is still no adequate explanation for them although it is surmised that they represent resonant fluorescence in the stellar radiation field in much the same way that comets fluoresce in the solar system. It is perhaps interesting to note that HD 44179 also presents a powerful spectrum of the still-unidentified infrared emission bands (at 3.3, 6.2, 7.7, 8.7, and 11.3 microns) (Fig. 2). Could these two sets of spectral features,

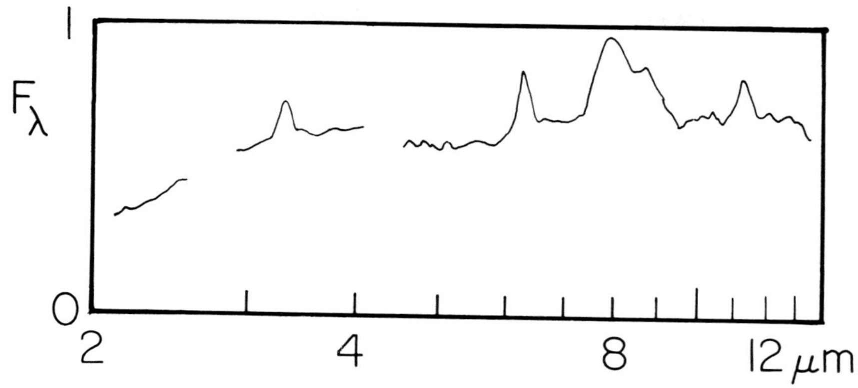

Figure 2. Infrared emission features in the spectrum of HD 44179.

one optical, the other infrared, be related? Thus far it has not been demonstrated that the infrared bands too arise solely from the nebula. This may have to await the time when narrowband, two-dimensional infrared imaging is routine, but it is surely a worthwhile experiment. Of the optical features one can only speculate but it could be that the sharp features represent molecular, gas-phase emissions while the broad orange structure arises from grain lattices that incorporate these same molecular species.

I am struck by one potential similarity between HD 44179 and GL 2688, which are rather unlikely bed-fellows, one with its strong carbon richness (GL 2688), the other without. However, GL 2688 too has yielded some vital information to nebular spectropolarimetry which, in turn, led to a set of observations bearing upon the existence of potential fluorescent optical bands in its spectrum. You may recall (Fig. 3) that certain features, common in carbon-rich atmospheres, seemed oddly lacking in the total flux spectra from the lobes of this nebula (Cohen and Kuhi 1977). Bipolar nebular light is often highly polarized, extremely so in the case of GL 2688. Nebular spectropolarimetry reveals that the C_2 molecular Swan bands arise from two totally different regions (for example, the (0,0) head at 5165A). Fig. 4 shows highly polarized C_2 absorption, from the stellar vicinity, superimposed on which is strong, unpolarized nebular C_2 emission. It is this combination that yields the curious weakness of these bandheads in GL 2688 (Fig. 3).

Where exactly does the C_2 come from, within the nebular lobes, and is it the sole molecule seen in emission? Fig. 5 presents a map of the two lobes of GL 2688 in which the weakness, or absence, of various molecular or atomic features is attributed to nebular emission that overlies reflected stellar absorption. You can see, mapped out in Fig. 5, the variations in strength of what is likely to represent resonance fluorescence in C_2, C_3, CH, SiC_2, and the Na D-lines. Once again, one

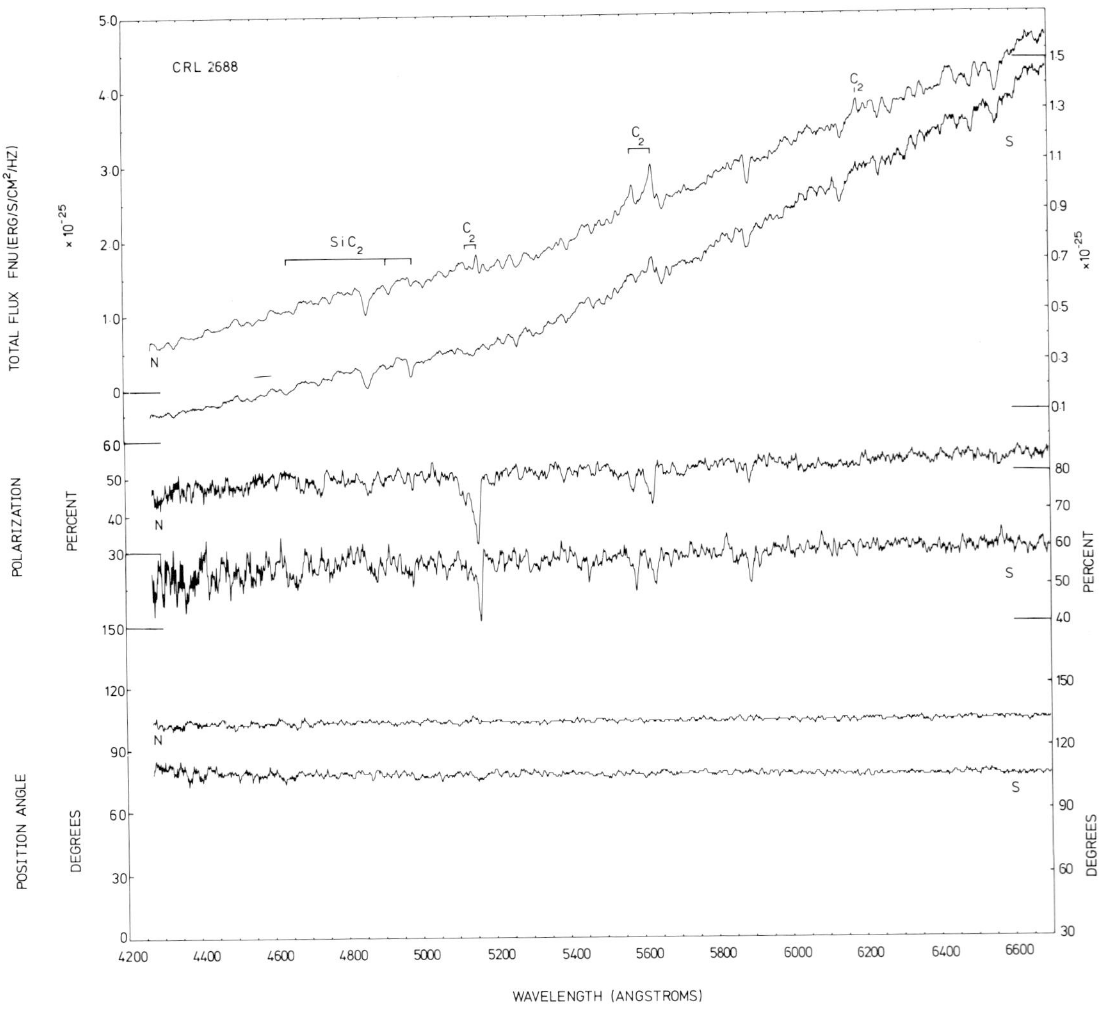

Figure 3. Optical spectropolarimetry of the north and south lobes of GL 2688. Shown are the total flux spectra (top), the degree of polarization (middle), and the position angle (bottom) for both lobes. Note the striking decrease in polarization through the swan bands at 5165, 5635, and 6192A.

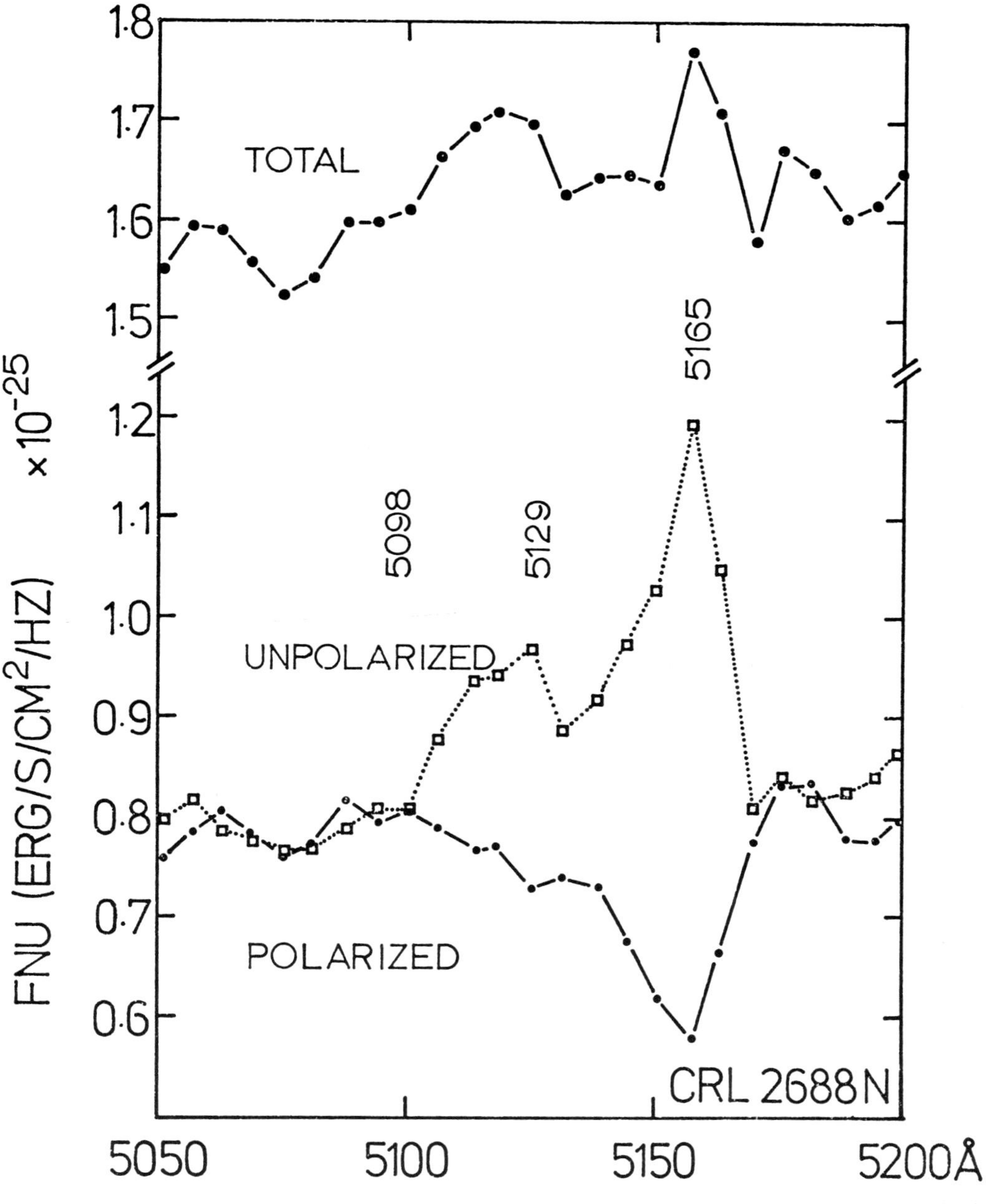

Figure 4. The resolution of the curious total flux spectrum in GL 2688N (shown in Fig. 3). We observe the superposition of polarized stellar C_2 absorption, and unpolarized nebular emission at 5165A.

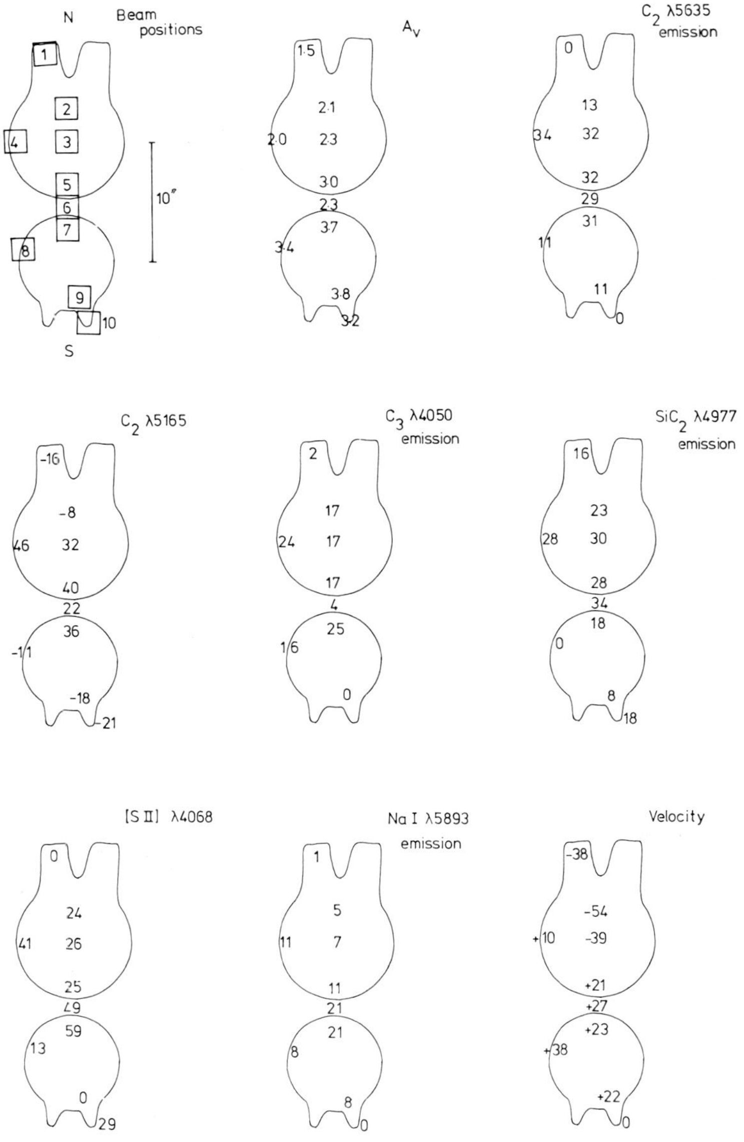

Figure 5. Distributions of the strengths of various molecular and atomic emission features across the lobes of GL 2688. Numbers are proportional to the total inferred emission strengths (from Cohen and Kuhi 1980).

is reminded of comets in sunlight. I find it easier to believe in fluorescence than to believe that the C_2 emission arises in those shocked regions recently mapped in GL 2688 by the distribution of near-infrared molecular hydrogen emission (Beckwith, Beck and Gatley 1984), principally because the H_2 and C_2 maps do not show the same extent or peak locations.

In this respect then, HD 44179 and GL 2688 seem akin to one another. It is striking that in no other bipolar (or other type of) nebula are any of these molecular bands seen in emission. Why should this be? Are the unidentified molecules in HD 44179 very short-lived? Do they require peculiar chemistry for their occurrence? Is it the extreme carbon-richness of the ejecta in the lobes of GL 2688 that yields sufficient carbon-bearing molecules for observable fluorescence?

2.2 OH 0739-14

Let me turn now to OH 0739-14 which contains an extremely late-type M-star, an M9 III in fact (Cohen 1981). As such it undergoes dramatic variations in brightness both optically and in its unusual OH maser structure. At times of maximum light, inferred from the strength of the OH maser, one can not only photograph the (brighter) south nebular lobe but even obtain its spectrum, whence the M9 III classification. Recently I obtained IPCS long-slit spectra along the two lobes, both of which were easily visible in April 1984 with the Anglo-Australian 4-meter telescope and integrating TV guider. If I can present you with data not yet fully reduced I think that you will find these interesting. What was obtained was a two-dimensional image, representing fully 4000A of spectrum, from 3500-7500A, for each of 50 pixels (2.3 arcsec along the slit per pixel) representing a different spatial location on the slit in the nebular structure. In Fig. 6 we see the spectra of both the south lobe (Fig. 6a) and the faint north lobe (Fig. 6b). Also seen are several very faint stars that happen to lie in or near the slit. However, also seen are two purely nebulous regions (delineated in spatial extent in these spectra by H-alpha emission), precisely along the major axis of the bipolar nebulosity, each lying beyond the relevant lobe. Let's examine the spectra of lobes and nebulous knots. The spectra represented in Fig. 6 are obviously those of a very late-type star, just as is expected for a reflection bipolar nebula. Each is separately classifiable as of type M9 III in April 1984, just the same as measured for the southern lobe during the maximum of December 1980. If you look at the overall spectra of the lobes you will see unmistakeable evidence for a blue continuum (don't worry about the odd shape of these continua: these reflect the quantum efficiency of the image-tube chain, not yet removed from the data; but the existence of the continua is not in doubt). These continua may be direct evidence for the existence of a hot stellar component in OH 0739-14, perhaps the putative binary postulated by Morris and co-workers (Morris 1981; Morris, Bowers and Turner 1982).

Now let's look at the spectra of the nebulous knots that lie far beyond the known reflection lobes (of order 1 arcmin each, from the nebular center). Both are H-alpha emitters (Fig. 7). What of their origin? Were they ejected from the M-giant during previous episodes of mass loss? We know from the tilt of the nebula that the south lobe material is approaching us, the north lobe receding from us. It is interesting to note that the radial velocities of the H-alpha lines in the knots also follow this pattern. The velocity difference between the

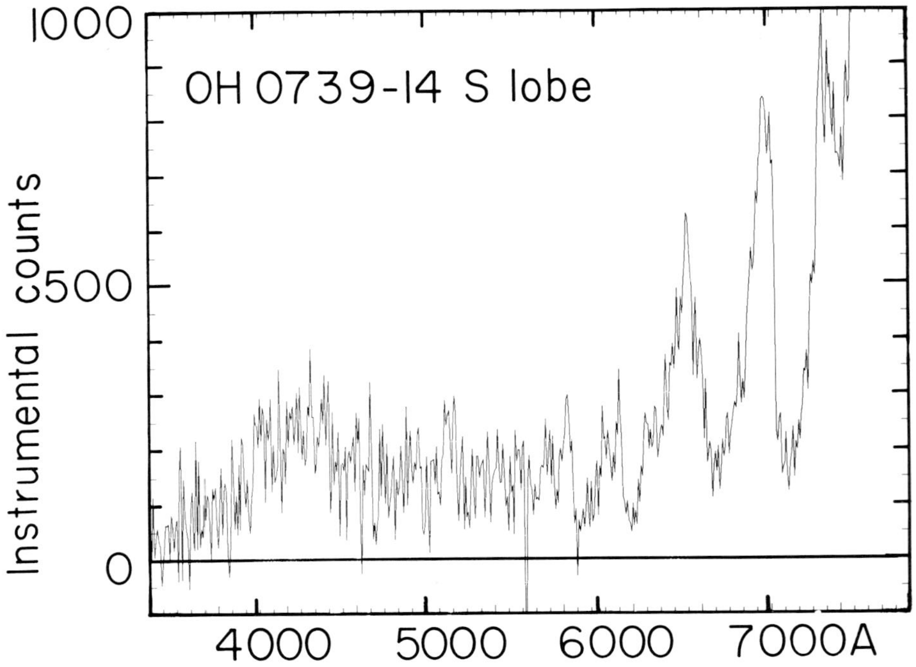

6a

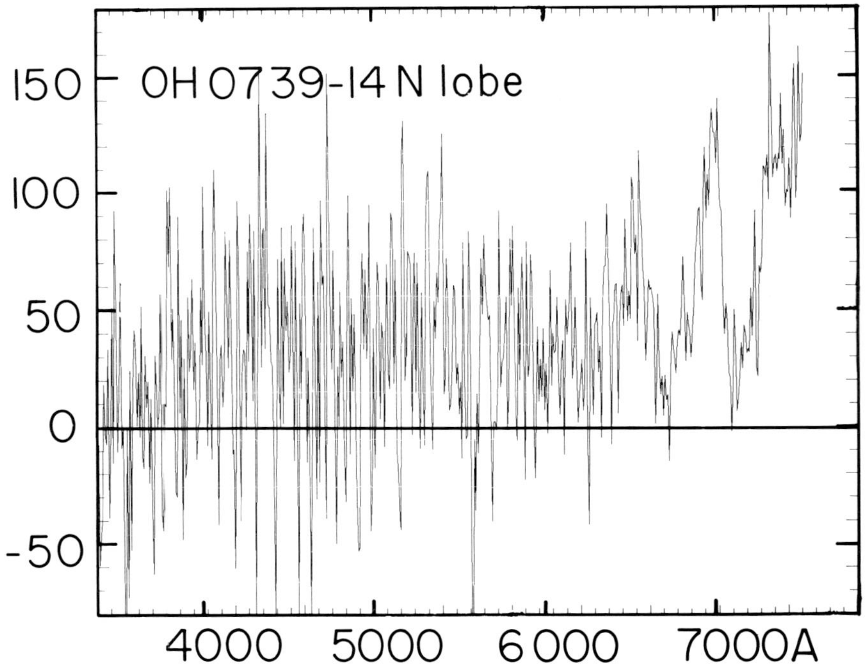

6b

Figure 6. a) Optical spectrum of the south lobe of OH 0739-14; b) as in a) but for the much fainter north lobe. Ordinates are proportional to flux density received from each lobe.

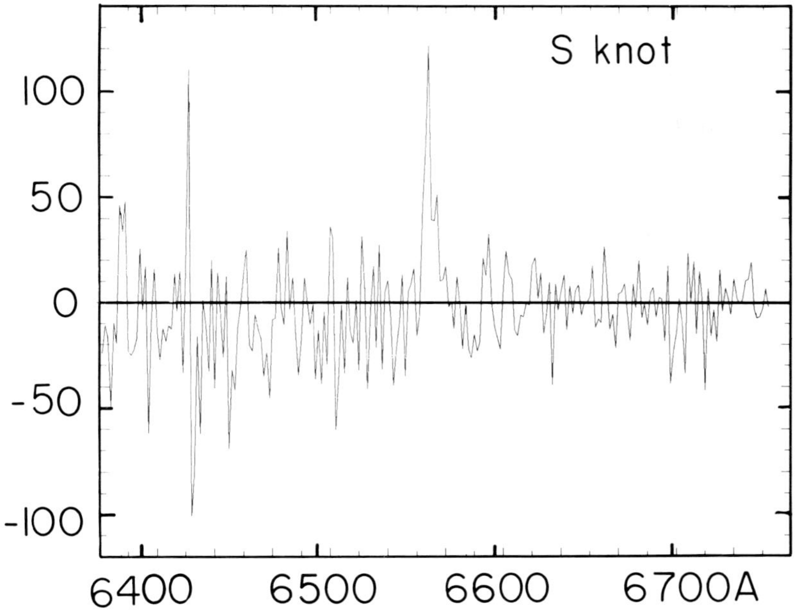

7a

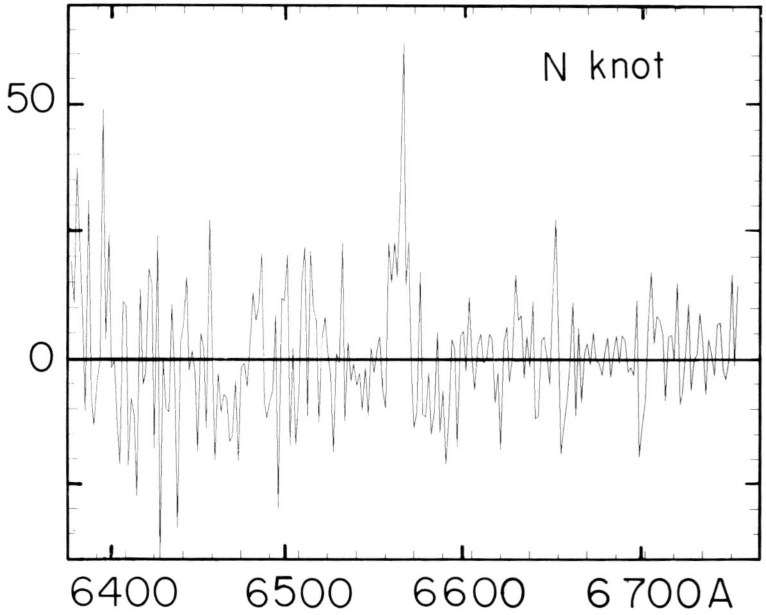

7b

Figure 7. a) Optical spectrum near H-alpha of the southern knot of OH 0739-14 that lies beyond the southern reflection lobe; b) as in a) but for the northern emission knot. Note the radial velocity difference between the two H-alpha emission peaks.

two emission knots is large: some 300 km/s, which speaks for very rapid motions, quite unlike those normally attributed to ejection from red giants. In fact, knowing the tilt relative to the plane of the sky of the entire bipolar system we can calculate that each knot must be moving away from the central star at $\pm$300 km/s! Does the hydrogen emission represent cooling of a region shocked by its high-velocity encounter with the ambient medium near OH 0739-14?

2.3 Extreme carbon stars

We move onto extreme carbon stars now. By this term I refer to carbon red-giant stars that are heavily enshrouded in dust of their own making. Again, spectropolarimetry has given us vital information on the geometry of the circumstellar dust shells around these stars. The three objects so far studied by this technique (Cohen and Schmidt 1982) all have dust disks: GL 2699 is the most highly linearly polarized star known (30%); IRC+30219 (CIT-6 to its friends) in particular has a bipolar structure with dust grains "over the poles" of the star that scatter light to us out of the circumstellar disk plane (Fig. 8), and starlight in the red coming to us by scattering in the plane of the

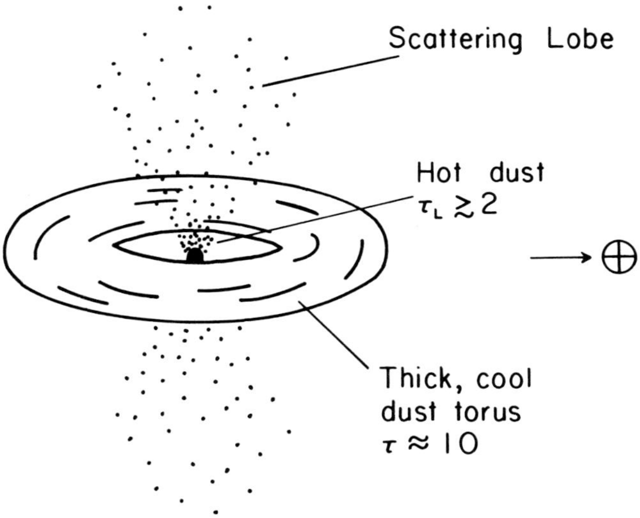

Figure 8. Bipolar morphology of the extreme carbon star IRC+30219.

dust disk. This orthogonally polarizing morphology produces a p(λ)-dependence that yields almost null polarization near 5200A (see Fig. 9) and an abrupt rotation of the plane of polarization through 90 degrees in only 200A.

The issue here is the conflict between this current geometry and the geometry of the ancient mass loss. HCN spatial interferometric mapping of, for example, IRC+10216 (Bieging, Chapman and Welch 1984) reveal the presence of a roughly ellipsoidal expanding envelope, yet we now see what is likely to be an equatorial ring shed by the central star with a bipolar, rather than spherical, flow direction. When does the initially spherical wind become bipolar, and why?

When thinking of velocities of ejection from red giants we should recall that CIT-6 has been seen to undergo regular (once per Mira period) shocks at which times emission lines of oxygen, nitrogen, sulfur and helium are prominent (Cohen 1980a, b). The velocity that best matches the character and line intensities of this shock spectrum is about 80 km/s (Cohen 1980a). This suggests that, at certain phases, red giants, both oxygen- and carbon-rich, are capable of much more rapid ejection velocities than the canonical 10 km/s.

If all three extreme carbon stars so far studied by spectropolarimetry reveal evidence of dusty disks, are these disks common to all these stars? Why then is OH0739-14 unique among M-giants in having a dust disk? What differences might there be between oxygen- and carbon-rich giants that could make disks in one group but not in the other? Might stars evolve from M-giants into carbon-giants along the red giant branch? It is clear that an observational program designed to study M-giants by spectropolarimetric means,in the detail in which these three extreme carbon stars have been studied, would be very worthwhile.

2.4 Parsamian 13

Finally, let me talk about Parsamian 13, a parabolic, but presumably truly bipolar, nebula apparently lying in a dark cloud. The central star, indicated in its position by a bright infrared object, lies at the apex of the parabola (Cohen et al. 1983). Optical spectra of this nebular lobe contain a uniquely curious aspect. Fig. 10 shows red spectra of the nebula in which oddly sharp, deep absorptions occur at the various TiO bandheads. The same features are detected in the yellow region too (Cohen at al. 1983). Clearly we have material that looks like it belongs in the photosphere of an M-star. However, to structure the TiO heads in this manner requires not only a high TiO abundance but also extremely low temperatures (of the order of 50K). Does this represent the consequences of an M-giant that has only very recently shed its outer layers to become a bipolar nebula (and then what?)?

3. GENERAL COMMENTS

Let me conclude by posing some simple and basic questions about the bipolar nebulae, as a class, the solutions to which must surely aid our understanding of mass loss in red giant stars.

Why are the lobes of bipolar nebulae usually so different in their brightness and color, their size, and sometimes their shape? Does the difference reflect the intrinsic asymmetry of mass loss from the two stellar poles? Is it a matter of viewing one lobe (the fainter and also the redder) through an overlying, large, tilted dust disk that represents an extension of the equatorial ring?

When does the initially apparently spherical wind from a red-giant star change and generate bipolar structure?

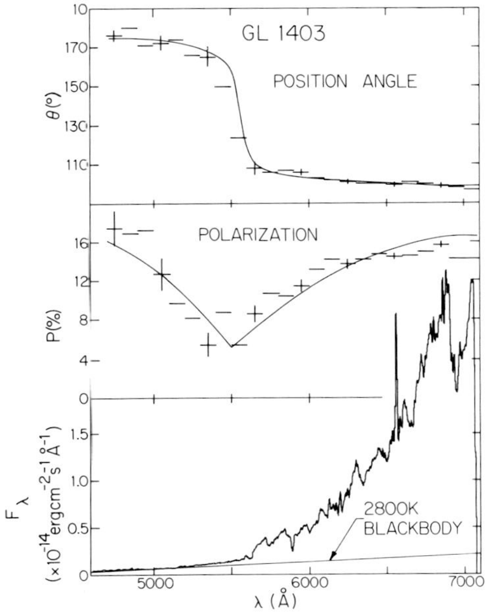

Figure 9. The spectrum, and wavelength dependences of degree of polarization and position angle of electric vector for CIT-6, the extreme carbon star that shows an abrupt reversal of position angle near 5200A.

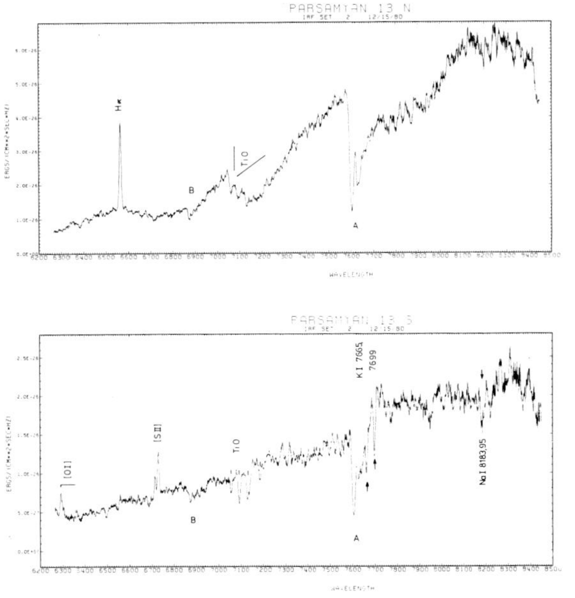

Figure 10. Red spectra of the nebulosity that is Parsamian 13 (P13S), and a nearby star in the same dark cloud (P13N), showing the different character to the TiO bandheads in P13S. Note other sharp absorptions due to neutral metals that are seen in this nebular lobe but not in a normal, cool stellar photosphere.

What generates the high velocities that are seen in IRC+30219 (CIT-6) (80 km/s) and in OH0739-14 (±300 km/s), both cool red giant stars?

Is Parsamian 13 the closest object that we can identify to the beginnings of a bipolar nebula/protoplanetary nebula?

Is OH0739-14 unique among extremely late M-stars in having a disk and bipolar structure, and, if so, why is it so unusual among M-giants?

Are carbon- and oxygen-rich giants radically different in the mechanisms and geometries of their mass loss?

REFERENCES

Bieging, J., Chapman, B., and Welch, W.J. 1984, Ap.J., in press 15 Oct

Beckwith, S., Beck, S., and Gatley, I. 1984, Ap.J., 280, 648.

Cohen, M. 1980a, Ap.J. Letters, 238, L81.

--------- 1980b, IAU Circular No. 3552.

--------- 1981, P.A.S.P., 93, 288.

Cohen, M. and Kuhi, L.V. 1977, Ap.J., 213, 79.

------------------------ 1980, P.A.S.P. , 92, 736.

Cohen, M. and Schmidt, G.D. 1982, Ap.J., 259, 693.

Cohen, M., Aitken, D.K., Roche, P.F., and Williams, P.M., 1983, Ap.J., 273, 624.

Morris, M. 1981, Ap.J., 249, 572.

Morris, M., Bowers, P.F., and Turner, B.E. 1982, Ap.J., 259, 625.

Schmidt, G.D., Cohen, M., and Margon, B. 1980, Ap.J. Letters, 239, L133.

INFRARED STUDIES OF THE BIPOLAR NEBULA OH0739

A.G.G.M. Tielens and M. Werner
NASA-Ames Research Center
MS 245-6, Moffett Field, CA 94035
R. Capps
Institute for Astronomy, University of Hawaii
Honolulu, HI 96822

The M9 giant, OH0739 is embedded in a bipolar nebula, as evidenced by far-red photographs and OH maser studies (Cohen and Frogel 1977, Morris et al. 1982). It is a bright infrared source, whose spectrum shows deep 3.1μm ice and 10μm silicate absorption. We have made infrared studies of this object with the aim of studying the structure and density distribution of the circumstellar disk. This paper presents some preliminary results of this study.

First, we studied the polarization across the 3.1μm ice band in an 8" beam centered at the 2.2μm peak. As expected for a bipolar scattering nebula, the degree of polarization increases with the extinction in the ice band. This is due to an increase in the contribution of the scattered, polarized light when the extinction to the central unpolarized source increases (Elsasser and Staude 1978). Assuming single, Rayleigh scattering, a tilt angle of the bipolar axis with respect to the line of sight of about 30° is inferred from the observed maximum polarization (40%). From the phase lag measurements of the OH maser (Bowers and Morris 1984) a distance of 2 kpc can then be inferred.

Second, we have made maps in the K band and in narrow bands (0.1μm) at 2.85, 3.1 and 3.5μm by rasterscanning OH0739 with a 2" beam and 1" separation (fig. 1). At each wavelength the scanning procedure was interrupted several times to peak up at the peak position at 2.2μm, in order to assure accurate relative positions (∿1"). In all maps the object is clearly elongated along the bipolar axis. Disregarding for the moment the map at 3.1μm, we note that the separation of the two lobes decreases with increasing wavelengths. This is probably due to a decrease in extinction, which enables us to see deeper into the envelope at longer wavelengths. We also note that the contrast between the two lobes is much less at the longer wavelengths, again, indicative of a decrease in extinction. The peak at 3.5μm is probably at the position of the central illuminating source. The trend of decreasing separation and contrast of the lobes with increasing wavelengths is reversed in the depth of the ice band (3.1μm). Again these differences can be attributed to an increase in extinction, this time due to the

M. Morris and B. Zuckerman (eds.), Mass Loss from Red Giants, 305–307.

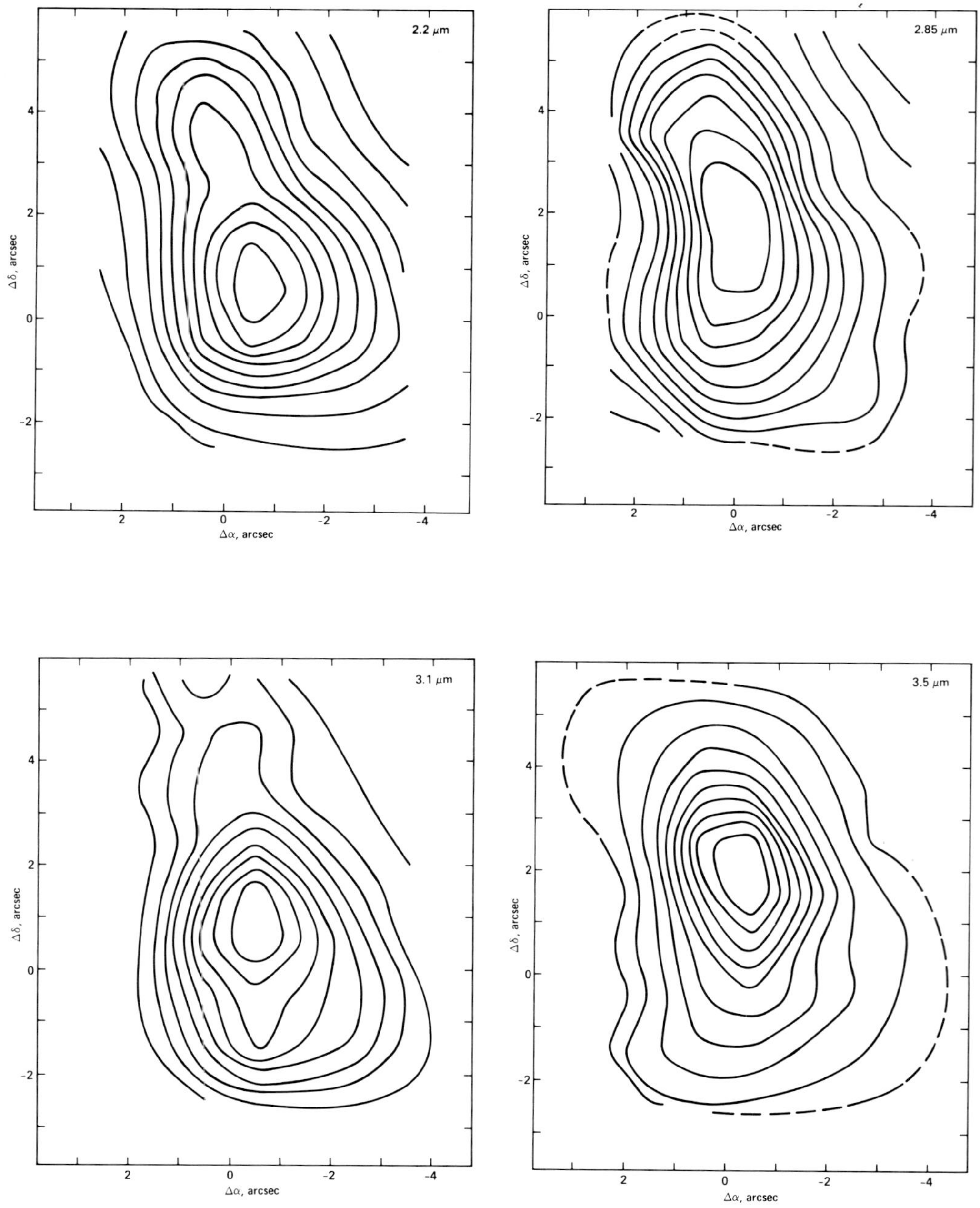

Figure 1: Near-infrared maps of OH0739. Contour levels are 100, 90...10, 5% of the peak flux. The lowest contour is 2 sigma.

ice particles. The structure in the north and south lobe at 3.1μm (cf. compare the maps at 2.2 and 3.1μm) suggests that the "ice envelope" becomes optically thin, rather abruptly, at about 2" from the central 3.5μm source. It should be emphasized that the south as well as the north lobe show a deep 3.1μm ice band absorption feature. All the data together seem to indicate that the ice particles are widespread through the nebula and not just concentrated to the disk. We do note, however, that the shape of the 3.1μm ice band in polarization and extinction may actually indicate that the ice particles do not dominate the scattering. Further IR spectropolarimetry studies are needed to clarify this point.

REFERENCES

Bowers, P.F. and Morris, M. 1984, Ap.J., 286, 646.
Cohen, J.G. and Frogel, J.A. 1977, Ap.J., 211, 178.
Elsasser, H. and Staude, H.J. 1978, Astr. Ap., 70, L3.
Morris, M., Bowers, P.F. and Turner, B.E. 1982, Ap.J., 259, 625.

OBSERVATIONS OF THE FAR-INFRARED EMISSION LINES OF OI AND CII IN PLANETARY NEBULAE

H. Benton Ellis, Jr.*, Michael W. Werner
NASA-Ames Research Center
Mail Stop 245-6
Moffett Field, CA 94035

We have made observations of the far-infrared, fine structure emission lines [OI] 63 μm, [OI] 145 μm, and [CII] 158 μm in the proto-planetary nebulae AFGL 2688 and AFGL 618, and the classical planetary nebulae IC 418 and NGC 7027. The observations were made with the UC Berkeley tandem Fabry-Perot spectrometer aboard the Kuiper Airborne Observatory in October, 1983. Apertures of 30" (63 μm) and 45" (145 μm and 158 μm) were used. We detected all three lines in NGC 7027, but obtained only upper limits to the 63 μm emission from the other three nebulae and the 158 μm emission from AFGL 618. The line fluxes in NGC 7027 are: [OI] 63 μm, $3.2 \pm 1.0 \times 10^{-17}$ W cm^{-2}; [OI] 145 μm, $1.2 \pm 0.4 \times 10^{-18}$ W cm^{-2}; [CII] 158 μm, $3.4 \pm 1.0 \times 10^{-18}$ W cm^{-2}. The [OI] 63 μm line has been detected previously in NGC 7027 by Melnick *et al.* (1981) but these observations represent the first detections of the [OI] 145 μm and [CII] 158 μm lines in a planetary nebula.

We interpret the observations of NGC 7027 in terms of emission from a warm, neutral atomic shell of gas located between the inner, ionized material and the extended molecular envelope observed in CO (Mufson, Lyon, and Marionni, 1975). We derive a characteristic temperature and density of 500 K and 7×10^4 cm^{-3} for this region using the line flux ratios 63 μm/145 μm and 63 μm/158 μm. To derive these values we have made the assumptions that, 1) the emission is excited by collisions with HI, 2) the OI and CII are coextensive, 3) the lines are optically thin, and 4) the C^+/O ratio is 1.7 (Kaler, 1981). From the [CII] 158 μm line flux and assuming a distance to NGC 7027 of 1.1 kpc, we determine that there is 0.25 $M_\odot$ of gas in the neutral shell. This mass estimate is relatively insensitive to the density and temperature because of the low critical density ($n_{crit} \simeq 10^3$ cm^{-3}) and excitation temperature (91 K) of the 158 μm transition. Comparing this mass estimate with the $\simeq 0.1$ $M_\odot$ of gas in the ionized region and the $\simeq 0.5$ $M_\odot$ in the molecular envelope, it is seen that a significant fraction of the total nebular mass may be in the neutral, atomic shell.

Details of these observations and analysis are presented in a paper submitted to The Astrophysical Journal. We would like to thank

M. Morris and B. Zuckerman (eds.), Mass Loss from Red Giants, 309–310.

C. H. Townes, R. Genzel, M. Crawford, and J. Lutgen at UC Berkeley for supporting these observations.

REFERENCES

Kaler, J. B. 1981, Ap. J., 249, 201-209.

Melnick, G., Russell, R. W., Gull, G. E., and Harwit, M. 1981, Ap. J., 243, 170-174.

Mufson, S. L., Lyon, J., and Marionni, P. A. 1975, Ap. J. (Letters), 201, L85-L89.

*NAS/NRC Research Associate

AMMONIA AND CYANOTRIACETYLENE IN THE EGG NEBULA

Nguyen-Q-Rieu
Observatoire de Paris-Meudon
92195-Meudon Principal Cedex, France

D. Graham
Max-Planck-Institut für Radioastronomie
Auf dem Hügel 69, 5300 Bonn 1
Federal Republic of Germany

V. Bujarrabal
Centro Astronomico de Yebes
Apartado 148, Guadalajara, Spain

We report the first detection of the NH_3 (1,1) and (2,2) inversion lines, and the rotational transition J=21-20 of HC_7N (at ~23.7 GHz) in the Egg Nebula (CRL 2688). The observations were carried out in

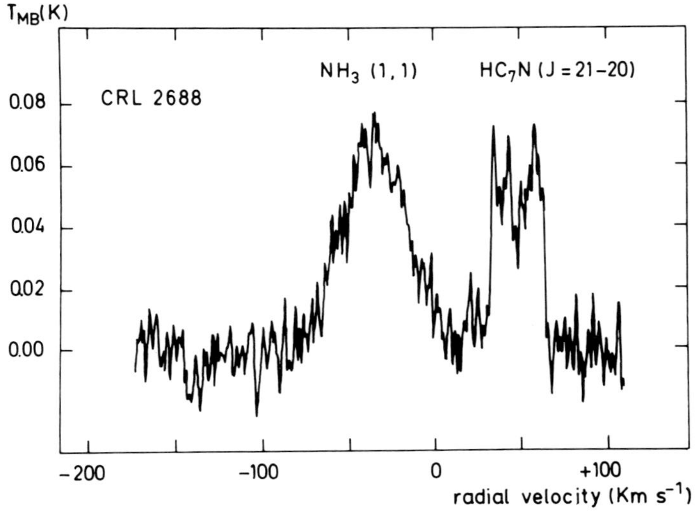

Figure 1: Composite spectrum containing the NH_3 (1,1) and HC_7N (J=21-20) lines. The velocity scale is only appropriate for the NH_3 line.

M. Morris and B. Zuckerman (eds.), Mass Loss from Red Giants, 311–312.

March and April 1983, using the Effelsberg radiotelescope. The beamwidth of the instrument at 23.7 GHz was ~40" and the velocity resolution was ~1.2 km s^{-1}.

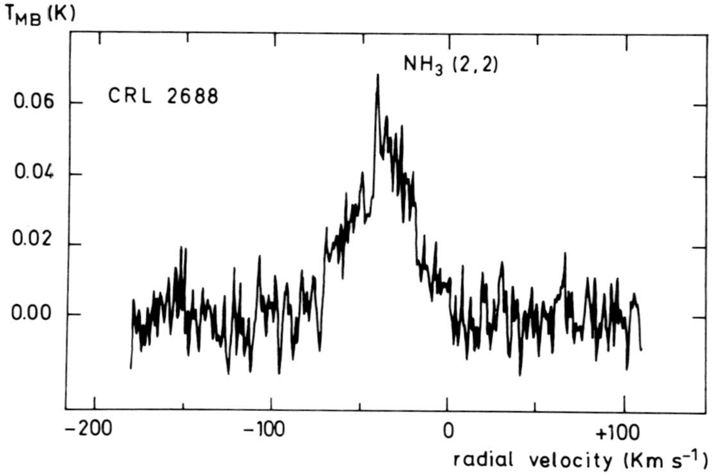

Figure 2: The NH_3 (2,2) spectrum.

The HC_7N spectrum is clearly double-peaked. If the molecular emission arises in a disk, a double-peaked profile does not necessarily indicate that the source is resolved. The presence of the wings can be due to the fact that the object is seen almost edge-on. The NH_3 lines consist of blended hyperfine components. These data have been interpreted in terms of a radiative transfer model from which the source size and the molecular abundance have been derived. Our calculations indicate that the column densities (averaged over the beam) of NH_3 (para) and HC_7N are ~10^{14} cm^{-2}. If we assume that the NH_3 abundance (relative to molecular hydrogen) ~2.8 10^{-8} as found in IRC+10216, we find a mass loss rate $\dot{M}$ ~10^{-3} $M_\odot$/yr. From $\dot{M}$, we can derive the abundance of HC_7N, ~10^{-8}. This mass loss rate seems to be too high. If NH_3 is more abundant than assumed above, $\dot{M}$ can be lower. However, if the Egg Nebula is a proto-planetary object, a high mass loss rate may be expected. More details can be found in our paper published in Astron. and Astrophys. 1984, vol.138, p. L5.

CONCLUDING REMARKS

Peter Goldreich
Division of Geological and Planetary Science
California Institute of Technology
Pasadena, CA 91125
USA

I would like to thank the organizers for inviting me to give this talk, and in particular for not putting any time limit on my contribution. I hope that you all have comfortable seats since I plan to speak for a long time.

The main point of this conference is that stars, like astronomers, tend to swell up as they age but unlike the case with astronomers, this swelling is invariably accompanied by mass loss. Aging astronomers also deliver summary talks. In this context, I note that this is probably the last talk I shall give without my spectacles unless I discover some way to stretch my arms.

The conference began with Icko Iben telling us that our subject is important because, in addition to providing us with some clues about the gas and dust in the uninteresting interstellar medium, it is of fundamental significance to the evolution of stars. Mass loss might cause stars with initial masses as large as 8 $M_{\odot}$ to end their lives as white dwarfs. Thus astronomers would be spared the burden of explaining away excessive production of heavy elements in supernovae. Next came a session in which we were presented with a slew of observational results.

I would characterize the observational situation as follows. The outer regions of the circumstellar envelopes of stars which are losing large amounts of mass are well-studied. There seems to be general agreement on the density of the gas and there is no question about how fast it is flying out. Remaining questions include the roles that photodissociation and photoionization play in altering the chemical composition. Of special interest to me were the talks by Baud, Bowers, Herman, Sargent, and Werner which dealt with the OH-IR stars. Theory and observation seem in reasonable accord for these objects which have yielded valuable insights about the structure of the outer envelopes.

Observations of the inner regions close to the stellar photospheres are at a much earlier stage. We learned from talks by Wannier on infrared spectroscopy and by Beckers and Beckwith on optical-infrared interferometry, and from discussions of the SiO masers, that in the future we can hope to untangle a lot of what is

M. Morris and B. Zuckerman (eds.), Mass Loss from Red Giants, 313–315.

going on but, at present, all we seem to have is some tantalizing clues. For example, the observations do not identify the mechanism responsible for the mass loss. We also don't have a very good idea of the gas temperature or of the composition, especially the abundance of the minor chemical constituents. So that is something for the future. Also for the future is the examination of the IRAS data set which should, at the least, improve our statistical understanding of stellar mass loss.

There is very little observational data on the intermediate regions which lie between the outer regions which are resolved by single-dish millimeter observations and the inner regions which are hot and bright enough to be studied in the near-infrared. Here we can look forward to contributions from the new millimeter interferometers just coming on line. Perhaps we can learn how the chemical composition changes as the gas moves away from the star to where it starts to interact with the interstellar radiation field.

Next I will discuss the mechanism of mass loss which surprisingly did not receive very much direct attention at this conference. Perhaps this reflects our poor understanding of the issue. Holzer's talk was the principal one dealing with the theory of mass loss. He identified four candidates for the acceleration of the gas away from the stars. The first is thermal expansion in which the gas is heated so that its scale height approaches the stellar radius and it asymptotically flows away from the star. This mechanism is probably the dominant one driving the solar wind and it is well-understood. The second mechanism involves shock acceleration. This is a natural one to consider for variable stars which provide ample observational evidence for strong periodic atmospheric shocks. A third mechanism is wave acceleration, in which the momentum and energy of waves are transferred to the gas and drive the expanding flow. The waves are perhaps by-products of the vigorous photospheric convection which occurs in red-giants. The fourth mechanism is acceleration by radiation pressure acting on dust grains which form in the stellar atmospheres.

Thermal driving of the mass outflow would require that a substantial amount of hot gas, T up to 100,000 K, be present in the inner envelope. The observational evidence, particularly with regard to the coolest stars for which the mass loss is greatest, is that hot gas is not present in the required amounts. As stressed, both by Linsky and Holtzer, we would have observed the emission from this gas if it were there.

Shocks also heat the gas but they don't necessarily keep it hot for very long. Thus shock acceleration appears to be more viable than thermal expansion. Shocks also dissociate and ionize the gas and thereby alter its chemical composition. An important, and to my knowledge unexamined question, is whether the current states of theory and observation are adequate to test the shock acceleration hypothesis. For example, would gas which was shocked strongly enough to be driven from the star be distinguishable from gas which was gently accelerated by radiation pressure on dust grains?

The correlation between mass loss and circumstellar dust is well-established but we don't know the cause and effect in this correlation.

It may be that the dust is required to scatter and absorb the radiation so that its momentum may be transferred to the gas, thereby driving the outflow. Alternatively, the conditions of low temperature and high gas density required to condense large quantities of dust may only be met around stars which possess large outflows. Very likely, both connections are important. Some problems associated with the hypothesis that radiation pressure acting on dust grains drives the outflows have been recognized for a long time and have received a good airing here. Over the past few years it had often been claimed that the momentum carried in many of the winds exceeded that which could be supplied by the radiation field. However, we heard from Knapp and Morris that most of these discrepancies seem to have disappeared. Indeed, the fact that momentum flux of the wind is frequently approximately equal to that of the central star's radiation field seems to support the notion that radiation pressure is reponsible for accelerating the wind. Another unresolved problem concerns the radial distribution of dust around red giants. Many OH/IR stars, among other objects, have been found to possess a significant opacity at near infrared wavelengths from dust whose temperature is rather low, typically of order 700 K. Nevertheless, the photospheres of these stars, with temperatures of order 2000 K, are seen at near infrared and often at visible wavelengths. This would not be possible if all the 700 K dust grains had condensed at the base of the outflow; the optical depth to the stellar photosphere would be much too high. On the other hand, some small quantity of dust could form within one stellar radius or so; there is nothing in the observations which contradicts this possibility. The infrared observations do suggest, however, that the bulk of the dust condenses considerably further from the star. Thus it seems to me that radiation pressure acting on dust grains, perhaps assisted in some cases by shock acceleration which carries the gas rather high above the mean photospheric level, is the best bet for the mechanism driving the winds from cool red giants.

To conclude, I shall mention a couple of secondary issues which caught my attention. The first is the possible importance of binaries. It seems plausible that there are some examples of binaries in which the companion star is enveloped by a red giant. These systems would be logical candidates for producing disk-like structures and possibly high outflow velocities. Perhaps, as also suggested by Cohen and others, binaries are at the heart of some of the more peculiar objects. The second issue concerns the symmetry of the envelopes. We were presented with evidence that spherical symmetry is a good approximation in the outer parts of the circumstellar envelopes but there were suggestions, from both theory and observation, that asymmetry may be common close to the central star. Alcock outlined a model for SiO masers which explicitly relied on large departures from spherical symmetry in the inner regions of the wind. He appealed for support to a paper written 10 years ago by Schwarzschild, who pointed out that the vigorous convection and large surface scale height of a red giant would lead to large photospheric temperature variations. Perhaps these result in uneven condensation of dust above the photosphere and asymmetrical mass outflow.

OBJECT INDEX

I. Optically Identified Stars

II. NGC + IC Objects

III. IRC Identifications

IV. AFGL (=GL=CRL=AFCRL) Identifications

V. OH Maser Sources

VI. Other Objects